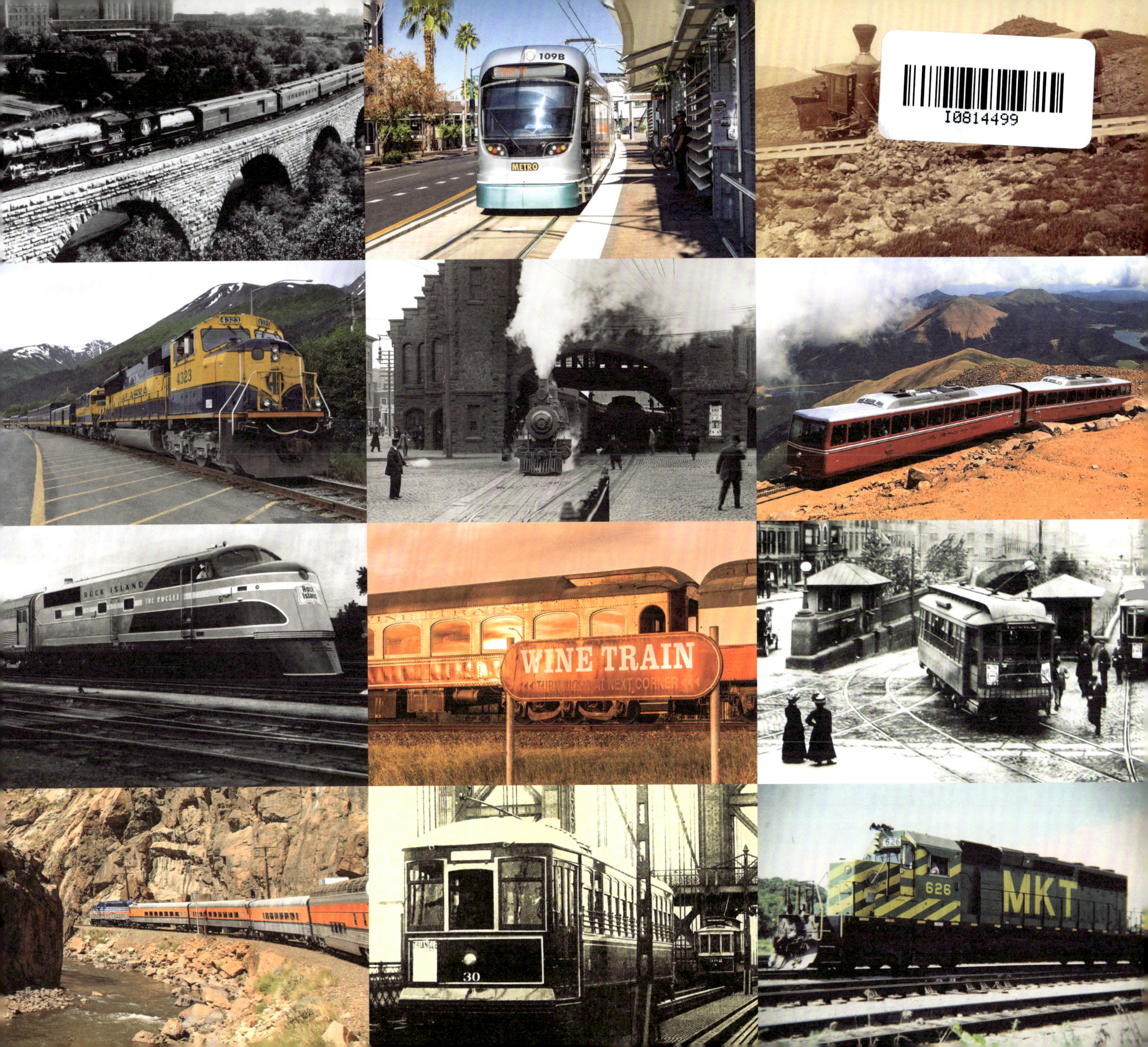
I0814499
109B
METRO
4323
ALASKA
ROCK ISLAND
Rock Island
WINE TRAIN
30
626
MKT

AMERICAN TRAINS

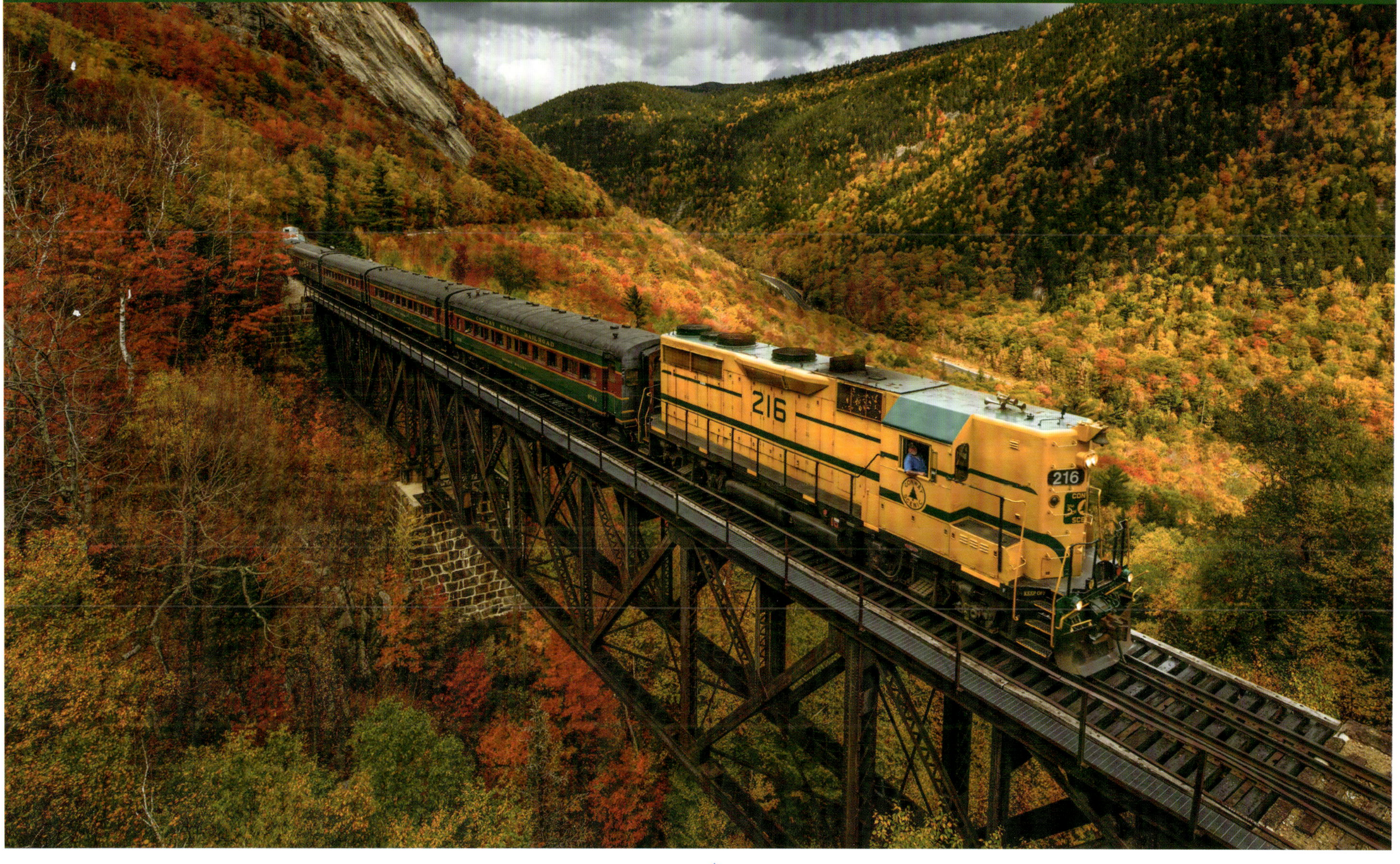

pil
Publications International, Ltd.

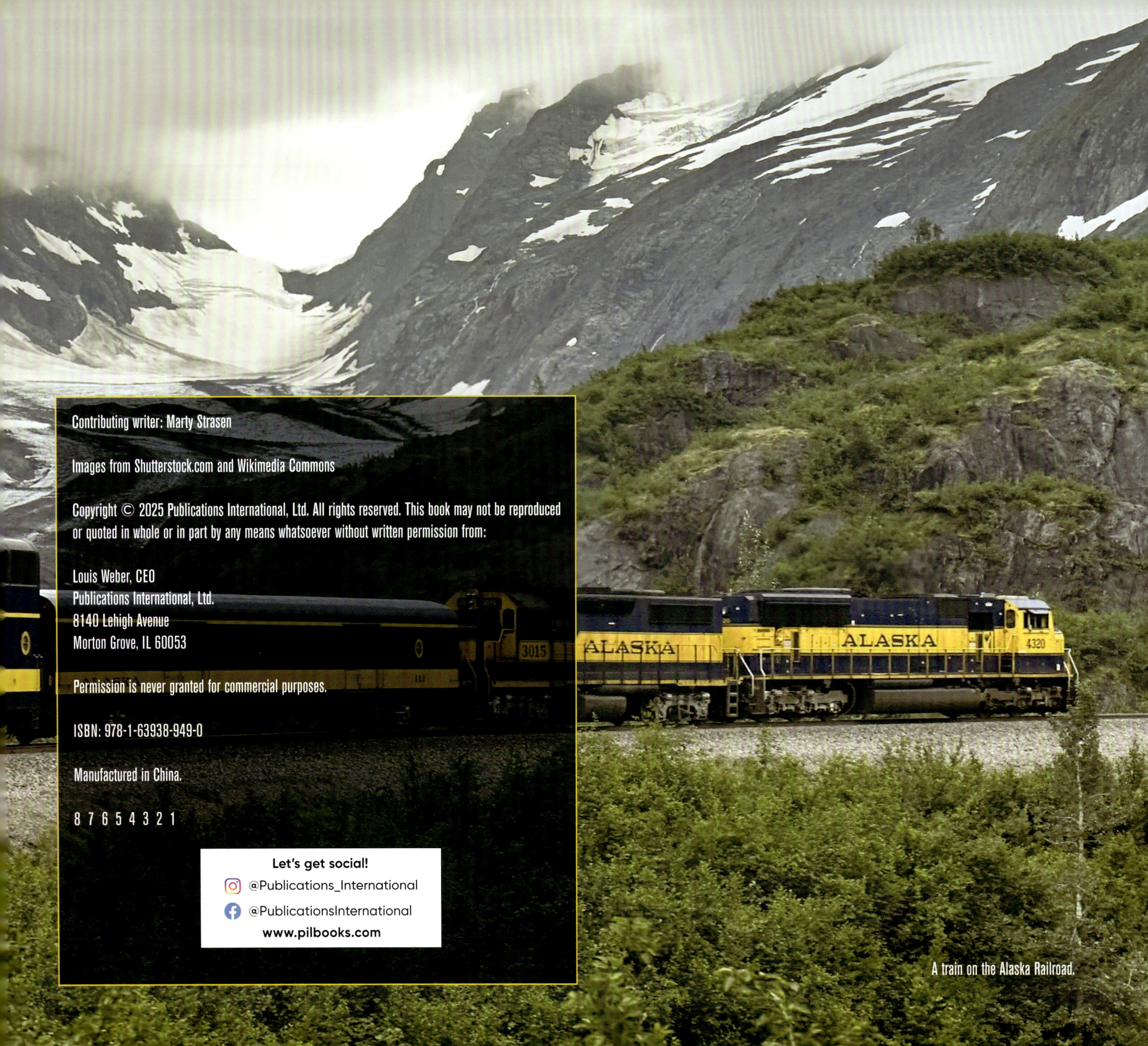

Contributing writer: Marty Strasen

Images from Shutterstock.com and Wikimedia Commons

Louis Weber, CEO
Publications International, Ltd.
8140 Lehigh Avenue
Morton Grove, IL 60053

ISBN: 978-1-63938-949-0

Manufactured in China.

8 7 6 5 4 3 2 1

A train on the Alaska Railroad.

TABLE OF CONTENTS

INTRODUCTION

Ride the rails with *American Trains* as it traverses the country to showcase the groundbreaking routes that have come to connect our coasts. *American Trains* covers some of the most fascinating and scenic routes from United States railway history, traveling along passenger and freight services that have evolved from horse-drawn rail carriages to locomotives powered by steam, diesel, and electricity.

There is perhaps no better example of the industriousness and fortitude of humanity as that of the railway. Hundreds of thousands of miles of track have been laid through dense, old-growth forests, treacherous river valleys and gorges, and the world's most drastic and difficult terrain with the brute force and drive of the human will. Tunnels have been bored through mountains and under bodies of water. Trestle bridges and viaducts have been erected over bogs, permafrost, and canyons. Trains and their railways have proven to be an indispensable technology that is not only a remaining vestige of industrialization but a beacon for us to follow in order to meet our growing and changing transportation needs.

Passenger service on historic and heritage tourist railways has grown in the past few decades. As railway construction boomed throughout the world in the mid-nineteenth and twentieth centuries, many of the shorter, older, and privately owned railways were either closed from competition or forced out of business due to shrinking freight demand. But as the ownership and operation of many railways switched hands between railway companies and governments over the years, the historic and scenic routes that once thrived have not been forgotten.

Routes like the Mount Washington Cog Railway in New Hampshire, the Grand Canyon Railway in Arizona, and the nation-spanning *California Zephyr* have all faced phases of heavy use or near abandonment, but they are thriving tourist attractions today. Light rails and electrified routes used for public transportation in the busiest cities in the United States have helped ease street congestion and allowed for faster movement across metro areas. *American Trains* covers these routes and many more to provide a glimpse into the vast and intriguing history of the Industrial Revolution's preferred method of transport and the ways it has been revitalized since.

THE EAST COAST

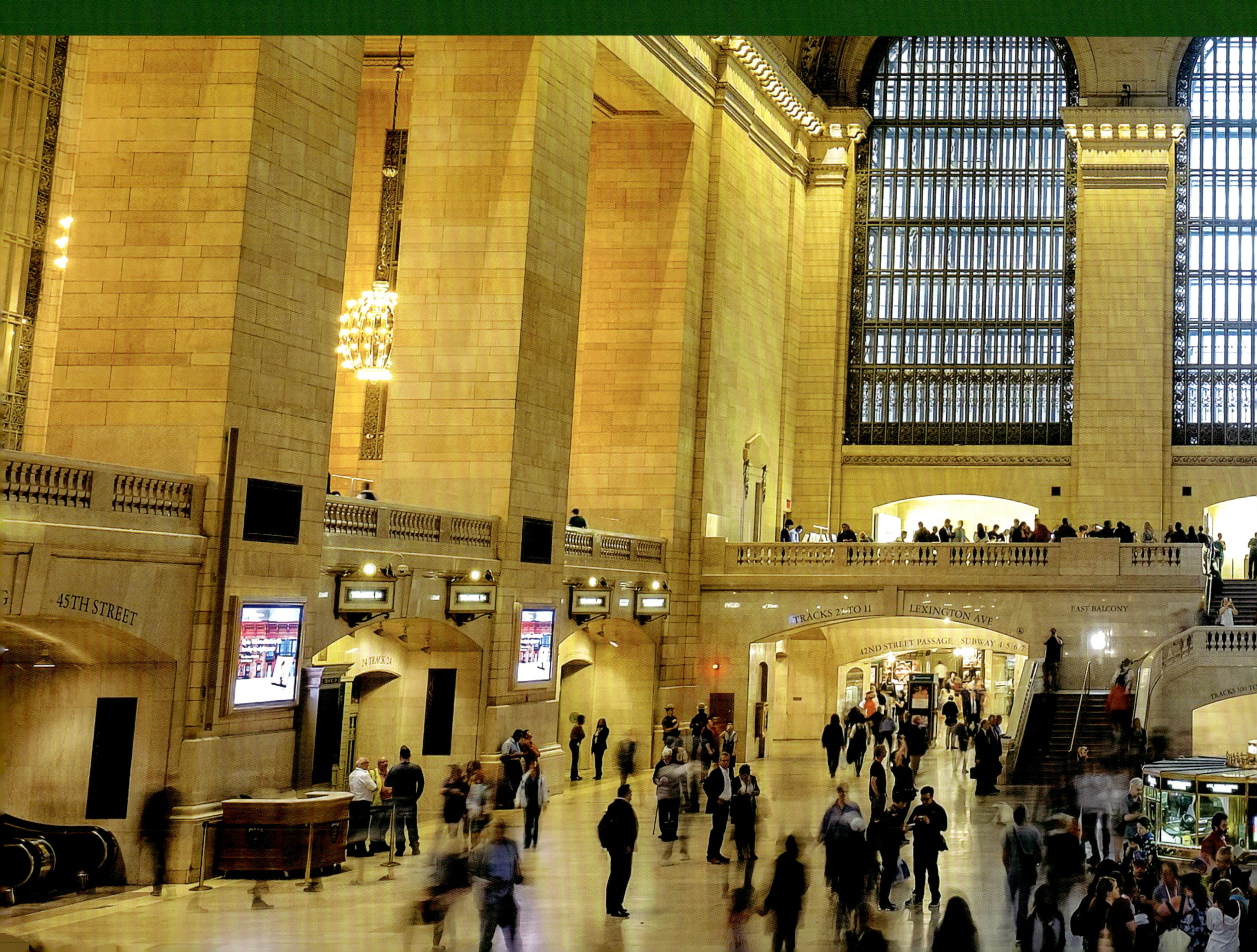

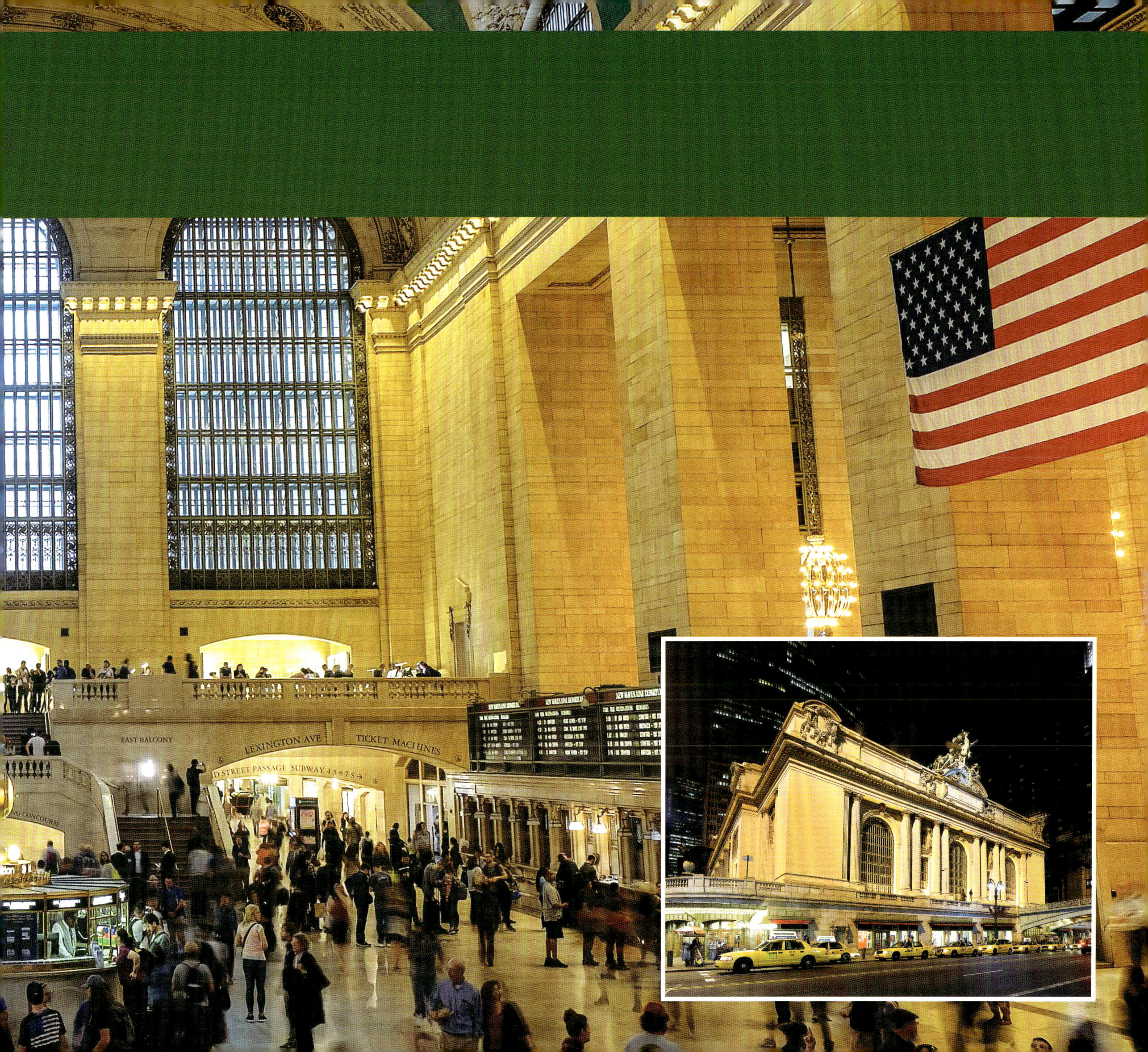
EAST BALCONY
LEXINGTON AVE
TICKET MACHINES

Port Authority Trans-Hudson

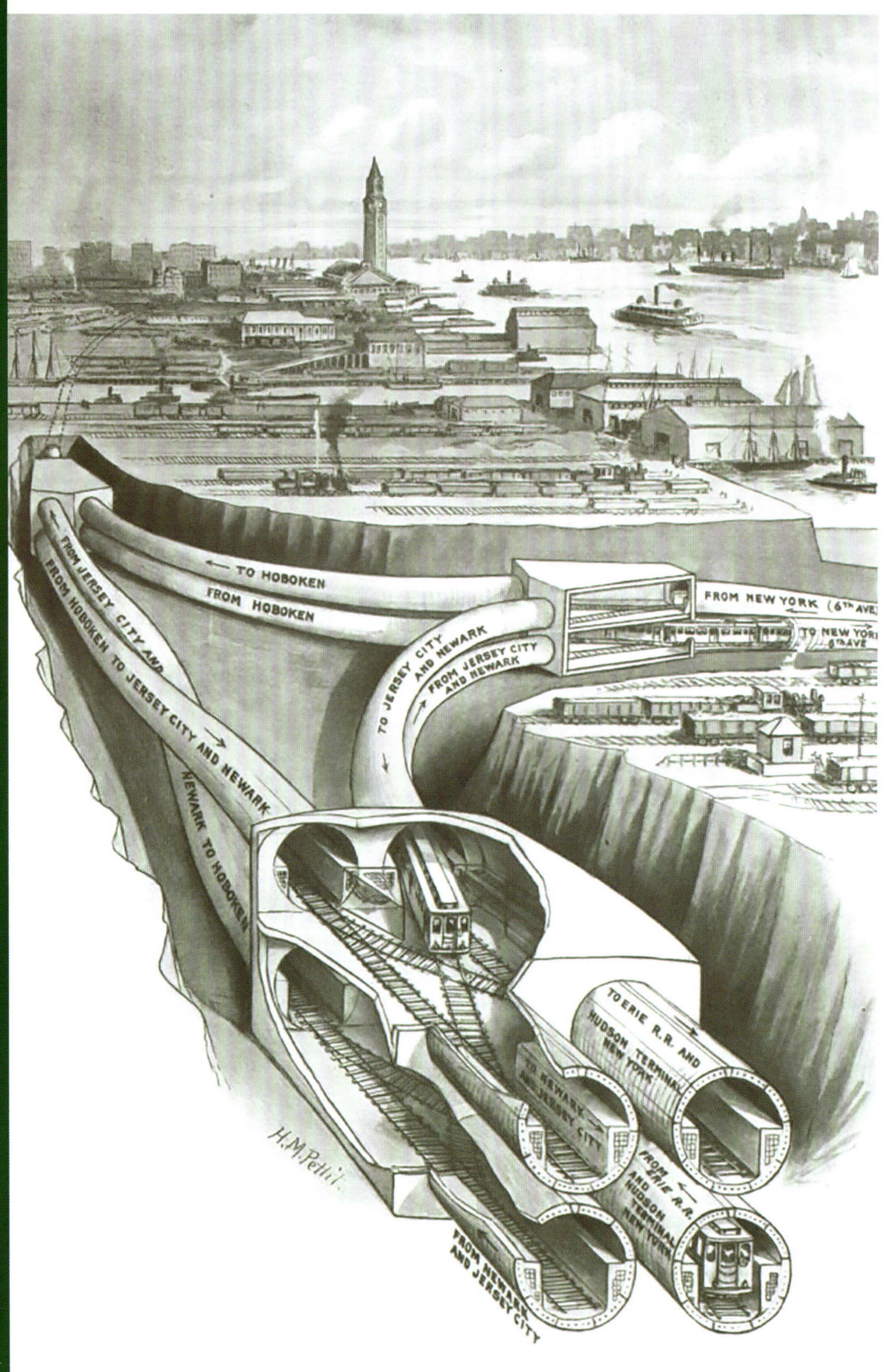

Though the Interborough Rapid Transit (IRT), which operates most of the subway lines in New York City, is perhaps better known, one might be surprised to learn that the Port Authority Trans-Hudson (PATH) from New York to New Jersey actually predates the IRT. PATH's roots date to 1874, when Dewitt Clinton Haskins raised enough money to create more than 1,000 feet of tunnel under the Hudson River. Unfortunately, it was not a quick, smooth, or happy operation. Haskins was forced to abandon his quest to connect New York to New Jersey via underground rail when an accident resulted in 20 deaths. The tunneling resumed in the 1890s, but this time financial troubles tabled the work.

William Gibbs McAdoo and Charles Jacobs picked up the effort after the turn of the century, heading a company called the Hudson and Manhattan Railroad (H&M), which later became PATH. This time, they took the work across the finish line. In December of 1907, the first test train made its run on tracks between New York and Jersey City through what people were calling the McAdoo Tubes or McAdoo Tunnels. The public began riding in 1908, and by the following year they could travel as far as Hoboken. Newark joined the route in 1911.

A diagram from 1909 showing the direction of rail traffic on the New Jersey side of the Hudson River.

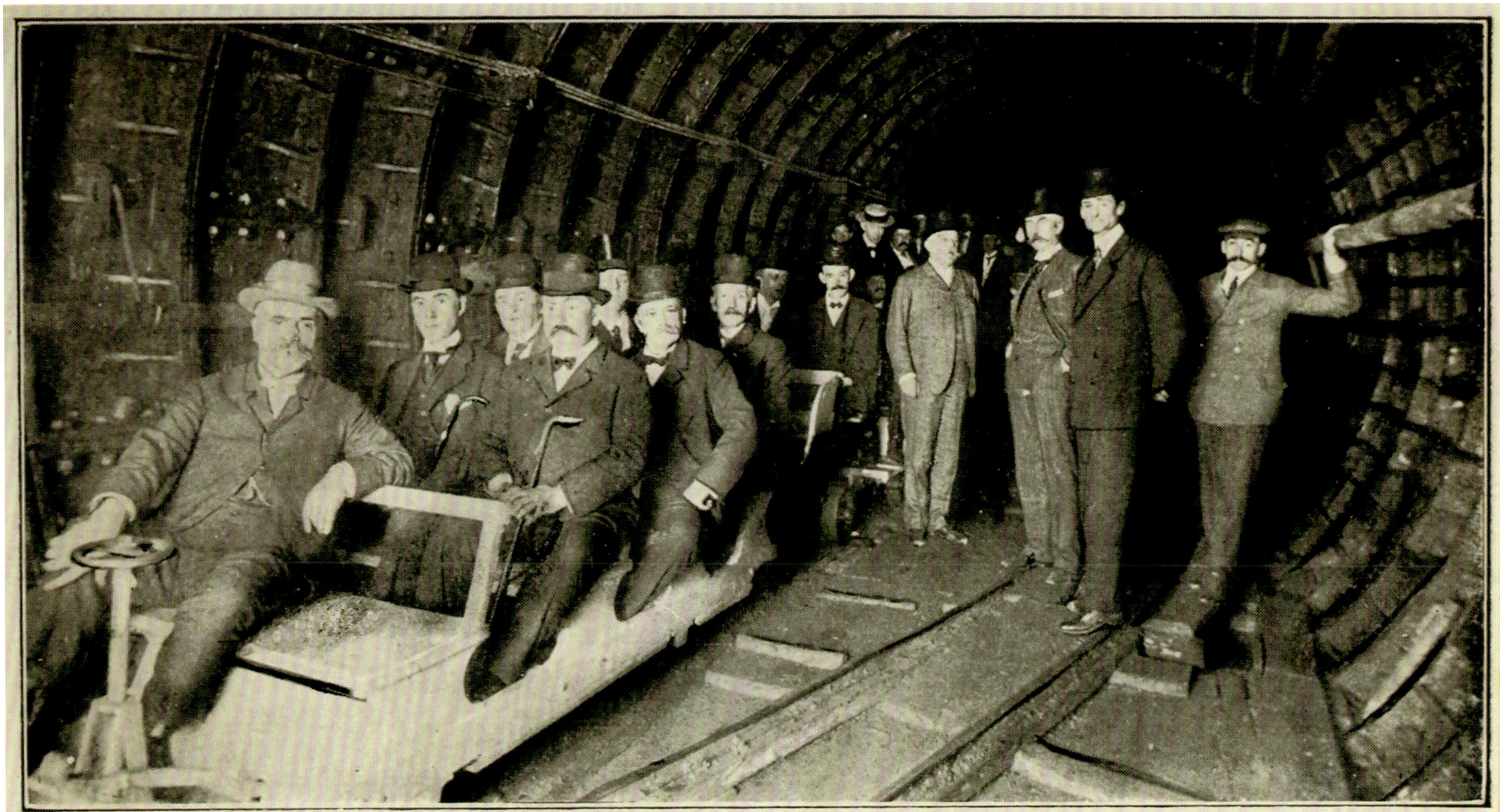

Here we look upon the giant minds who have made possible the tunnel: McAdoo, Jacobs, Davies and associates.

The line enjoyed great success in its first two decades. H&M ridership peaked in 1927, with some 113 million passengers. The Great Depression soon followed, however, as did the addition of the Lincoln and Holland Tunnels, which took business away from McAdoo's group. H&M fell into bankruptcy in the 1950s. The following decade, the New York Port Authority agreed to buy and take over the lines in exchange for the right to build the World Trade Center towers. The Port Authority Trans-Hudson was born.

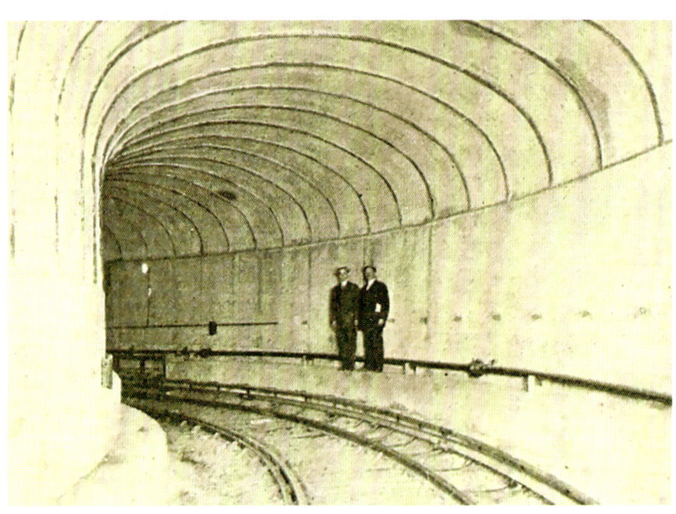

Top: A photo of the investors who made the McAdoo Tunnel possible.

Bottom: Charles Jacobs and William Gibbs McAdoo in the McAdoo Tunnel after it was completed in 1908.

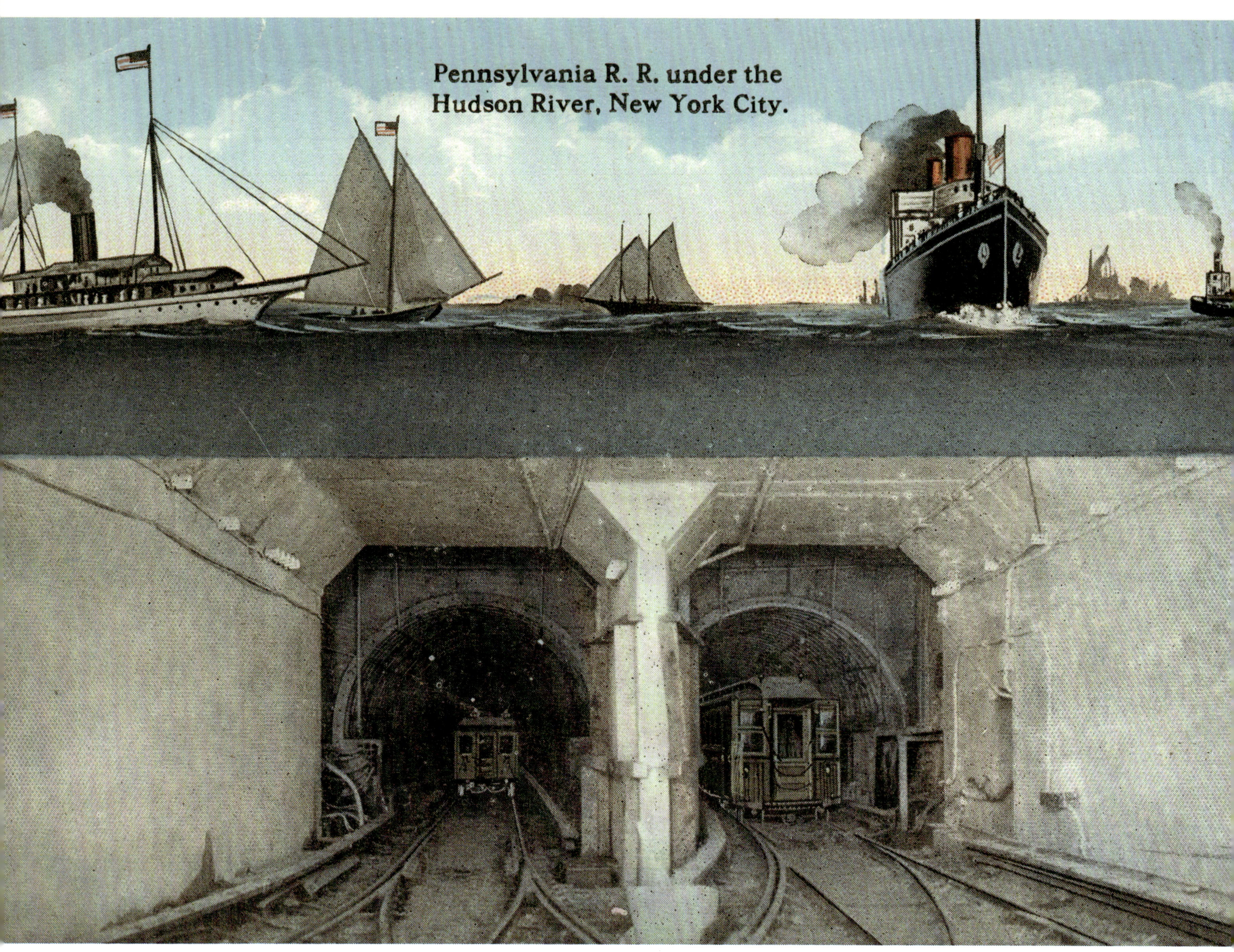

Although the postcard claims that it depicts the Pennsylvania Railroad under the Hudson, the design of the tunnels and the rolling stock in the photo tell us that the Hudson and Manhattan Railroad (now the PATH) is pictured.

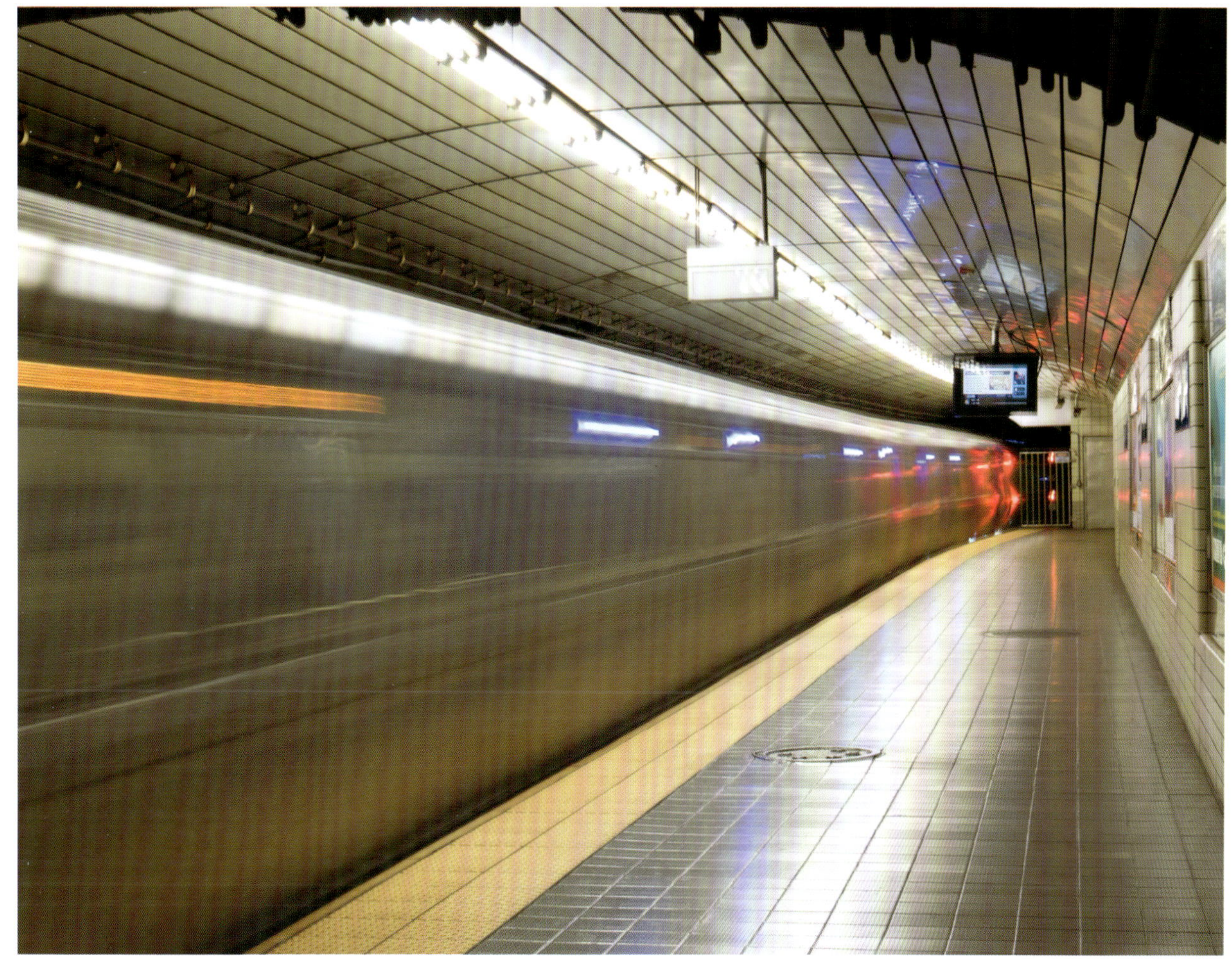

A train at the Exchange Place Station in Jersey City, New Jersey.

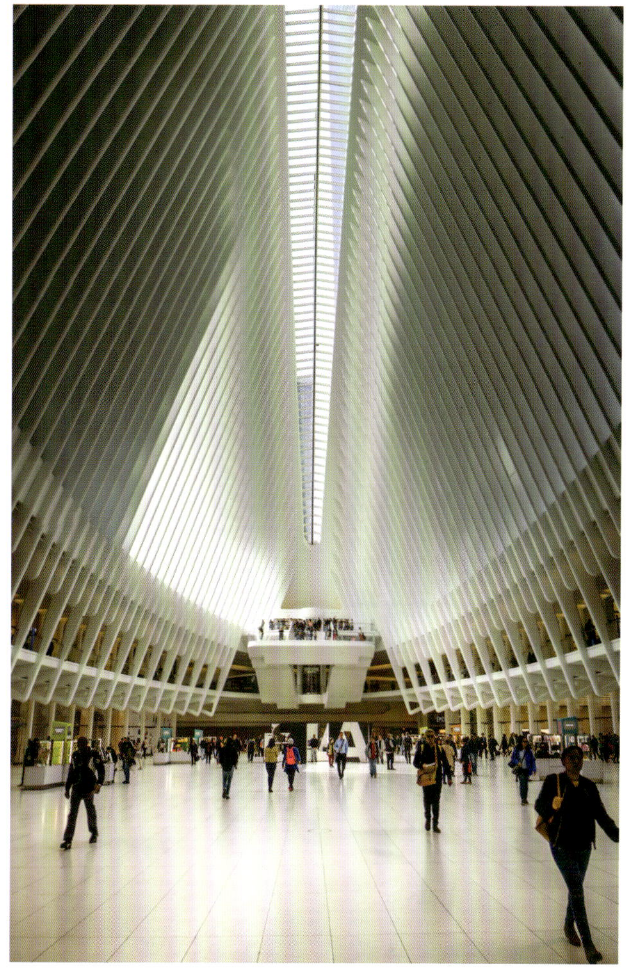

The Oculus at the World Trade Center Transportation Hub.

Though the line has evolved over the years, including influxes of new cars in the 1960s, late 1980s, 2000s, and most recently in 2023, the purpose of connecting popular Manhattan spots with key points in New Jersey remains central. There are Blue and Yellow Lines running from 33rd Street in New York to Hoboken and Journal Square, respectively. Green and Red Lines run from the World Trade Center area to Hoboken and Newark, respectively. Trains run 24 hours a day, seven days a week over the nearly 14 miles of rail. In 2024, the Newark to World Trade Center line rolled out nine-car trains to increase capacity during peak ridership times. In all, there are 13 stations serving the more than 200,000 people who ride every weekday.

On September 11, 2001, all passengers in the World Trade Center station were evacuated before the first of the towers fell in the worst terrorist attack in United States history. Much of the money that has gone into upgrading the PATH system has gone into the new terminal at the World Trade Center site. A $253 million temporary station was opened in 2003 to restore service to the area, but that was just a fraction of the billions of dollars that went into the impressive station that welcomes guests now.

The World Trade Center Transportation Hub rises from the ground like a giant white whale's tail (some say a bird in flight), offering four platforms and a seemingly endless white interior. In the center of it all is the Oculus, a glass and steel structure designed by Spanish architect Santiago Calatrava to serve as the headhouse. Construction on the station began in 2008 and the fourth and final platform was completed in 2016, when the hub officially opened to the public.

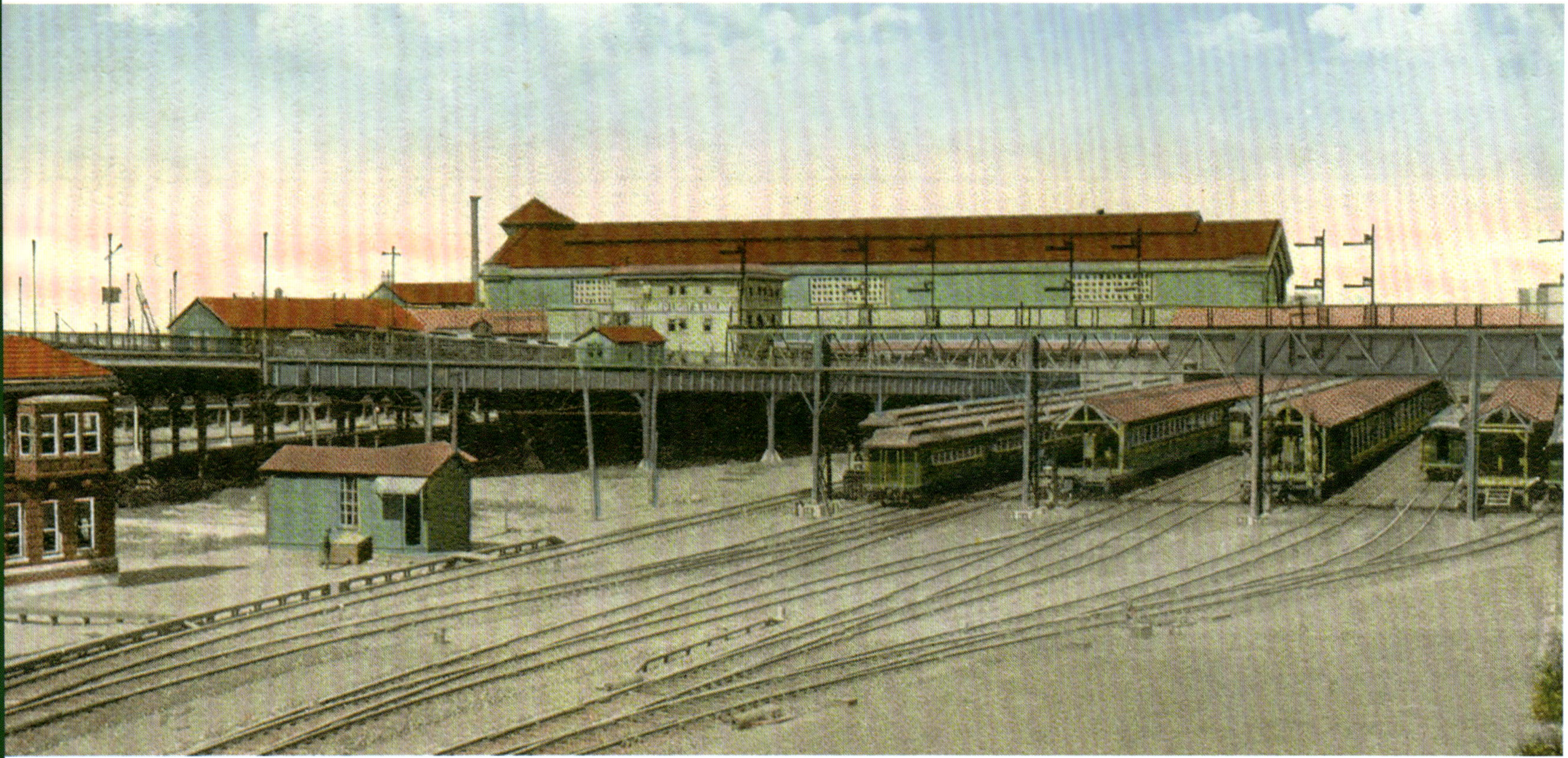

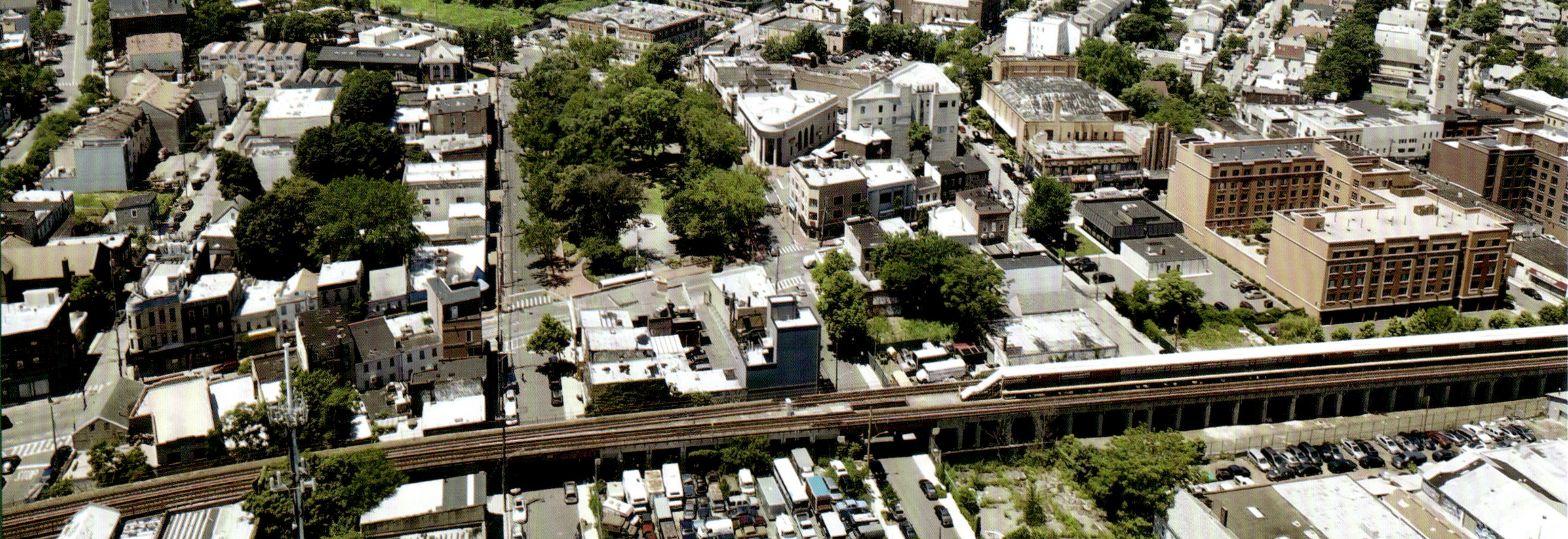

Top: A photo of the St. George station on Staten Island. This station was later destroyed by a fire in 1946.

Bottom: An aerial view of the Tompkinsville station.

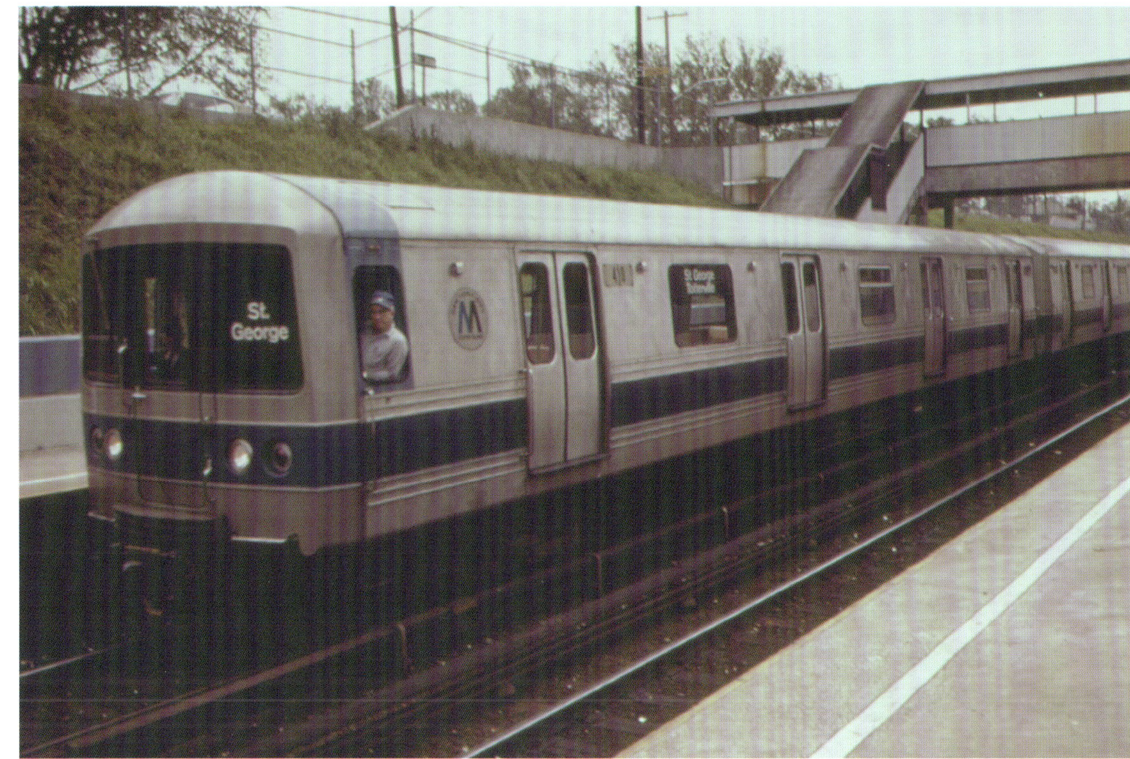

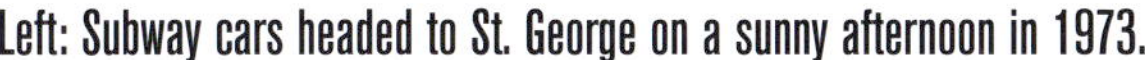
Left: Subway cars headed to St. George on a sunny afternoon in 1973.

Right: A Staten Island Rapid Transit map from 1885.

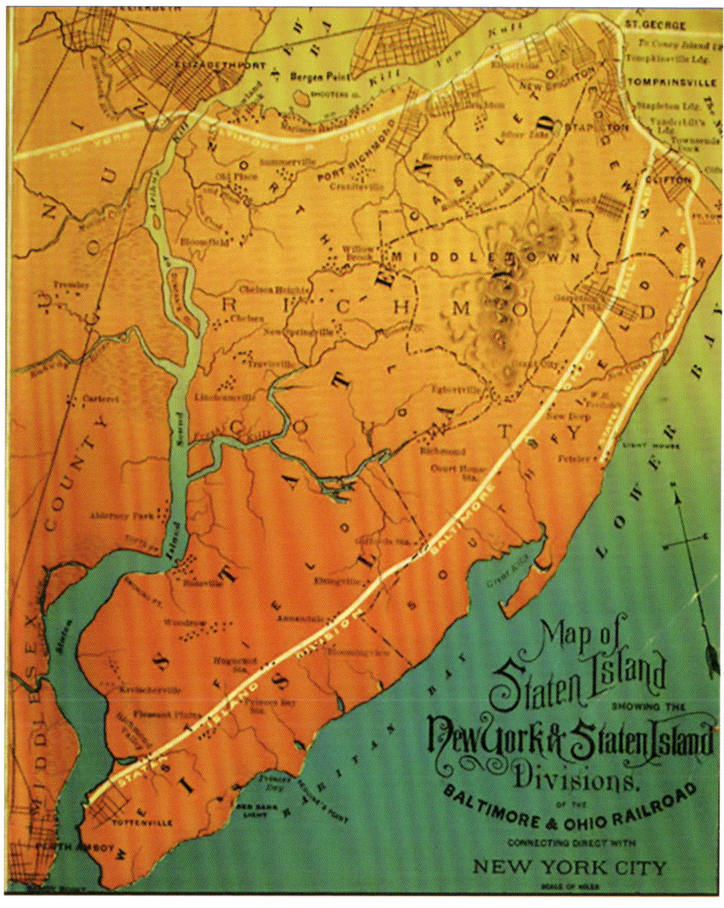

Like many railways in the Northeast, the Staten Island Railway—the one and only commuter railway in the New York City borough of Staten Island—was backed financially by Cornelius Vanderbilt. Vanderbilt was actually born on Staten Island, so this particular project was close to the tycoon's heart. The railway first started moving passengers in 1851, making it one of the longest-running railways in the country.

Originally called Staten Island Rapid Transit, the line was serving more than 10 million passengers per year by the time World War II broke out. Lower-fare buses in the 1940s stole traffic from the trains, and for much of its existence the operation was not a big money maker. By 1971, the famed Baltimore and Ohio (B&O) Railway that had been leasing the Staten Island line for years threatened to end service.

That's when the Metropolitan Transportation Authority (MTA) stepped in and purchased the line. The railway is now run by the Staten Island Rapid Transit Operating Authority (SIRTOA). It's a division of the New York City Transit Authority, which still operates under the MTA umbrella. More than six million riders hop aboard the Staten Island trains annually.

The line starts on the northern end of the island at the St. George Terminal, just a few nautical miles from Manhattan. Trains run all the way across Staten Island to Tottenville at the southern tip, and service is 24 hours. The 14-mile, 21-stop route takes 42 minutes in all, but express trains cruise nonstop from New Dorp to the St. George Terminal and from St. George to Great Kills each morning, and from St. George to Great Kills in the evening. Trains arrive every 15 minutes during rush hour for both express and local service. The evening and overnight hours offer local service only, with trains running every half hour.

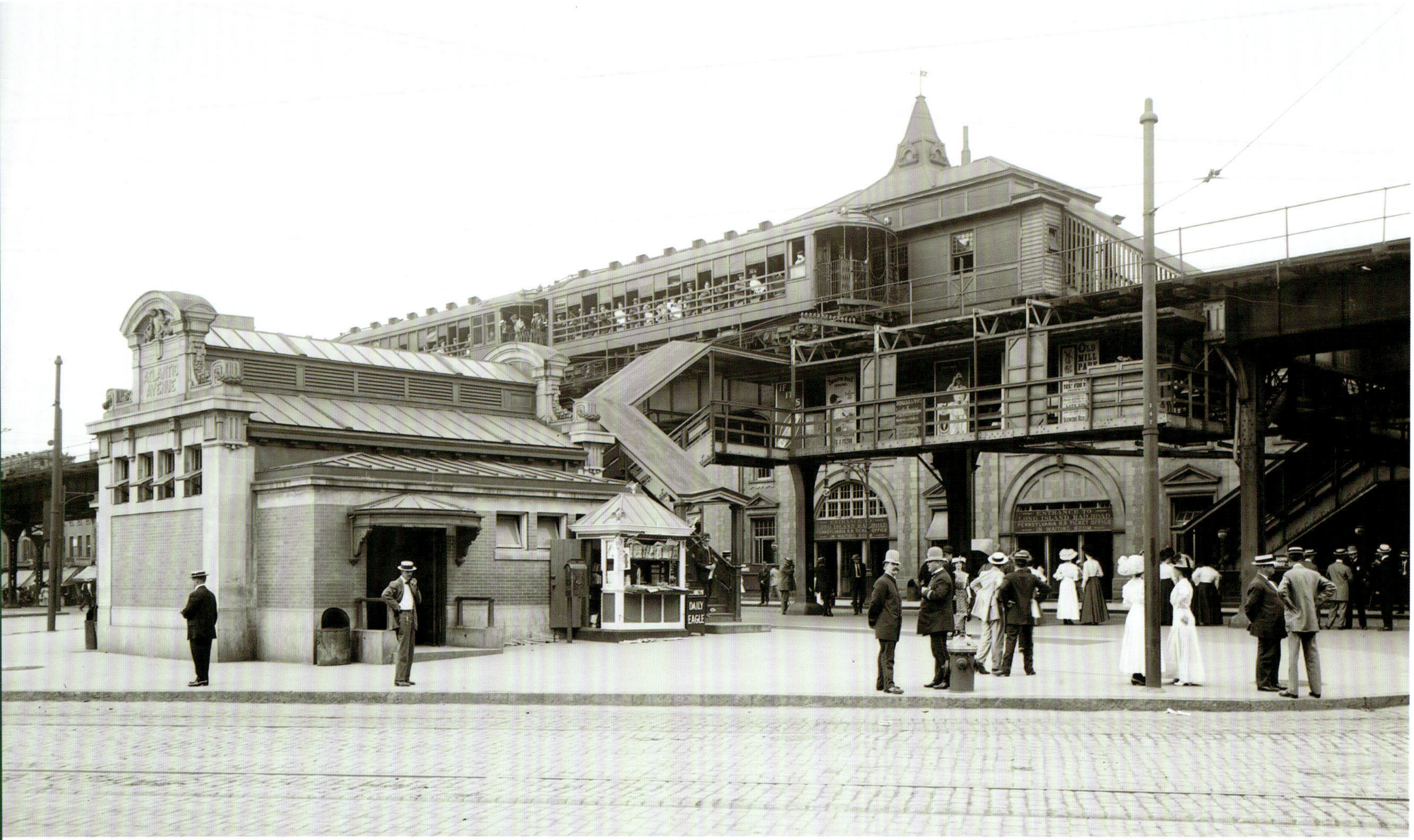

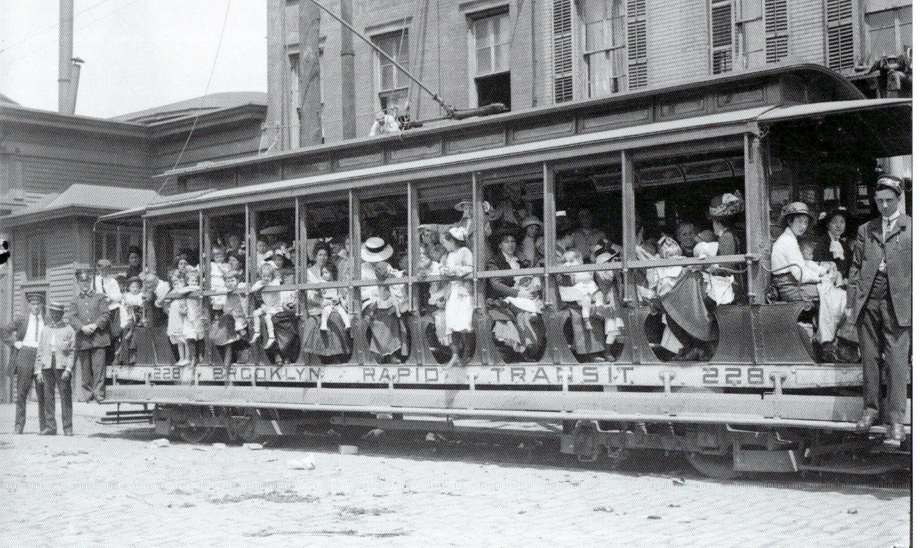

Top: A photo of the Atlantic Avenue station in Brooklyn circa 1910. The station is located at the intersection of Fourth, Atlantic, and Flatbush Avenues and is used today by IRT's Eastern Parkway Line.

Bottom: A BRT trolley car transporting families to Sea Breeze, Coney Island, in 1913.

Right: A train of the BRT system operating on the upper deck of the Manhattan Bridge.

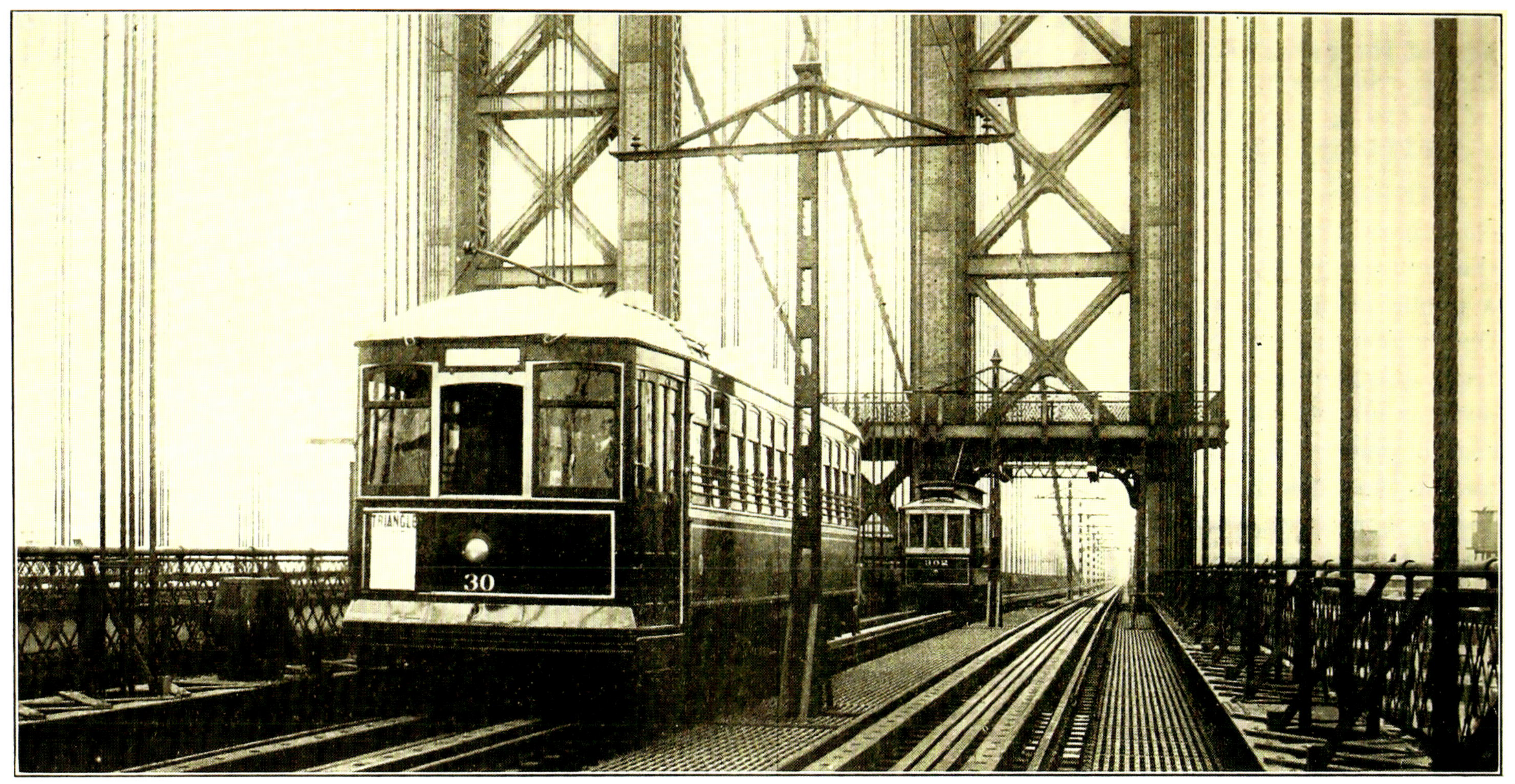

Though it has been more than a century since its dissolution, the Brooklyn Rapid Transit Company (BRT) is still remembered as an early leader in America's rail industry. In fact, it earned a *B* on the New York Stock Exchange—a prestigious designation—before fading into the history books. Some of the major hotels in Brooklyn were looking for ways to get customers to their check-in desks, and rail was the way to go. It was the Brooklyn Union Elevated Railroad Company that built the first elevated lines in the New York borough. Trains began spanning the Brooklyn Bridge to Park Row in 1883, connecting Brooklyn residents to New York's commerce hub.

Several smaller companies jumped into the race to bring rapid transit and street car service to Brooklyn. Seeing this, Timothy S. Williams created the BRT in 1896 as a holding company that quickly began buying up these smaller entities. The BRT took over the Sea Beach Railway, Nassau Electric Railroad, and Brooklyn Union Elevated Railroad. By the turn of the century, all but one Brooklyn line fell under the BRT umbrella. The BRT oversaw the electrification of Brooklyn's rail lines and a big expansion along the "Els." It made the popular move to begin running trains to Coney Island. It also operated a four-track subway leaving Chambers Street, running under Centre Street, and heading east under what today are Kenmare and Delancey Streets.

The BRT did it all in a very progressive way, allowing its subsidiary companies to make many of the operational decisions—even to design and acquire rolling stock. With World War I-era inflation and a tragic accident on Malbone Street that killed at least 93, the BRT came upon hard times. The BRT entered bankruptcy in 1919, and was reorganized four years later as the Brooklyn–Manhattan Transit Corporation.

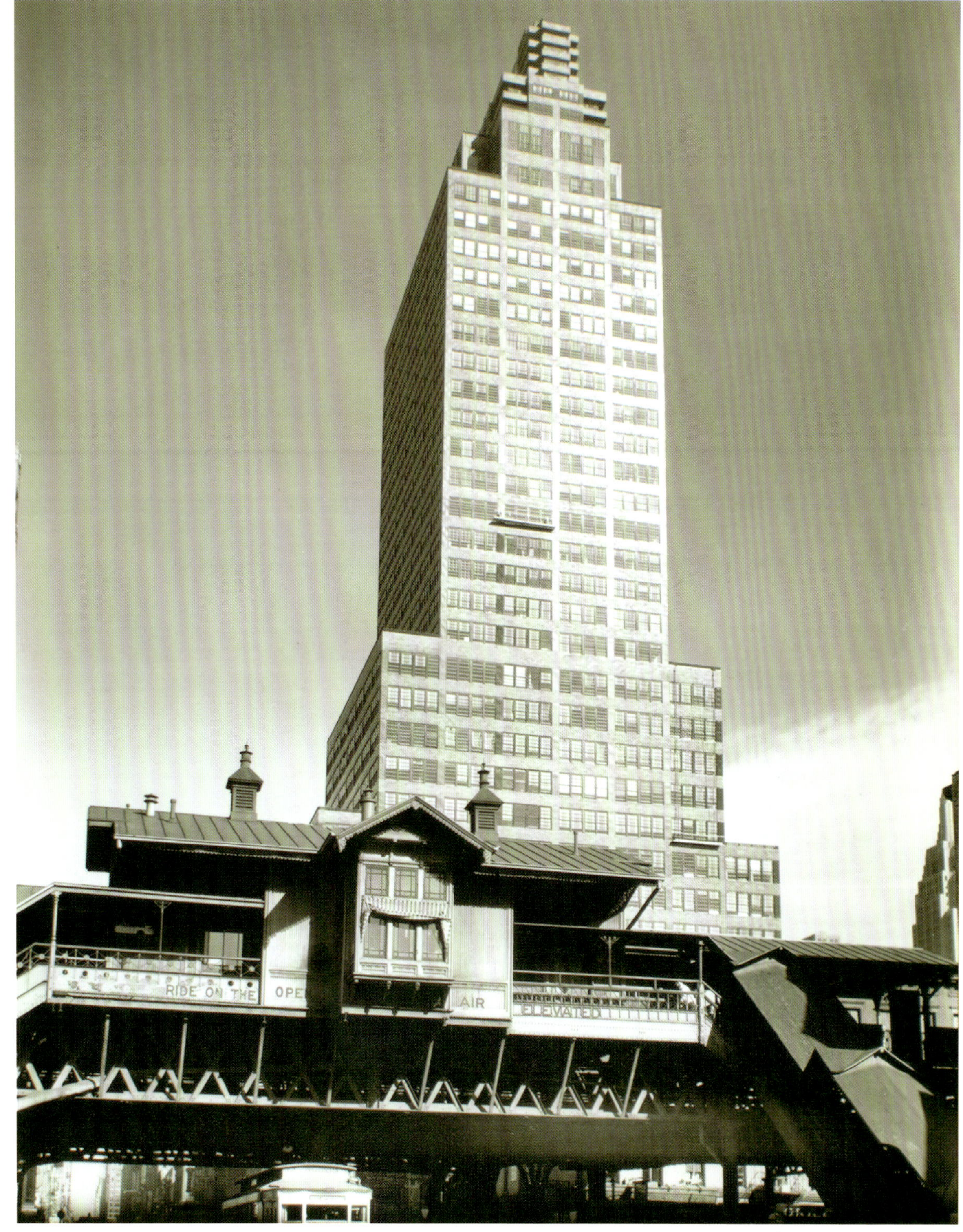

The McGraw Hill Building towering above the 42nd Street station in 1936. The station was closed in 1940.

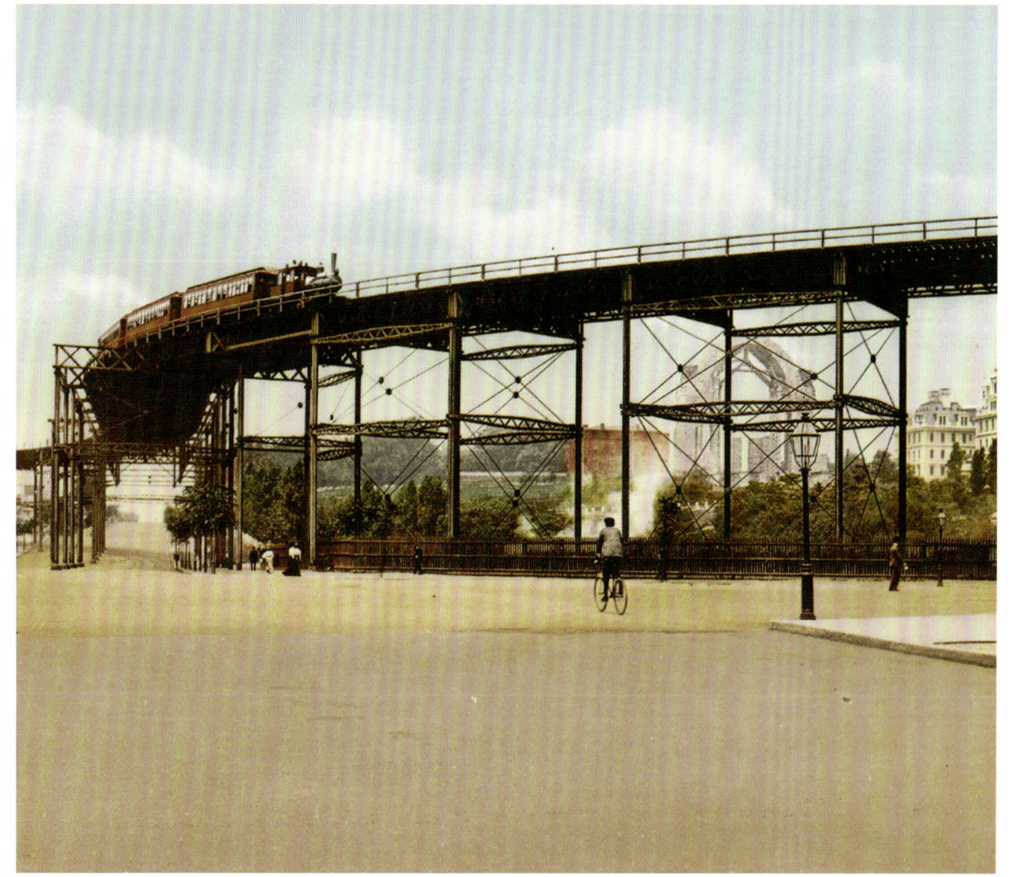

The IRT Ninth Avenue Line traveling over what was known as the "Suicide Curve" near 110th Street.

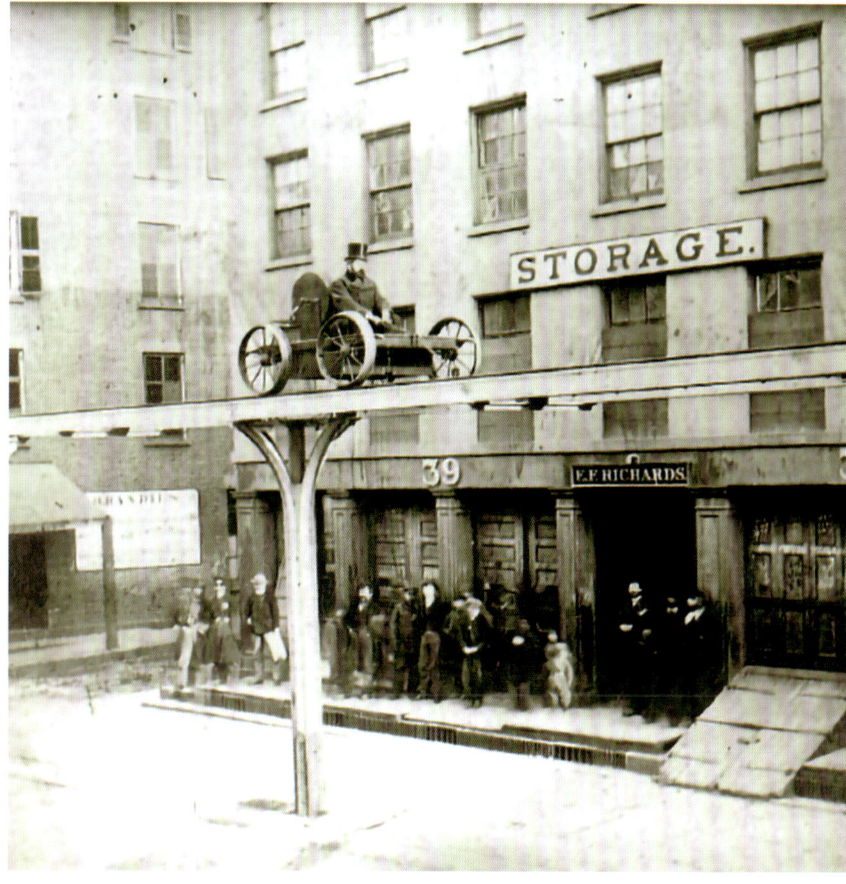

Charles Thompson Harvey testing the track along a portion of Greenwich Street in December of 1867.

For a while, New York City's mass transit system ran well above ground, rather than beneath it. The Interborough Rapid Transit (IRT) Ninth Avenue Line, often called the Ninth Avenue El, was New York's first elevated rail. It opened in 1868 as the West Side and Yonkers Patent Railway, an experimental, cable-powered single track that ran from Battery Place on the south end of Manhattan up Greenwich to Cortlandt Street.

"The car ran evenly from Battery to Cortlandt," *The New York Times* reported, "starting at a rate of five miles an hour, and increasing to a speed of ten miles." The train ran using one-mile cable loops powered by steam engines in the basements of buildings next to the track. There were several issues with the cars grasping the cables, and as time went on a better steam system was implemented, followed by the arrival of electricity in 1903.

That's also the year the IRT took over the Ninth Avenue Line and several other Manhattan Railway lines. Expansion soon followed. The Ninth Avenue Line was extended to 167th Street in 1918, using a bridge built by the New York and Northern Railroad. The next year, service was stretched to the Woodlawn station. One of the most famous things about the Ninth Avenue Line was its five-cent fare, which was in effect for almost the full run of the line (save for a failed increase to 10 cents in the late 1800s).

There was no surviving the Great Depression for the IRT, however. The end of the line for the company came in 1932, although its trains ran until 1940 and one of them—the Polo Grounds Shuttle—survived even longer. It was a huge success in getting baseball fans to New York Giants games, and it remained in use until the Giants moved to San Francisco in 1958.

IRT Broadway–Seventh Avenue Line

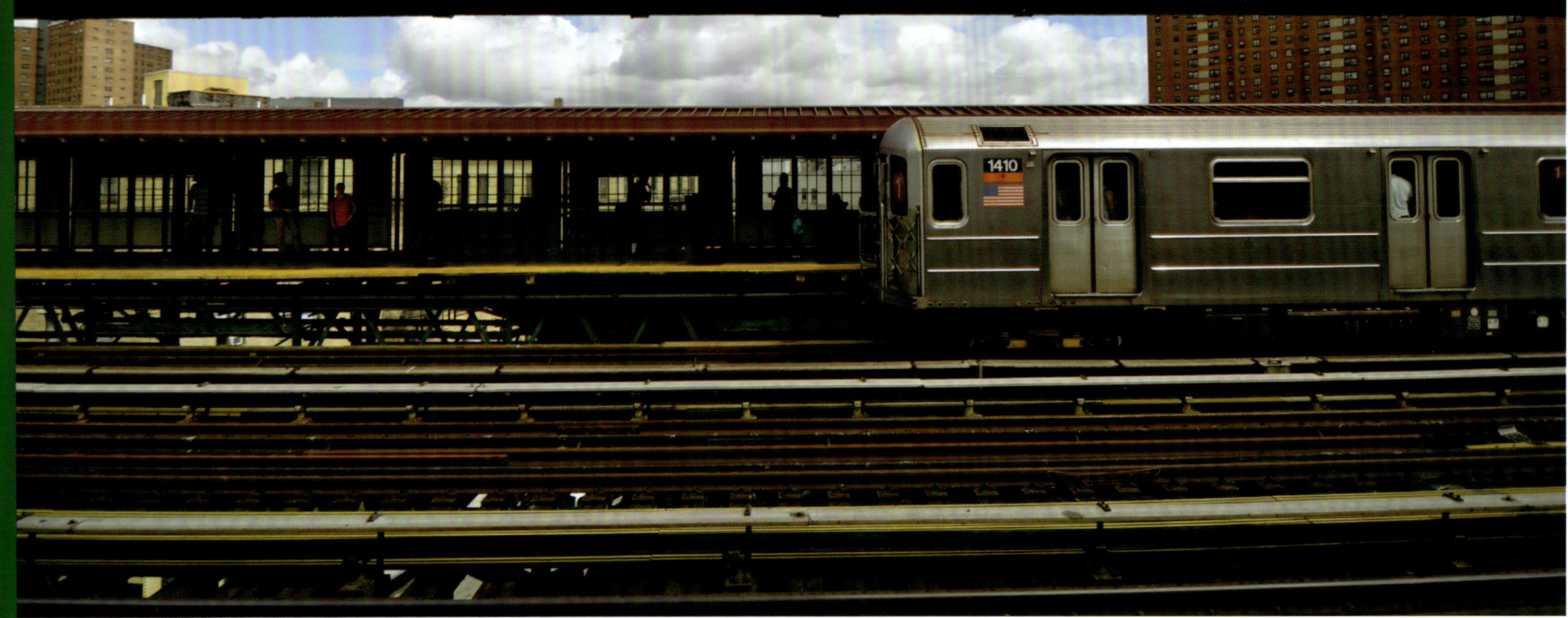

More than one million people hop aboard the IRT Broadway–Seventh Avenue Line every weekday in New York City. Also known as the West Side Line because it traverses Manhattan's west side, it is among the best-known and most ridden lines in New York, which has the busiest mass transit system in the country. It runs from South Ferry in Lower Manhattan to Riverdale, Bronx, giving New Yorkers and visitors access to some of the most popular spots in the city.

Another distinction held by the IRT Broadway–Seventh Avenue Line is that it still uses the oldest stretch of track in New York. The first part of the line, north of 42nd Street in Times Square, was built between 1904 and 1908 as part of the first subway line in the city. In fact, it was called the "First Subway" plan, drawn up in 1898 by chief engineer William Barclay Parsons.

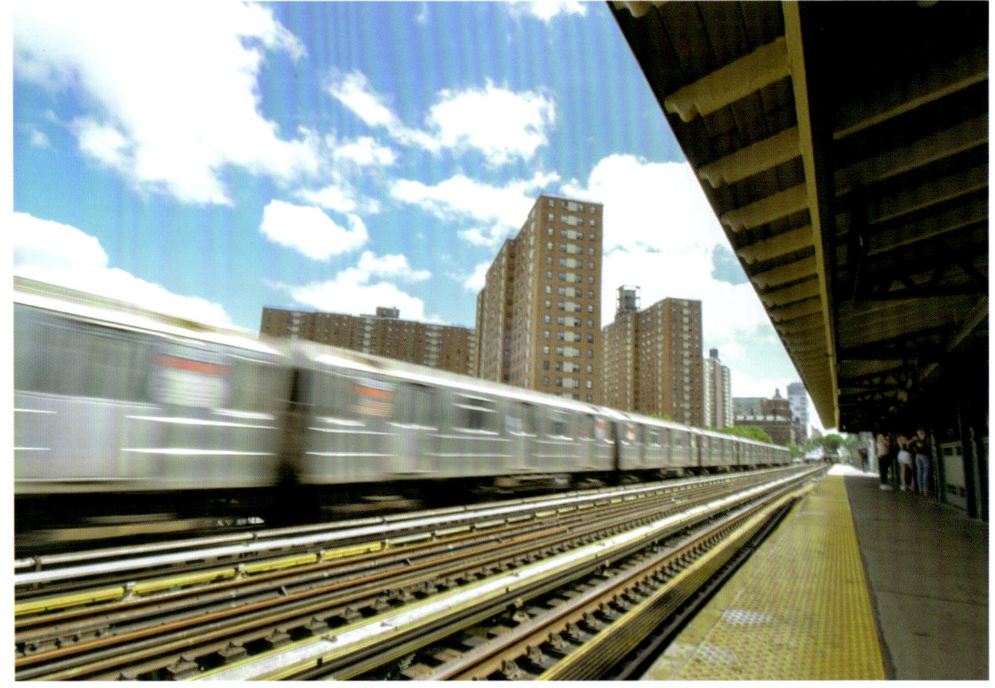

Top: The IRT Broadway–Seventh Avenue train at the station on 125th Street.

Bottom: The IRT Broadway–Seventh Avenue train running north from the station at 125th Street.

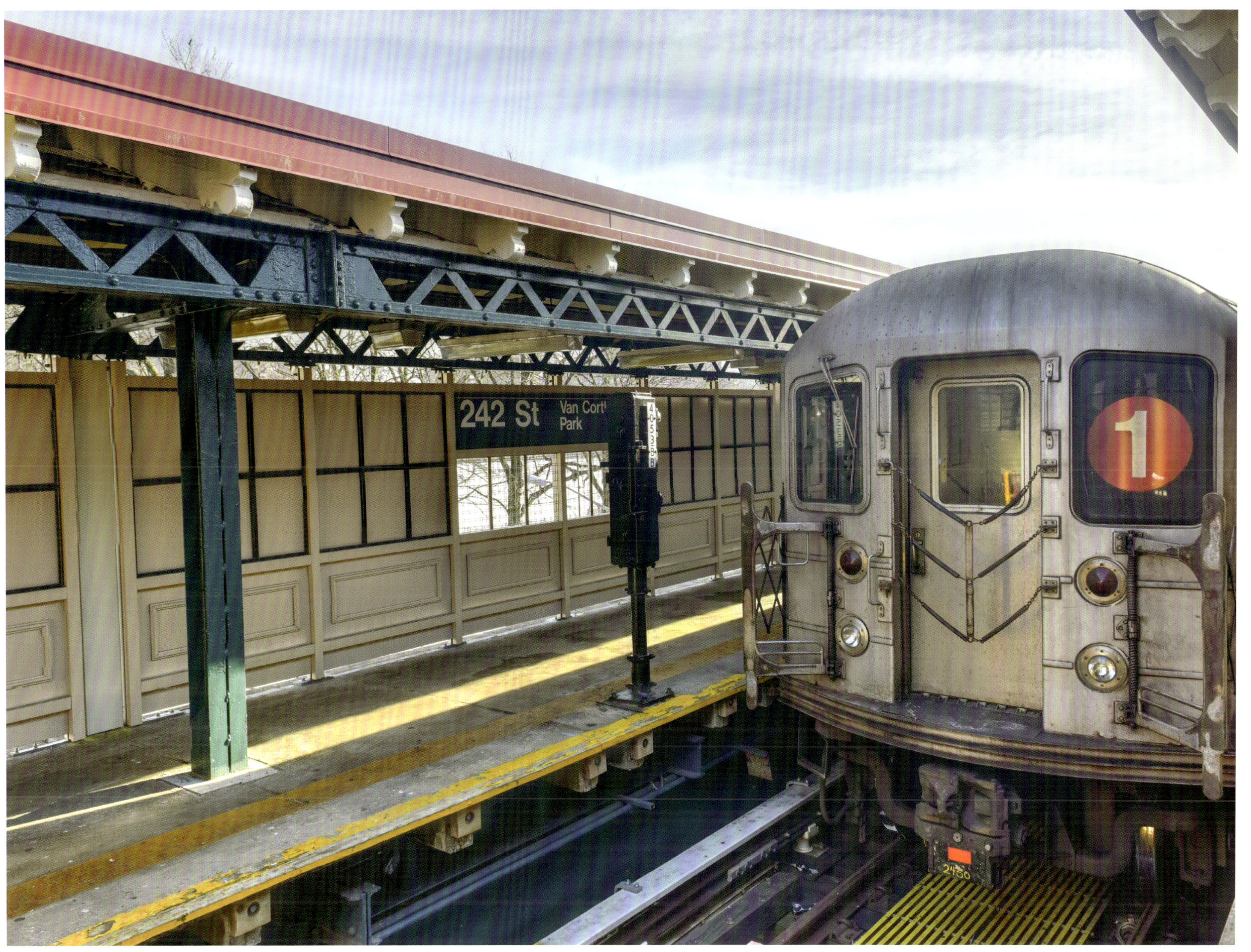

The northernmost station on the line, Van Cortlandt Park, 242nd Street, is a wonder of Victorian Gothic architecture with its red control house and ornamented stairs. It was designed by the same firm that designed the Cathedral of St. John the Divine.

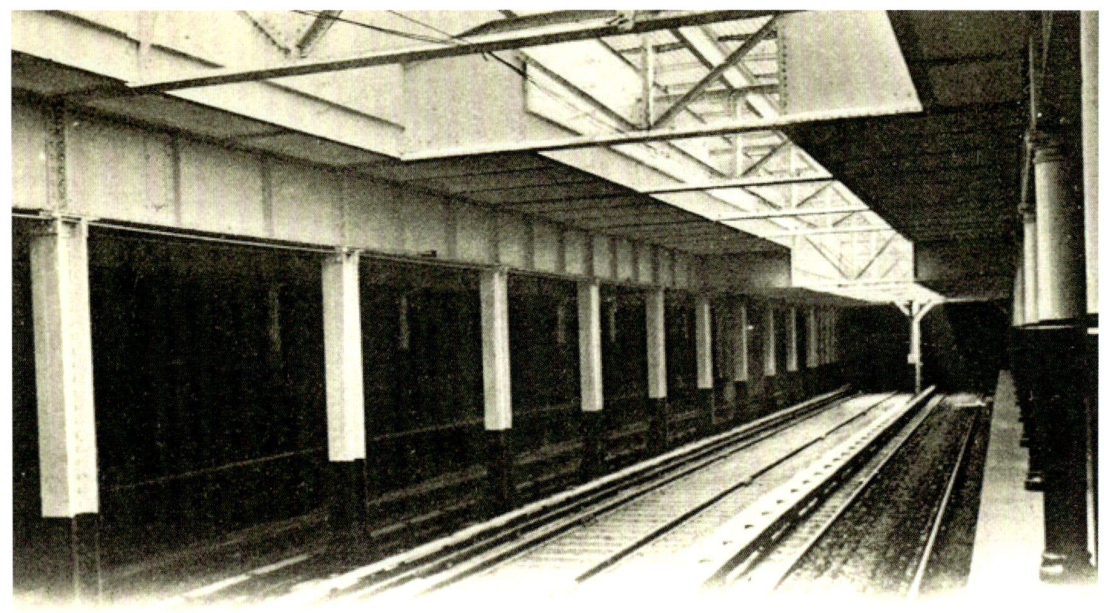

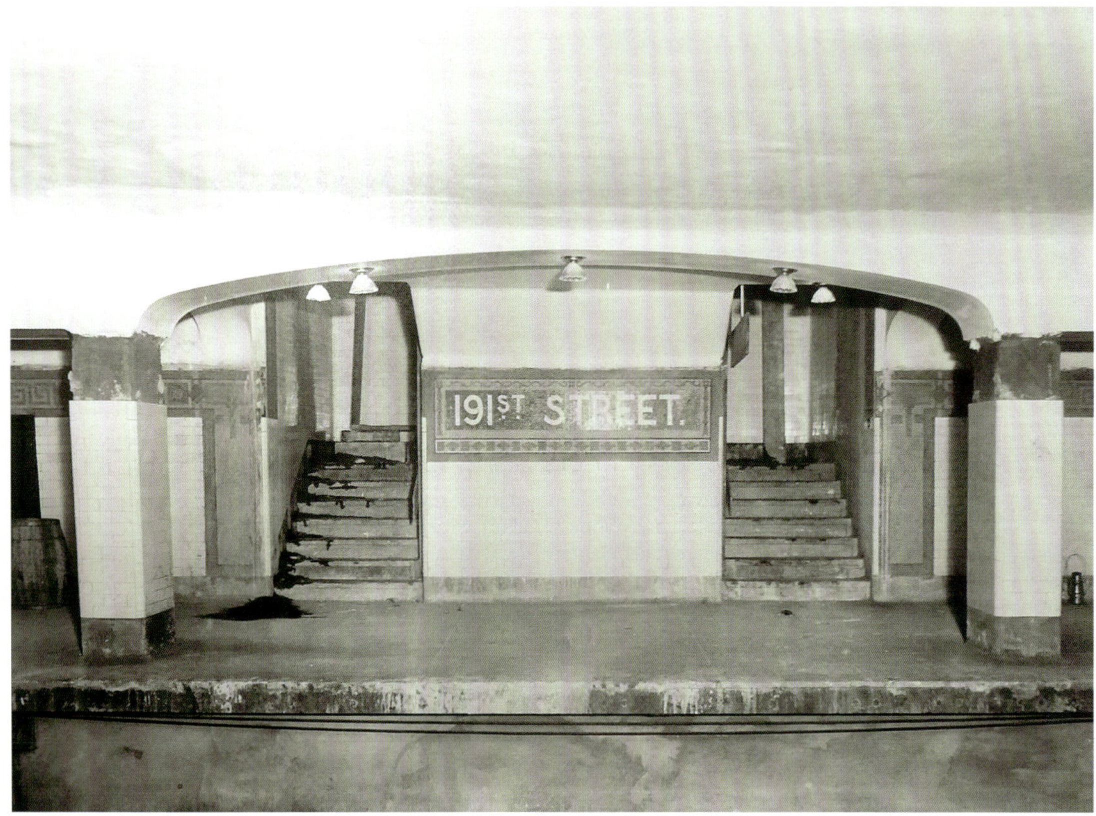

The IRT Broadway–Seventh Avenue Line owes its roots to two main construction contracts from the early twentieth century. In addition to that First Subway deal of 1904 and several extensions to it (there were 15 subsections), the Broadway–Seventh Avenue Line underwent massive expansion and reconstruction between 1910 and 1920 thanks to what became known as Dual Contracts, a deal between the IRT, the Brooklyn Rapid Transit-owned New York Municipal Railway, and the City of New York.

The Dual Contracts section included the Seventh Avenue section south of Times Square and a Brooklyn section known back then as the Park Place, William and Clark Street Subway. It put stations at Wall Street, Fulton Street, and Park Place. By 1919, the West Side Line stretched to Atlantic Avenue in Brooklyn via the Clark Street tunnels that connected to the Wall Street station.

From those beginnings emerged a line now known for its tomato-red color coding and the numbers 1, 2, and 3—1 for local service, and 2 and 3 for express trains. There have been several changes to the line since its original structure, of course, including some rerouting to address overcrowding issues, the lengthening of several platforms, and the accommodation of bigger, better cars.

One set of upgrades came about as an unexpected and tragic need. The Cortlandt station was destroyed in the September 11, 2001, terror attacks on the World Trade Center. It was rebuilt as the WTC Cortlandt station, which offers a connection to the PATH World Trade Center station, helping to get visitors to One World Trade Center and the 9/11 Memorial and Museum.

Top: A postcard showcasing the 116th Street station the year after it opened in 1905.

Bottom: A photo of the 191st Street station in 1911, featuring the iconic mosaic tiling that is still used today to call out the stops along the line.

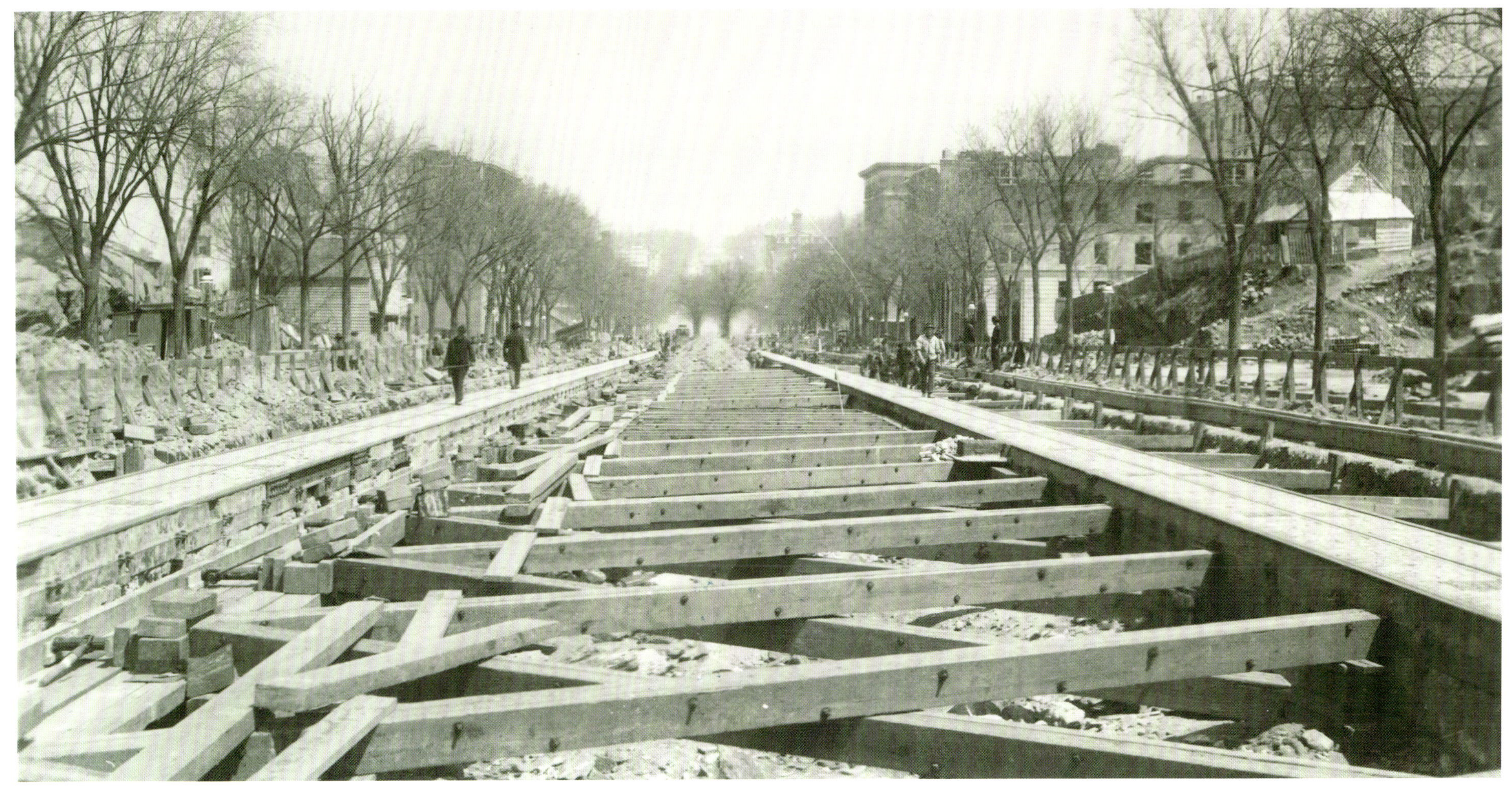

Though mostly a subway, the Broadway–Seventh Avenue Line also features the only elevated stations in Manhattan. The 125th Street station is elevated, as are the rails north of Inwood. There are 44 stations in all along the route from South Ferry to 242nd Street, delivering people not only to and from their jobs but also to and from some of the most popular spots in the city. Times Square, Lincoln Center, One World Trade Center, and the Brooklyn Bridge are just a few.

Construction of the subway in 1901 along Broadway at 122nd Street.

Some of the stations themselves are also worth a look. The 191st Street station, for example, sits further below street level—almost 200 feet—than any station in New York. It also contains a pedestrian tunnel to Broadway, where walkers are treated to bright and colorful mural art along the walls.

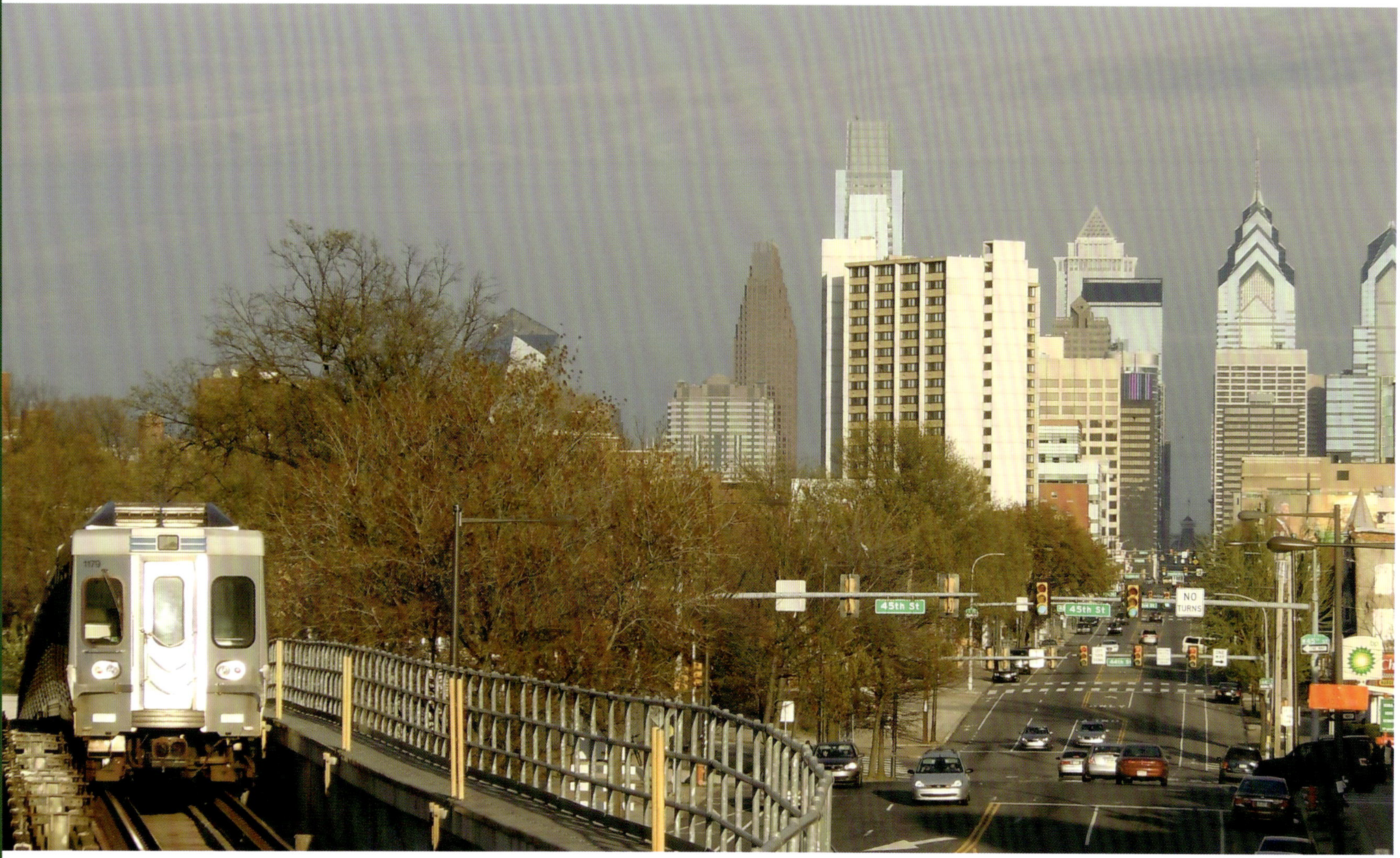

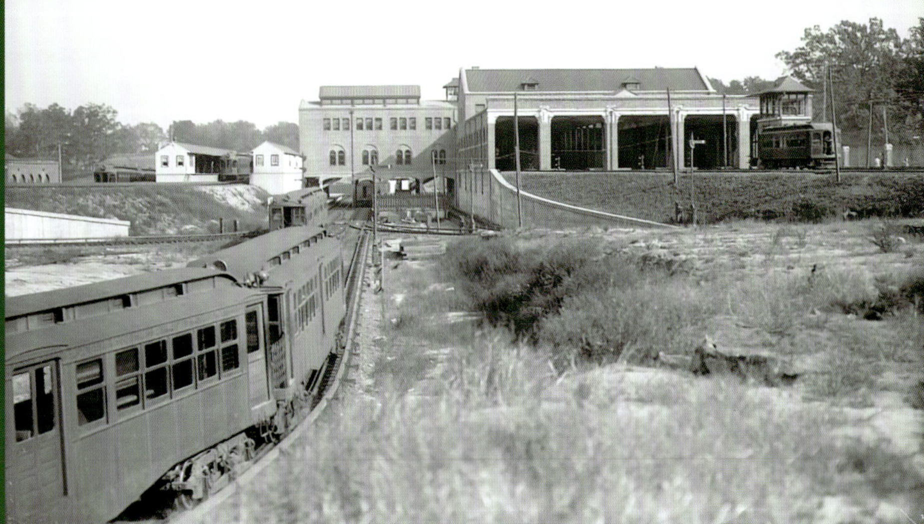

Top: The Market-Frankford Line leaving the 46th Street station with a view of Philadelphia's Center City.

Bottom: A photo of the 69th Street terminus from 1908.

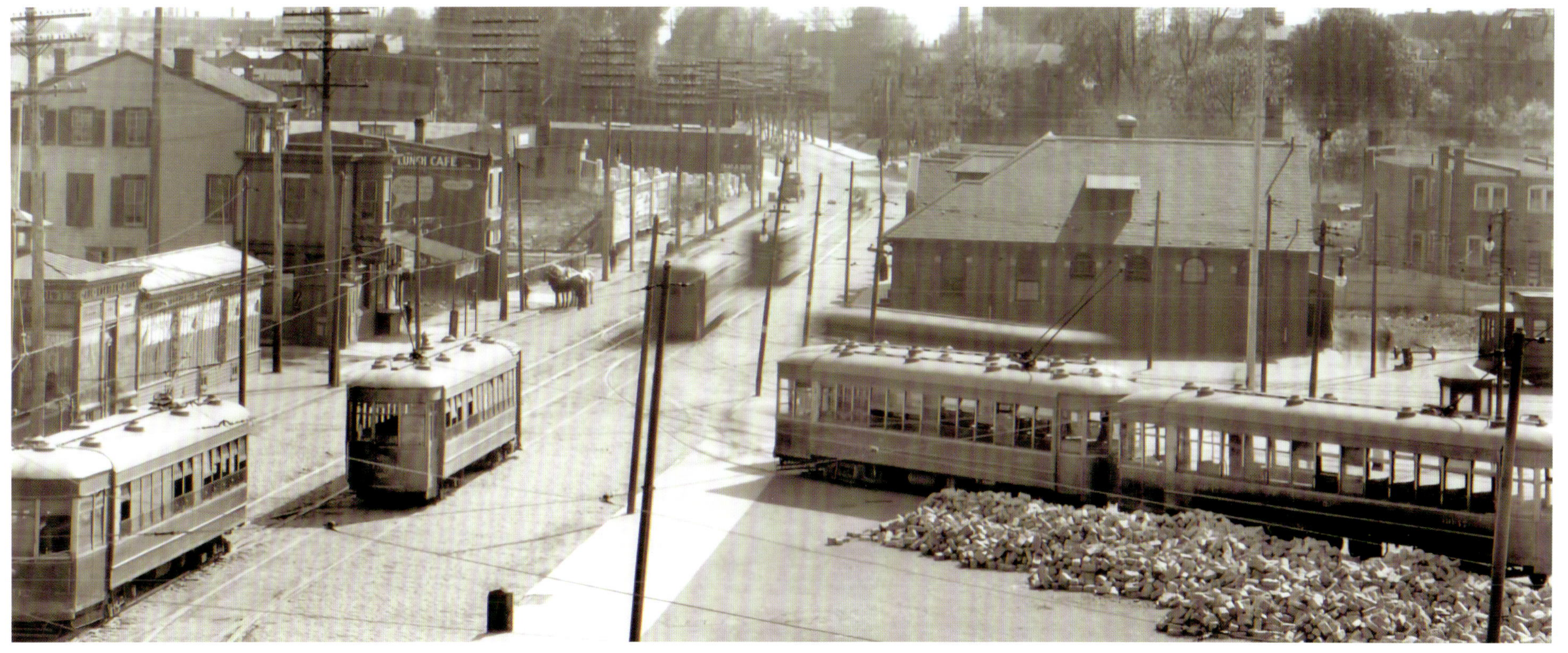

The Frankford terminus in 1918 before the elevated track was built.

Philadelphia-based 1970s band The American Dream put out a catchy tune entitled, "Can't Get to Heaven on the Frankford El." True enough, but more than 100,000 Philly commuters use the rail system, formally called the Market–Frankford Line, to get to and from work on weekdays.

Philadelphians have dubbed it the "L" for years, though the line dips underground between 2nd and 46th Streets and runs above ground the rest of the way. In all, the busiest line in the SEPTA (Southeastern Pennsylvania Transportation Authority) system winds 13.6 miles from the 69th Street Transportation Center in Upper Darby, just outside West Philadelphia, to the Frankford Transportation Center in Near Northeast Philadelphia. It has 28 stations.

The Market–Frankford Line contains some of the oldest rail stretches in American subway history. The tunnel from City Hall to the portal at 23rd Street and the bridge over the Schuylkill River were built between 1903 and 1905. The line evolved over the years, of course, keeping up with urban growth and commuter demand. It was completely overhauled from 1988 to 2003, at a cost of almost $500 million to SEPTA, between Frankford Transportation Center and the 2nd Street portal.

The line's rolling stock has been continually replaced over the years, with air-conditioned cars containing LCD signage and automated announcements hitting the tracks in the late 1990s. SEPTA launched some of the first commercially deployed battery storage systems powered by regenerative braking at the Letterly and Griscom Substations in 2012 and 2014, respectively. In 2016 SEPTA received the Pennsylvania Governor's Award for Environmental Excellence for implementing the wayside energy storage systems, and the transit authority continues to plan more sustainability improvements.

One update that won't happen is the conjunction of the Market–Frankford Line with either the Broad Street Subway or Norristown High Speed Line. That's because the "L" is unique in its use of thicker, 5'2 ¼" trolley gauge rails compared to the more standard 4'8 ½" gauge. The "L" also uses a bottom-running electric third rail for its operation.

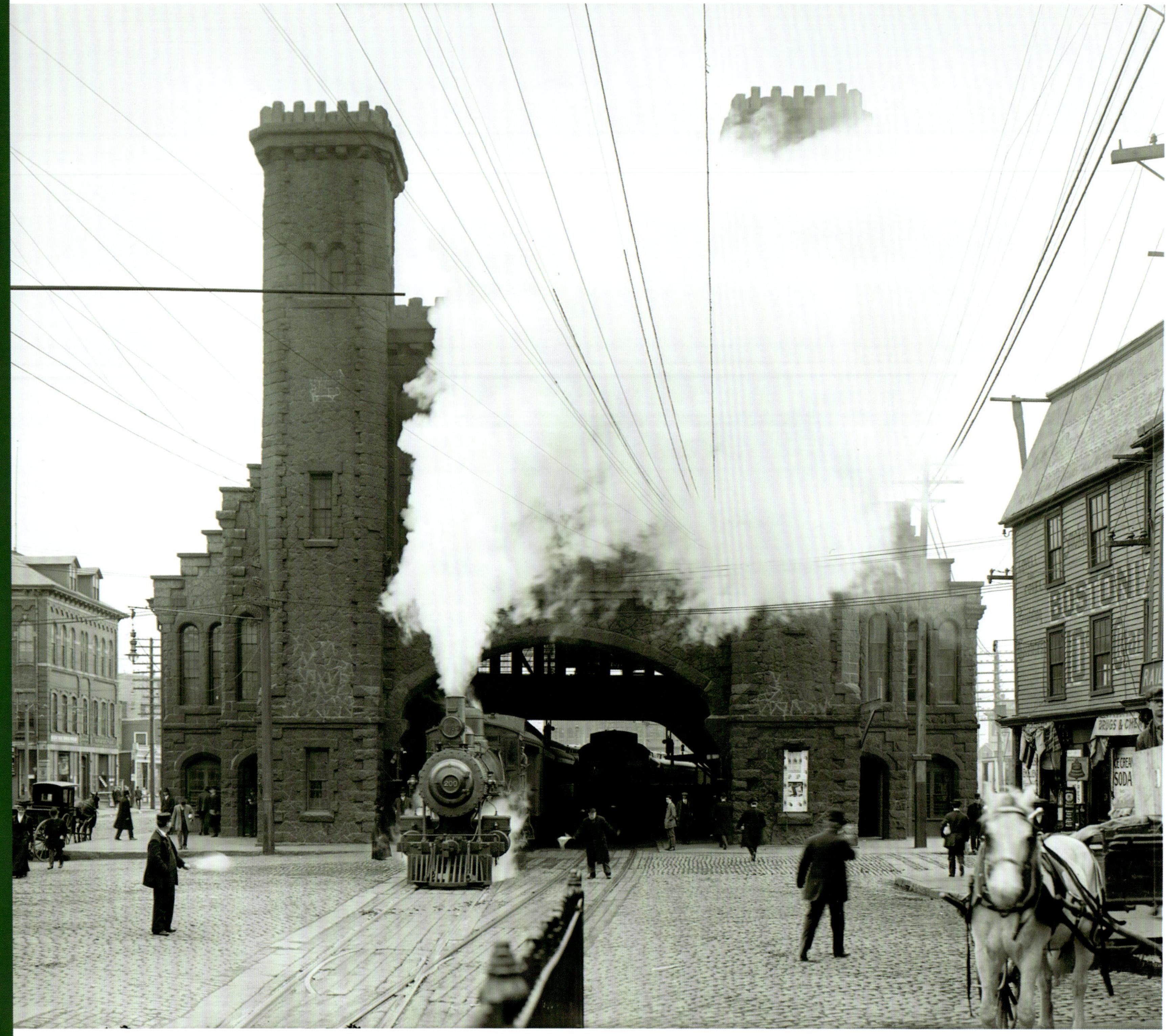

A train leaving the Salem station on the Eastern Line in 1910.

The Boston and Maine Railroad (B&M) was a fixture in New England and beyond for more than a century. At the peak of its dominance in delivering passengers and freight all over the northeast, the B&M maintained more than 2,300 miles of track, 1,200 steam locomotives, and a workforce made up of 28,000 employees. And it all started with a simple mission: make it easy for people to get from Boston to Portland, Maine.

The B&M opened its first segment in 1840 and began buying up several smaller railroad companies to fuel its massive growth. The B&M merged with the Maine, New Hampshire and Massachusetts and the Boston and Portland Railroads in 1842, retaining the B&M name. Its early rival, the Eastern Railroad, was also eyeing a line from Boston to Portland, but the rivalry ended when the B&M leased Eastern in 1883 and ultimately purchased it in 1890. So, not only did the B&M win the race to Portland, it also became a force in the industry, connecting hundreds of cities and towns in New York, Maine, New Hampshire, Vermont, and Massachusetts with its steam trains.

Over the years, the B&M proved itself to be a pioneer in diesel power, switch and signal mechanics, automatic stopping technology, and centralized traffic control, among others. It earned acclaim for building the Hoosac Tunnel in western Massachusetts. At the time of its 1875 completion, the almost five-mile-long tunnel was the second-longest in the world, and it remains in use today.

The B&M was also ahead of its time when it came to marketing. It touted things to do and places to visit along its routes in its company magazine and newspaper advertisements, helping popularize New England resorts, scenic getaways, and even winter sports. In this way, the railroad was an active partner in the success of many small towns. Like most railroads, the B&M struggled during the Great Depression, and with the growing popularity of the automobile thereafter. It survived longer than most, however—until 1983, to be precise. That's when the company was purchased out of bankruptcy by Guilford Transportation Industries. It later became Pan Am Railways, which still operates over many of the B&M rails.

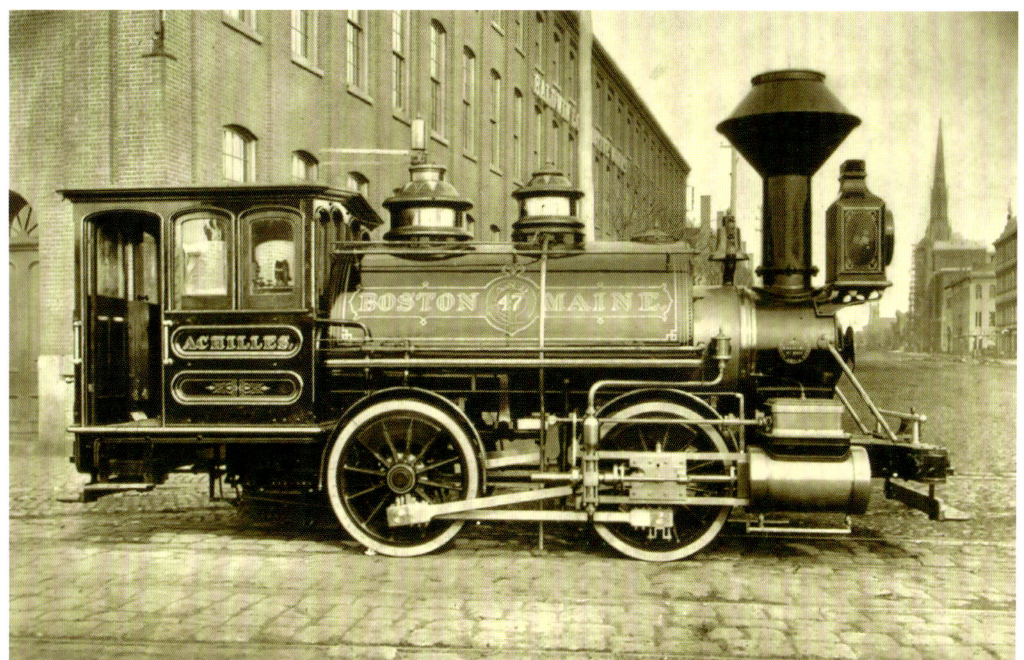

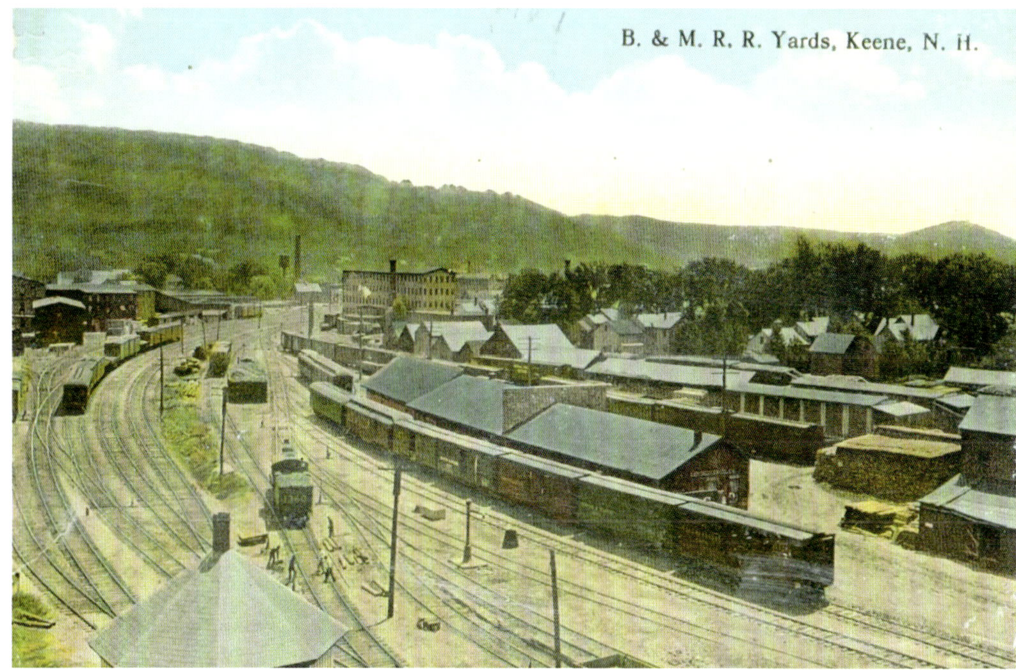

Top: The B&M was chartered as a company in New Hampshire on June 27, 1835, as a successor to the Andover and Wilmington Railroad.

Bottom: A postcard of the B&M rail yard in Keene, New Hampshire.

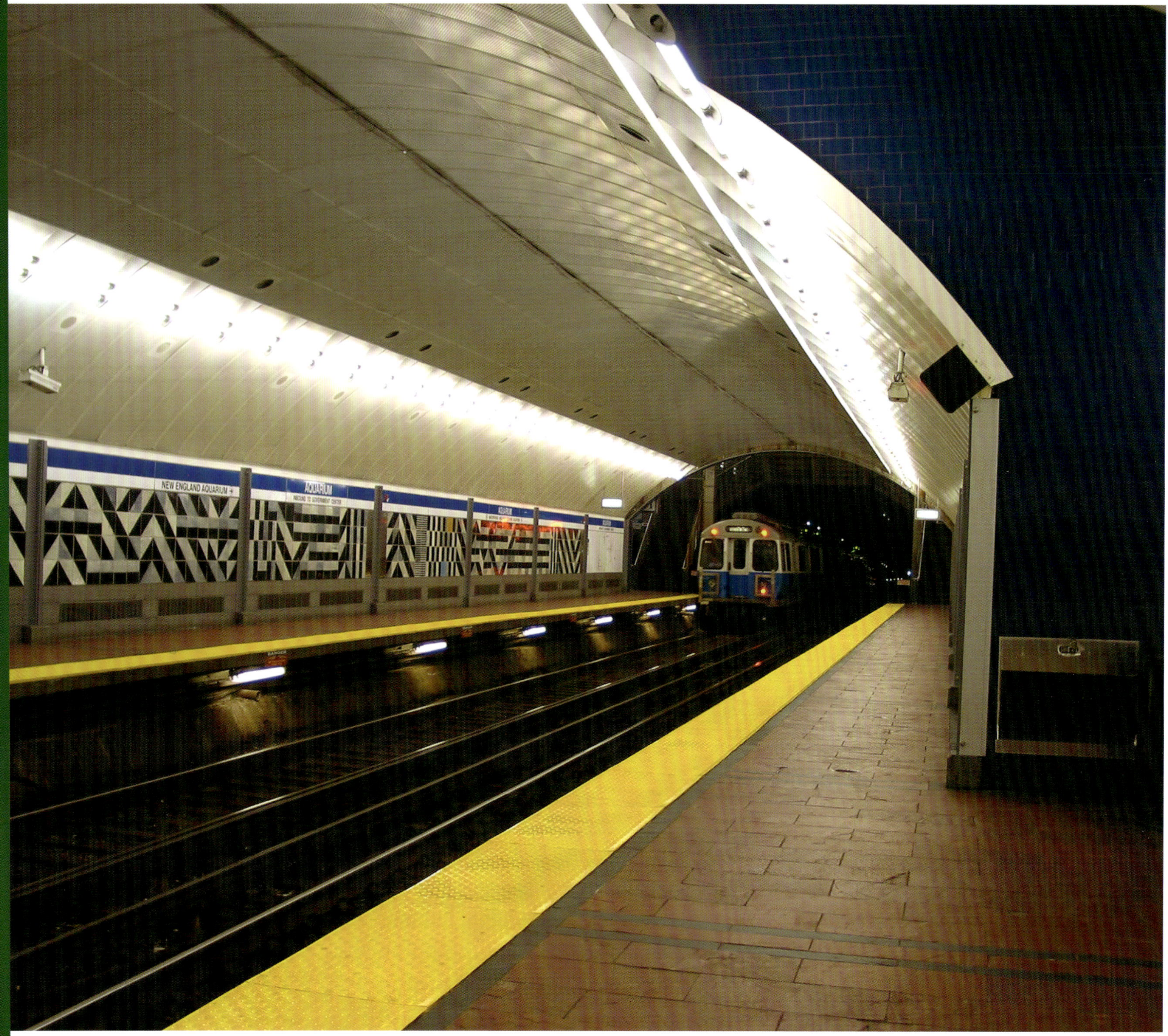

The Blue Line emerging from the East Boston Tunnel, which passes under the Boston Harbor, at the Aquarium station.

When the Massachusetts Bay Transportation Authority (MBTA) went to color-code its transit lines in 1967, the Blue Line got its name because it passes under the blue waters of Boston Harbor. Some 63 years earlier the East Boston Tunnel was constructed as a streetcar tunnel in 1904. The East Boston Tunnel would later become a part of the Blue Line's route.

While the sections of the Blue Line tunnel came about in that manner, the surface lines owe their history to the narrow-gauge Boston, Revere Beach and Lynn Railroad in the 1940s. That part of the route was initially expected to reach Lynn, but it was stopped short at Wonderland, which remains the northernmost stop on the MBTA Blue Line (though expansion to Lynn is under discussion).

The Blue Line employs standard gauge heavy rail just like the Orange and Red Lines, but it is unique among the Boston "T" lines in that its vehicles—shorter and narrower than those on other lines—use both third rail power and pantograph electrical current pickup from overhead catenary wires. The switch occurs at Airport station. The pantograph was a weather-induced addition, as it's not uncommon for the third rails to ice over during the winter months.

The Blue Line is the MBTA's shortest line, at just six miles, and is the only rapid transit line that terminates in downtown Boston at the Bowdoin station—one of 12 stations along the Blue Line. The line has not been expanded since the 1950s. In addition to talk of extending the northern terminus to Lynn, there have also been discussions about stretching the Blue Line from the Bowdoin stop to the Charles/MGH station on the Red Line.

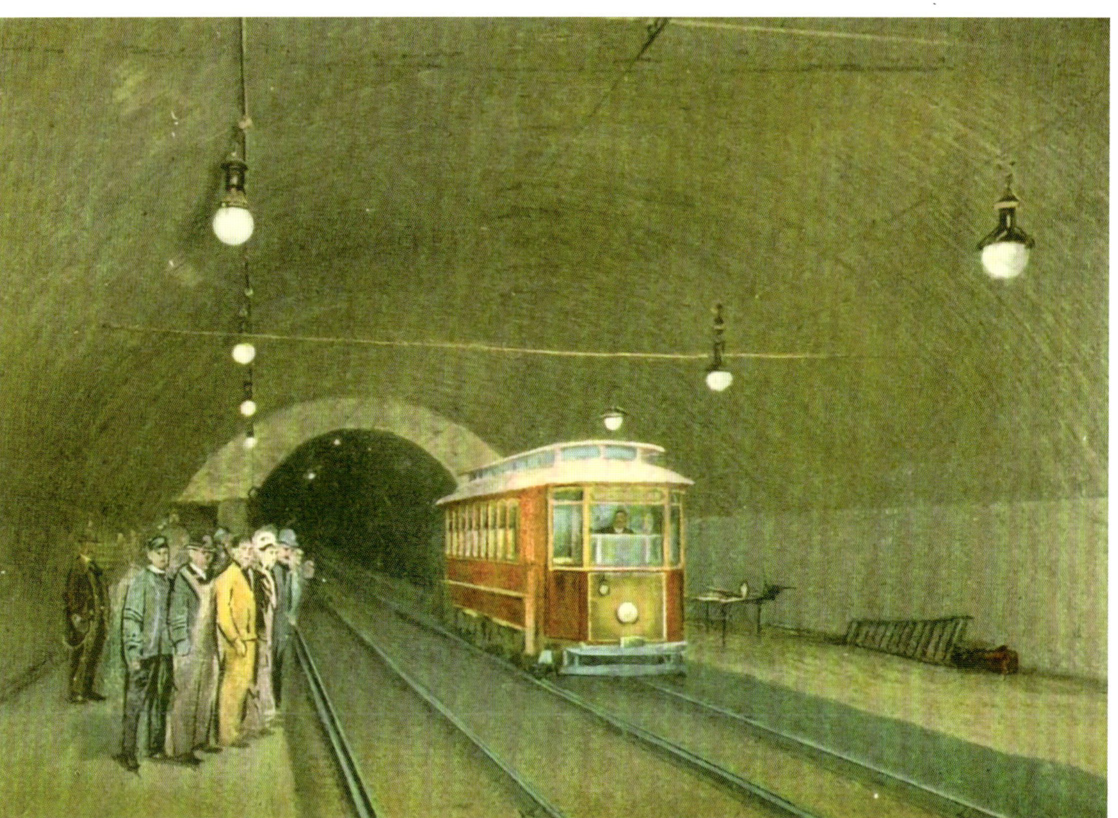

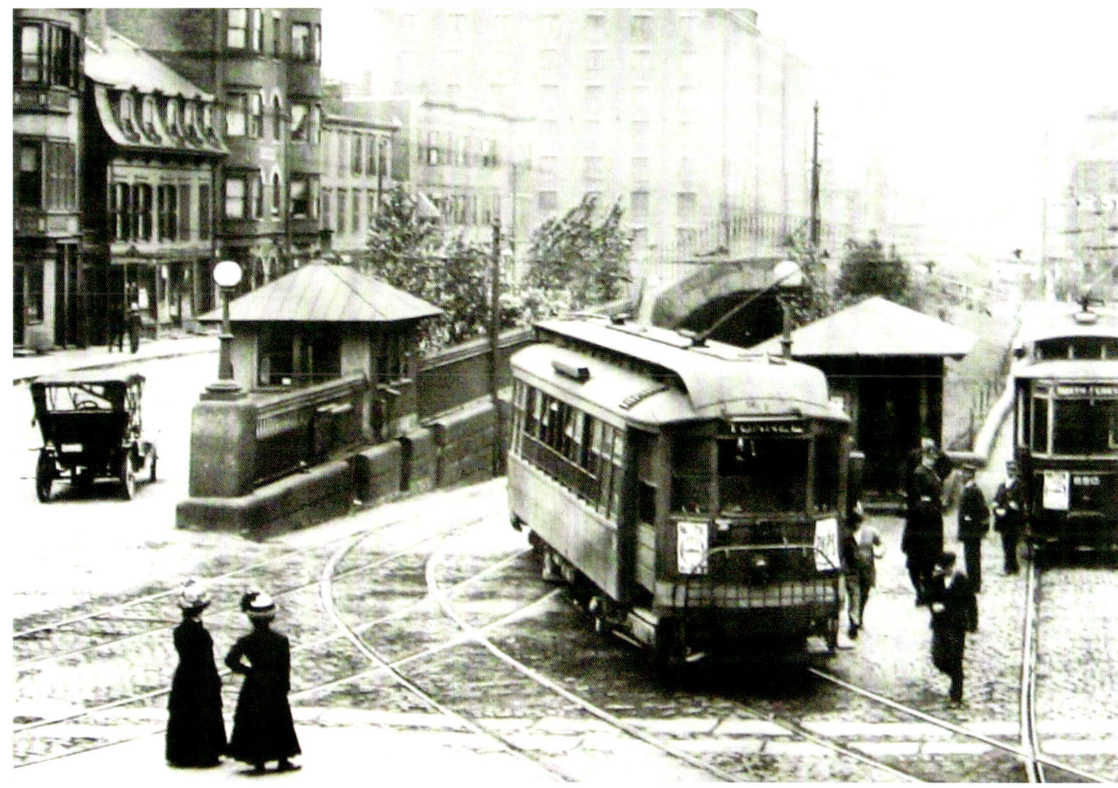

Top: A postcard from 1906 of a trolley at the Aquarium station (then known as the Atlantic station) with the East Boston Tunnel behind it.

Bottom: A photograph taken in 1918 of a trolley emerging from the East Boston Tunnel at the Maverick Square portal. It was not until 1924 that a station was built at Maverick Square.

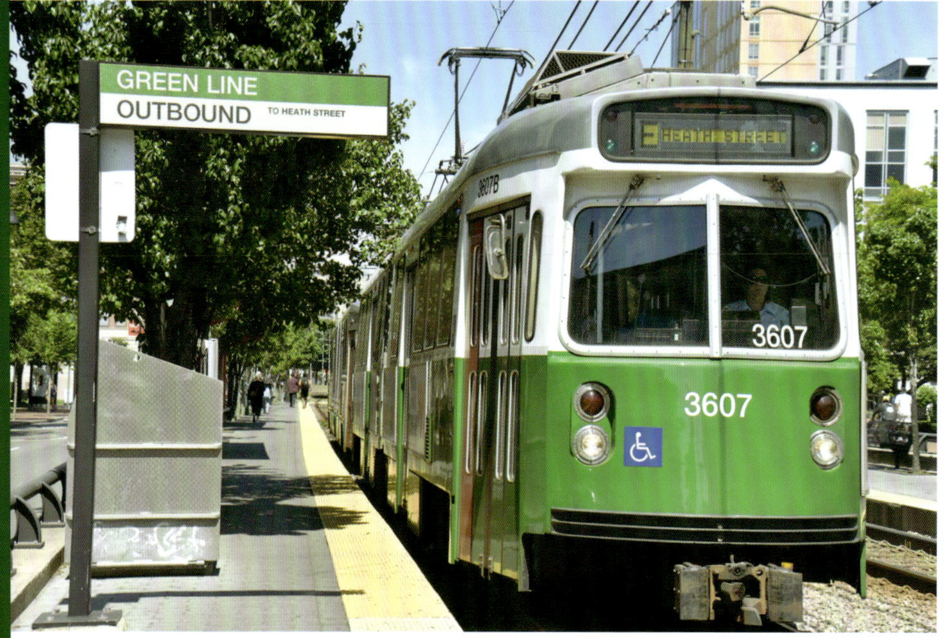

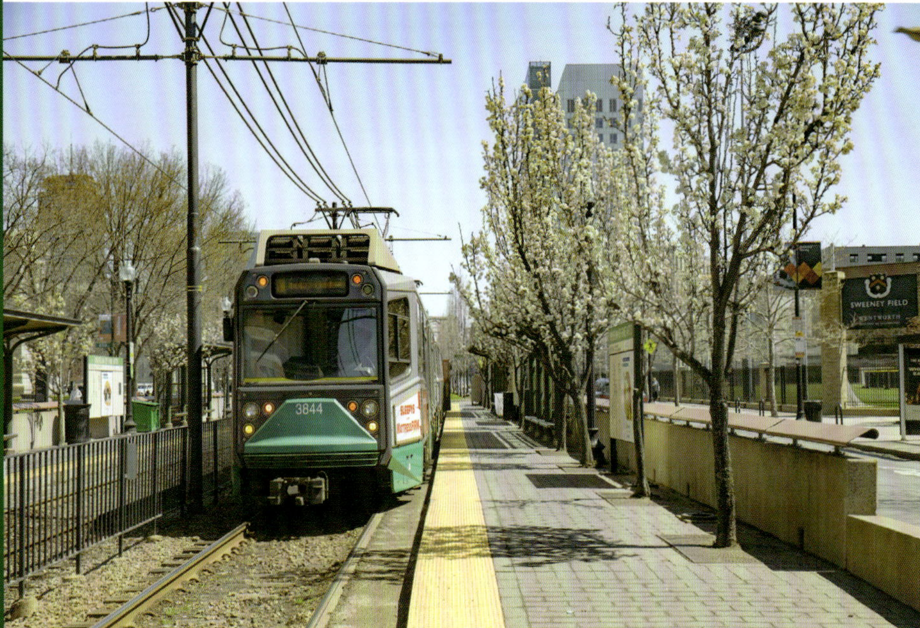

Top: The Green Line "E" branch at the Northeastern University station.

Bottom: The Green Line "E" branch at the Museum of Fine Arts station. The Green Line travels on elevated tracks, street-level tracks, and underground tracks.

Boston and New York were stepping on the gas pedal in the late 1800s to build America's first subway system. More accurately, they were cranking up the electricity. The steam trains that had been spewing ash and soot in London's underground transportation system were not the envy of the U.S. developers, so the invention of the electric motor in 1886 was truly a game changer.

Boston's need for an underground rail system was twofold. First, its narrow, colonial-era streets were seriously overcrowded due to the city's swelling population. Second, winter weather could all but cripple the city's commuters. Together, those two issues called for a solution. Boston leaders looked underground for just that. For the record, Boston won the race with New York when the Tremont Street Subway opened to the public on September 1, 1897. It was a two-track line between Park and Tremont Streets and a four-track line between Park and Boylston Streets. Portions of those tunnels—the first subway system in America—remain in use along the Massachusetts Bay Transportation Authority (MBTA) Green Line, making it the oldest line in America.

There are four branches on the Green Line with an average weekday ridership of around 100,000 as of 2023, making it among the most-used light rail systems in the country. Those branches date to the Cambridge Horse Railroad in 1856 and later became the Boston Elevated Railway. Some might find it strange that the routes begin with the "B" branch. That's because the letter A was given to the stretch that was used to run out to Watertown but was closed in 1969, before the letter system went into effect. The northern terminus of the Green Line used to be at Lechmere in East Cambridge, but the line was extended to Medford/Tufts in 2022. Trains run south through the old Tremont Tunnel and under downtown to Boston Common. Just west of Copley, the additional branches split off and run south from the main line.

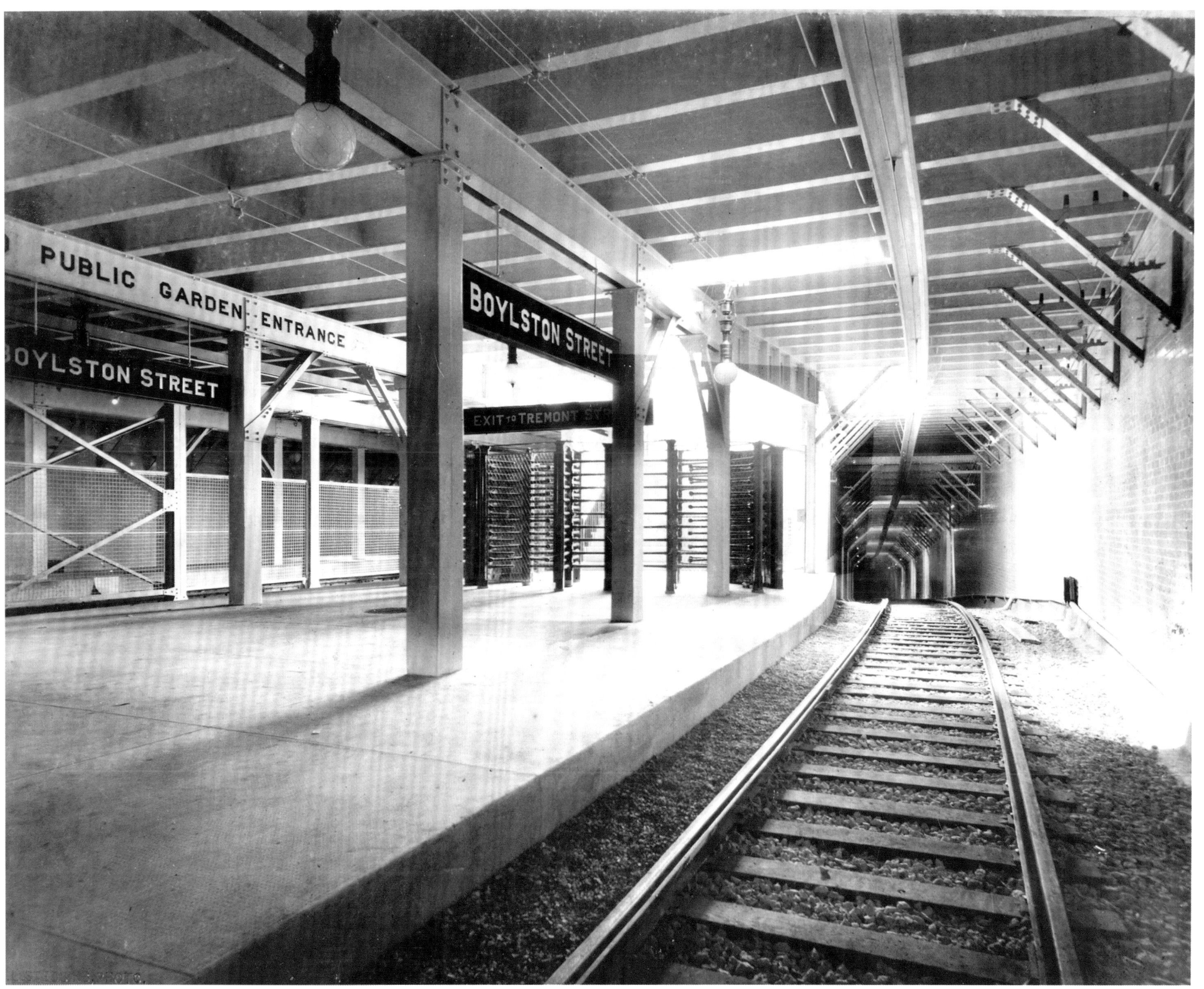

The Green Line's Boylston station in 1915.

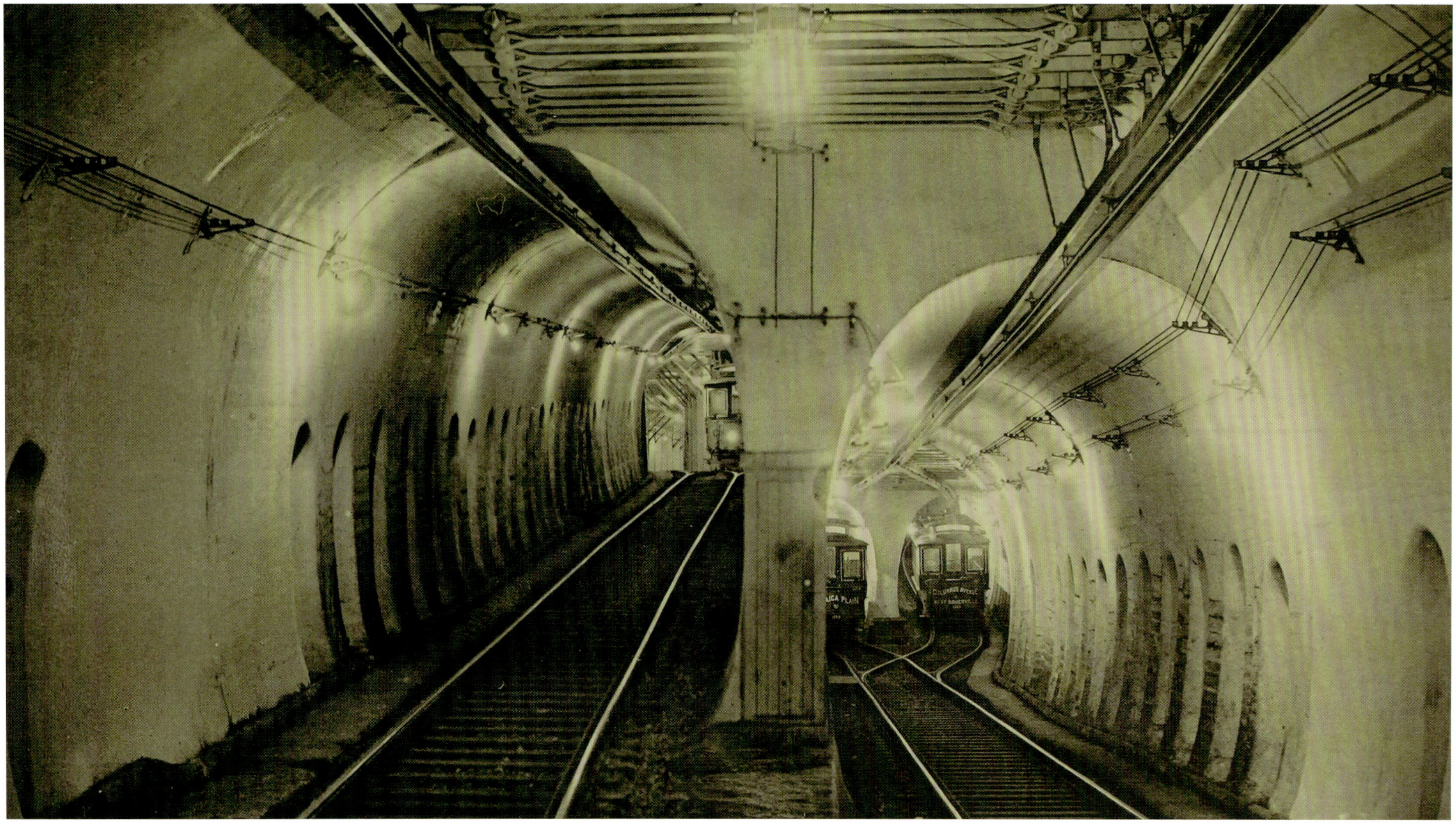

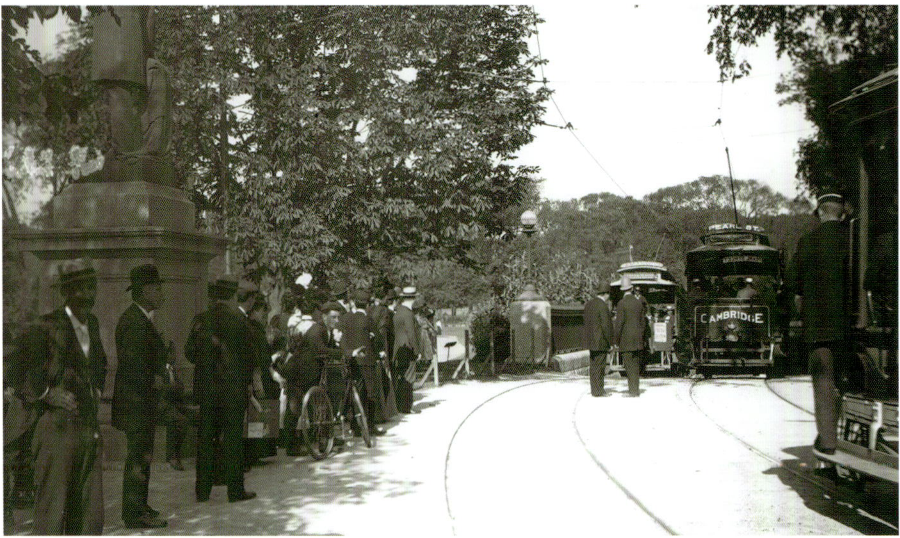

Top: The Tremont Street Subway in 1897. The Tremont Street Subway was the first subway system built in the U.S. and is now used as part of the MBTA's Green Line.

Bottom: The Tremont Street Subway at the Public Garden portal, now Boston Common, on the subway's opening day. Although the portal has been closed, this section of the subway is still used by the Green Line.

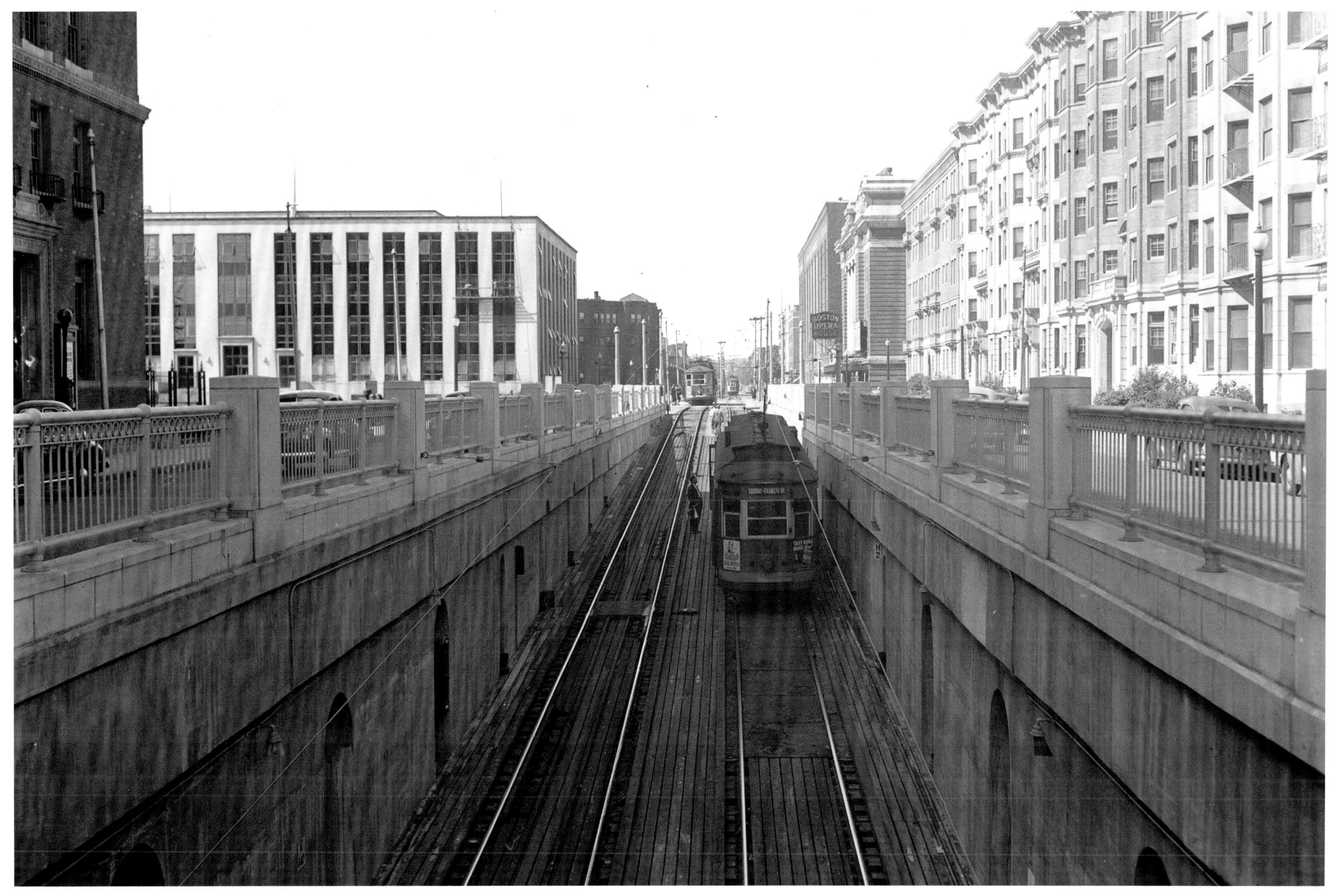

A trolley entering the Huntington Avenue portal outside of Northeastern University. The portal was opened in 1941.

The "E" runs to Heath Street, near the V.A. Medical Center. The "D" makes its first southern stop at Fenway, to the delight of Red Sox fans, and runs a little more than 10 miles south of the main line to Riverside. The "C" heads to Cleveland Circle on the Chestnut Hill reservoir. The main line continues west to Boston College. In all, there are 70 stations along 26.7 miles.

Bostonians know the system as the "T," which rolls off the tongue a little better than the Massachusetts Bay Transportation Authority. Before the Massachusetts Transit Authority (MTA) came about in 1947, public transportation in Boston was run by independent operations. The MBTA replaced the MTA in 1964, and it's now a division of the Massachusetts Department of Transportation.

Boston, Revere Beach and Lynn Railroad

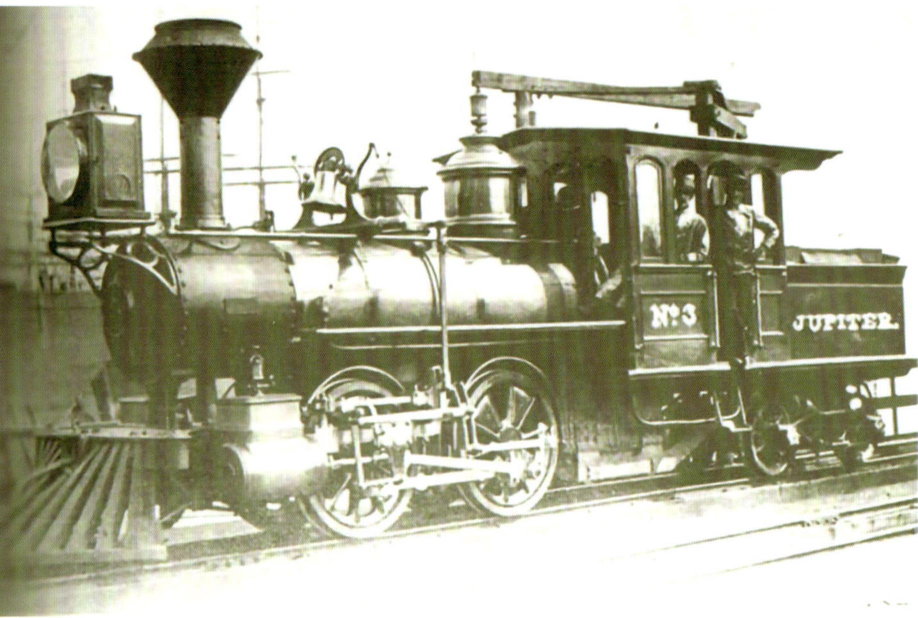

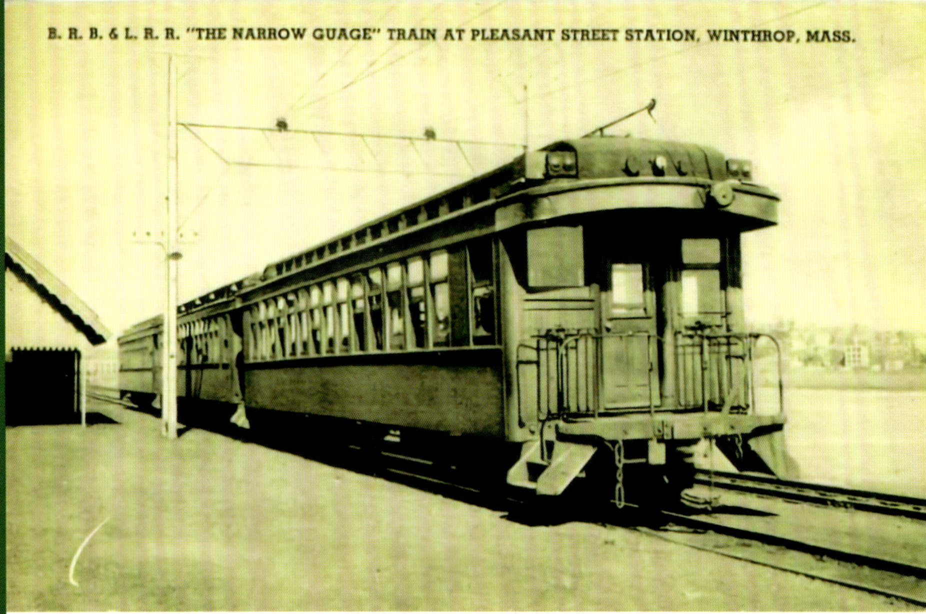

Top: The *Jupiter*, one of the first three locomotives operated by the BRB&L, photographed in 1875.

Bottom: A postcard of the electrified BRB&L at the Pleasant Street station. The railway was electrified in the late 1920s.

The Boston, Revere Beach and Lynn Railroad (BRB&L), one of the most successful narrow-gauge railroads in the country, emerged from some contentious transportation and political battles in the late 1800s. The Eastern Railroad, the predecessor to the Boston and Maine, had not made many friends in Lynn, Massachusetts, due to the location of its depot there and what residents deemed to be poor service. The hard feelings led to what locals called the "Depot War" in the 1870s. It influenced a mayoral election and ultimately spawned a new railroad: the Boston, Revere Beach and Lynn Railroad.

The new three-foot, narrow-gauge railroad was chartered in 1874 and opened to passenger traffic on July 29, 1875. The day before opening to the public, three "Shore Line" trains with three cars apiece made ceremonial runs between Lynn and Boston, a stretch of some 10 miles. According to a local newspaper account, "The road and equipment being entirely new, and the trip purely experimental, the trains were run slowly, the first one occupying an hour and a half in the passage. No accident occurred during the trip, and the excursionists were well pleased with the road. Large crowds of people gathered at the ferry houses in Boston and East Boston to see the excursionists start, and they were received in this city with marked demonstrations."

Orion, *Pegasus*, and *Jupiter* were the new company's first three locomotives. The southern terminus at East Boston linked passengers to a ferry from Rowes Wharf. The rail followed the coastline northeast through the resort area of Revere Beach and on to a new depot in Lynn, with a branch at Orient Heights that took people through Winthrop. The heavy traffic from Boston to the seaside resorts gave the BRB&L a level of success that few railroads its size would ever enjoy.

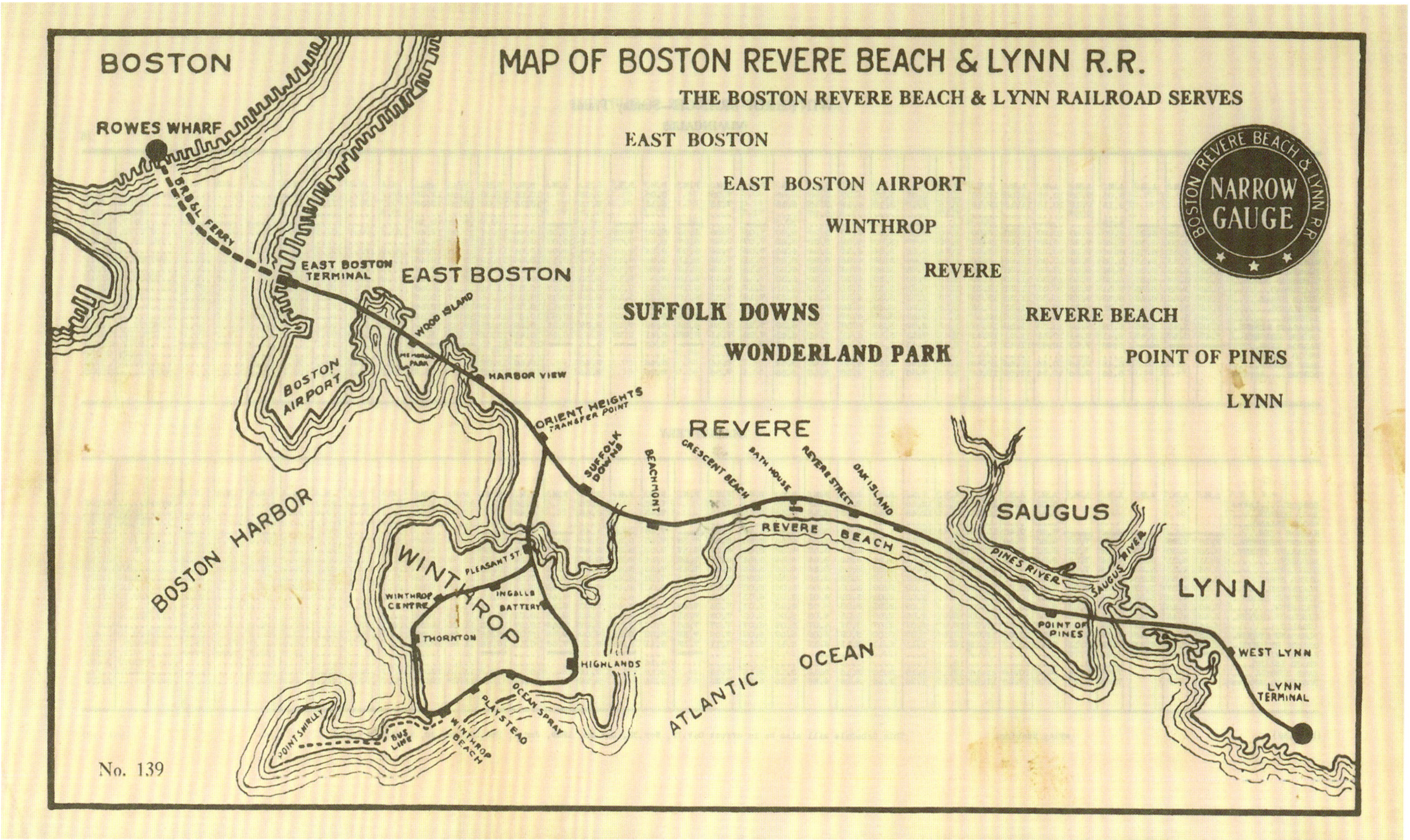

MAP OF BOSTON REVERE BEACH & LYNN R.R.

THE BOSTON REVERE BEACH & LYNN RAILROAD SERVES

EAST BOSTON

EAST BOSTON AIRPORT

WINTHROP

REVERE

SUFFOLK DOWNS

REVERE BEACH

WONDERLAND PARK

POINT OF PINES

LYNN

BOSTON REVERE BEACH & LYNN R.R. NARROW GAUGE

BOSTON
ROWES WHARF
BRB&L FERRY
EAST BOSTON TERMINAL
EAST BOSTON
WOOD ISLAND
MEMORIAL PARK
HARBOR VIEW
BOSTON AIRPORT
ORIENT HEIGHTS TRANSFER POINT
SUFFOLK DOWNS
BEACHMONT
REVERE
CRESCENT BEACH
BATH HOUSE
REVERE STREET
OAK ISLAND
REVERE BEACH
SAUGUS
PINES RIVER
SAUGUS RIVER
LYNN
POINT OF PINES
WEST LYNN
LYNN TERMINAL
BOSTON HARBOR
WINTHROP
PLEASANT ST
WINTHROP CENTRE
INGALLS
BATTERY
THORNTON
HIGHLANDS
OCEAN SPRAY
PLAYSTEAD
WINTHROP BEACH
BUS LINE
POINT SHIRLEY
ATLANTIC OCEAN

No. 139

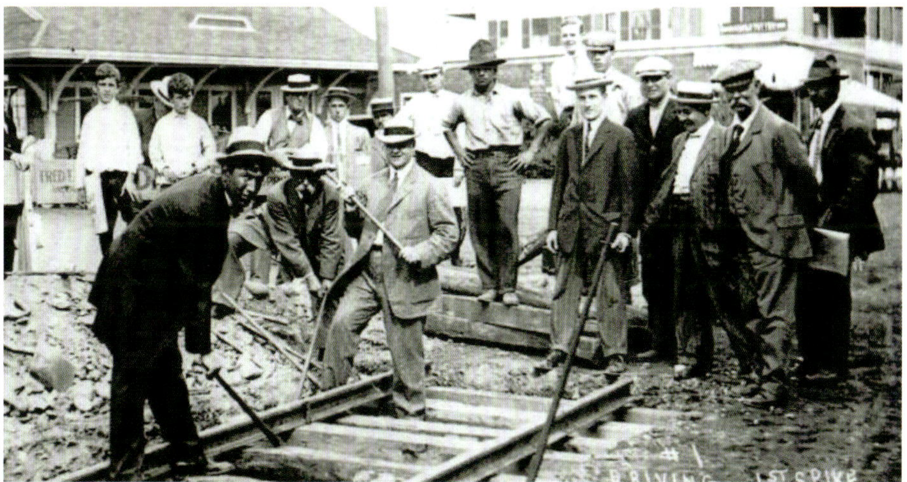

Top: The BRB&L's system map in 1939 just a year before it was decommissioned.

Bottom: A photo of ceremony attendees driving the first spike for the Point Shirley Street Railway at the Winthrop Beach station in 1910. The BRB&L acquired the line in 1912.

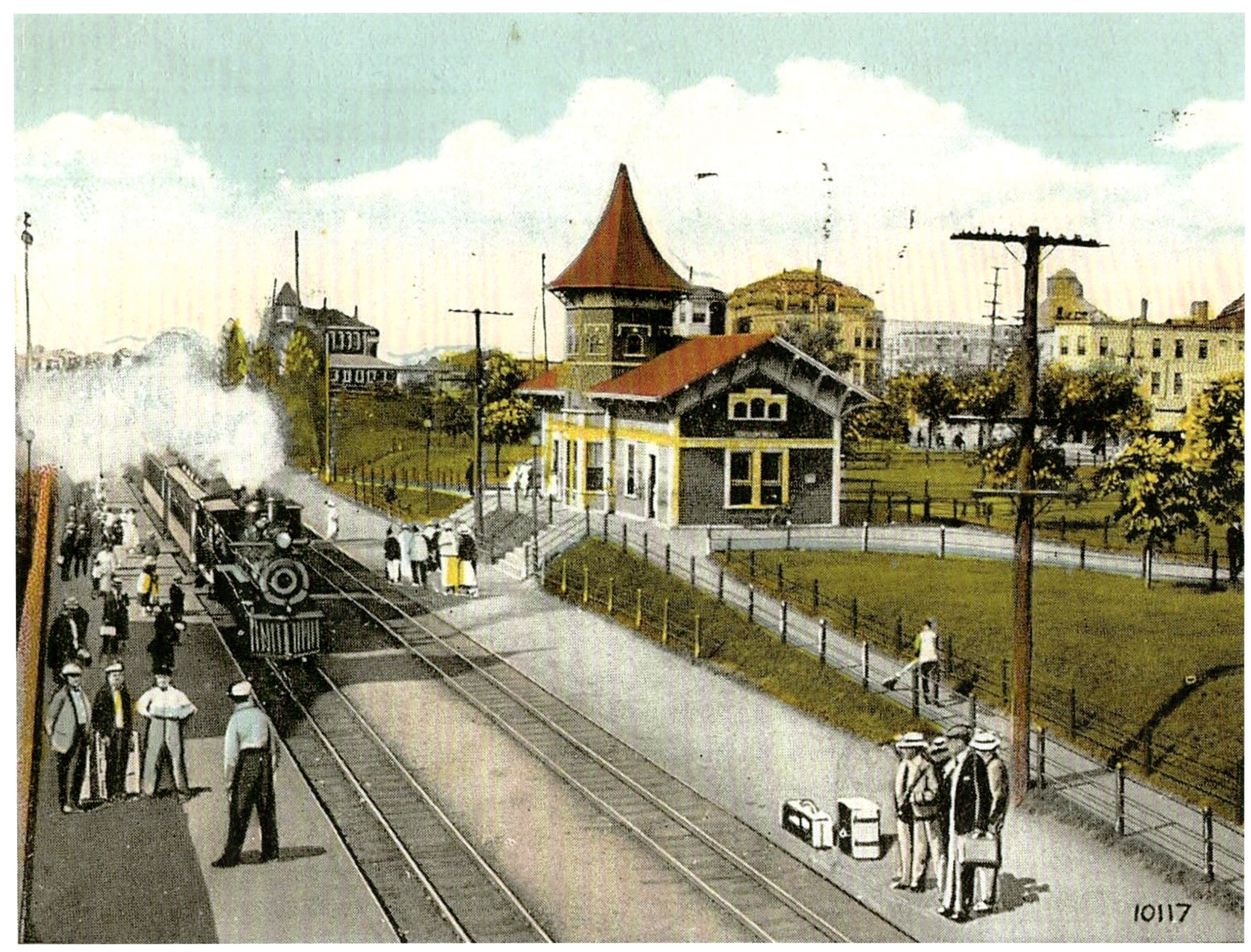

A postcard depicting the Crescent Beach station in 1924.

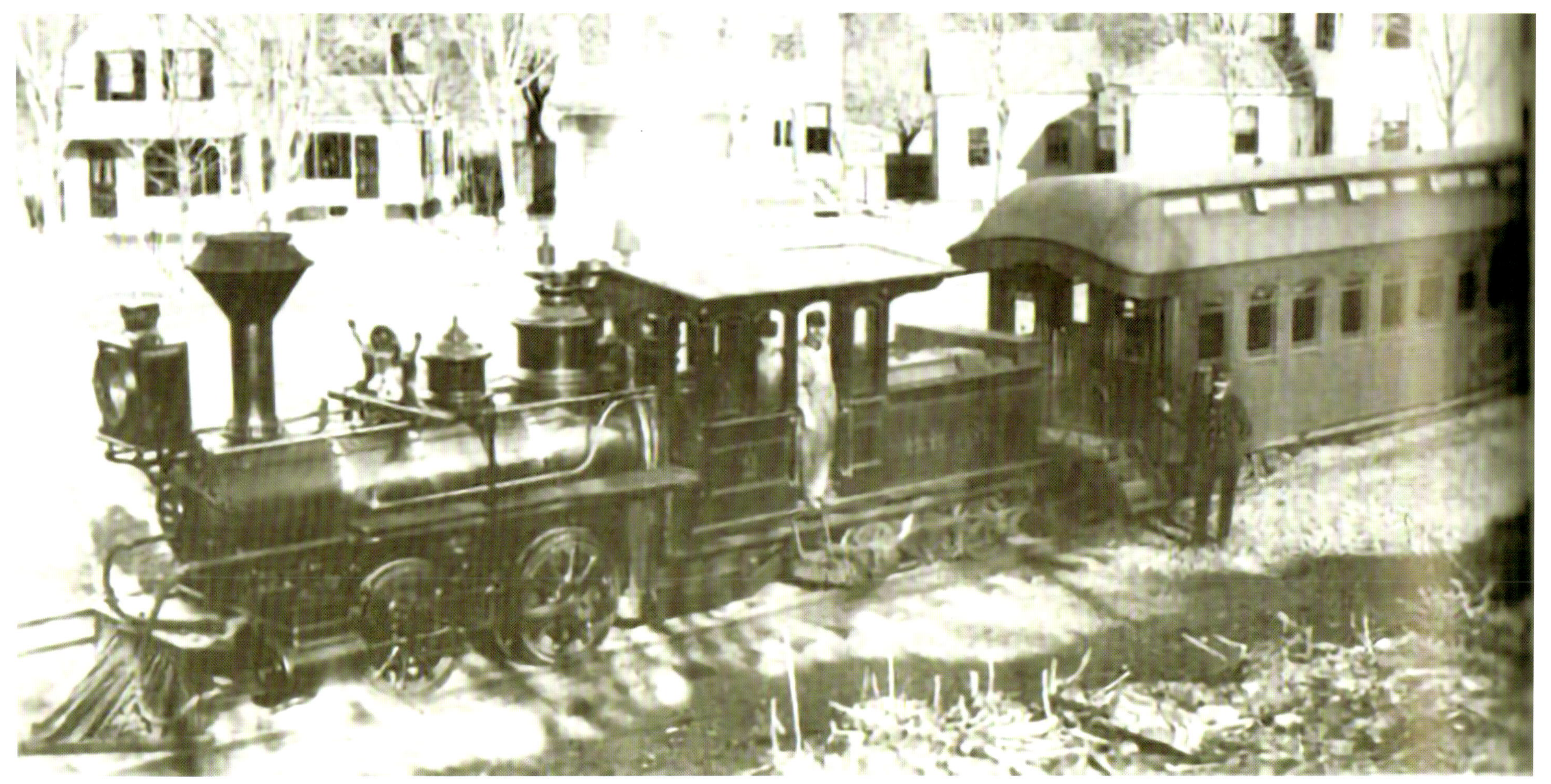

The *Pegasus* in 1888.

More than seven million passengers per year were being transported by the mid-1910s, giving the BRB&L one of North America's most traveled stretches of railroad. Its steam locomotives had been scrapped by the end of the 1920s, as the money coming in was enough to support electric motors, trolley poles, and control stands—full electrification of the system. The railroad's logo even touted the fact the trains ran on narrow gauge, a distinction that seemed to further endear the BRB&L to its growing throng of passengers.

The city of Revere still credits the railroad as the largest factor in its development and growth. Lynn residents were thrilled with what they considered vastly superior service. The great ride of the BRB&L, however, would be hit hard by the Great Depression and the popularity of the automobile. The 1930s were hard times for the railroad, which filed for bankruptcy in 1937. It continued to operate for three more years, but ridership dwindled and management filed for abandonment in 1939. The following year, the BRB&L made its final run. The railroad that was worth about $2 million after its electrification was sold to a scrapyard for just over $100,000.

An iconic *Boston Globe* photo of the last BRB&L train leaving East Boston on January 28, 1940, shows the full gamut of emotions around the event. Passengers, many of whom were wearing the lifejackets they kept from their final ferry run from Rowes Wharf, were having a great time as they smiled for the camera. Meanwhile, train conductor John J. McCarthy wore the thoughtful look of a man who would be hunting for a job for the first time in 33 years.

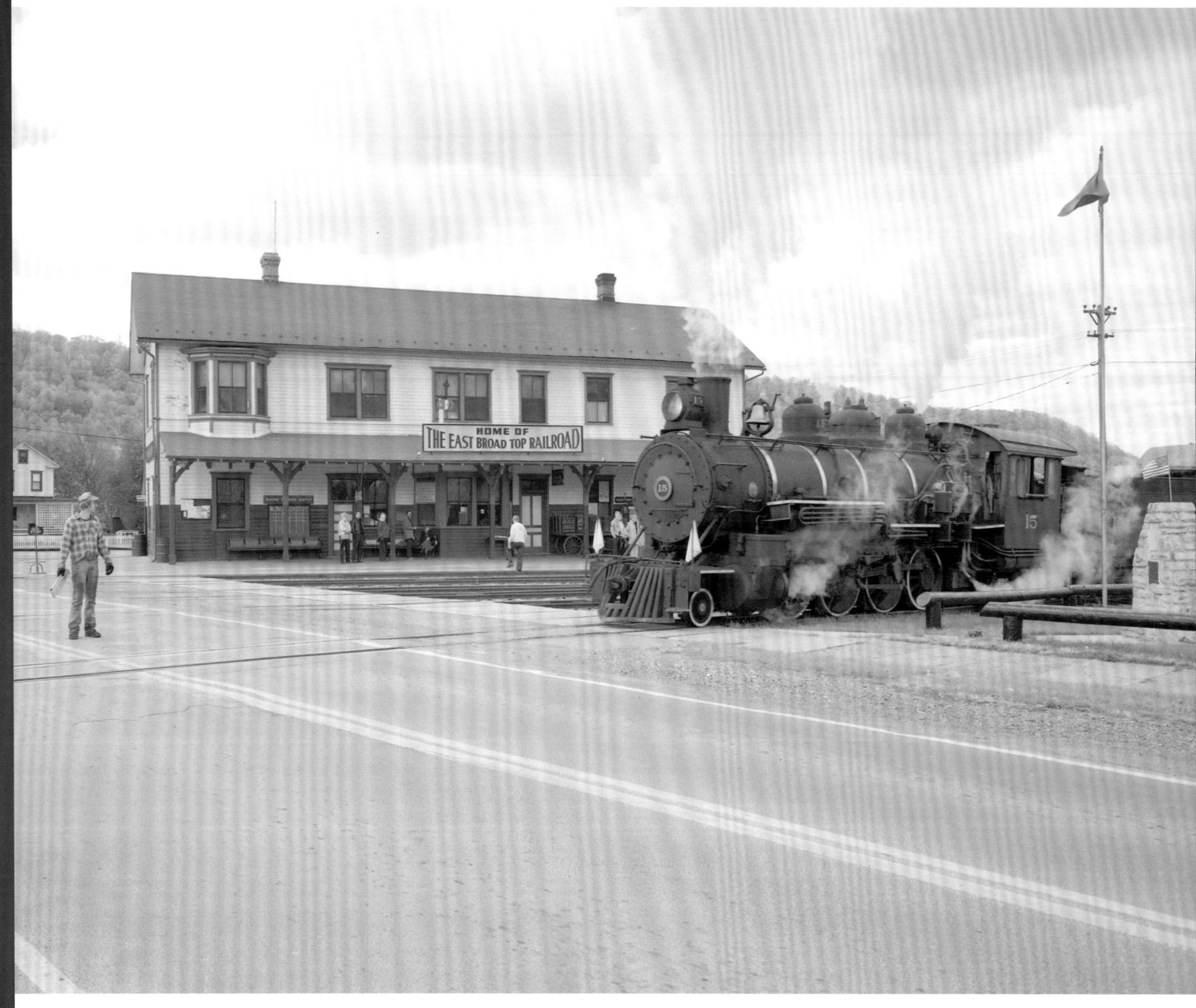

The East Broad Top Railroad after it was revived in the mid-twentieth century. Here Locomotive #15 is seen operating in Orbisonia, Pennsylvania, in 1986.

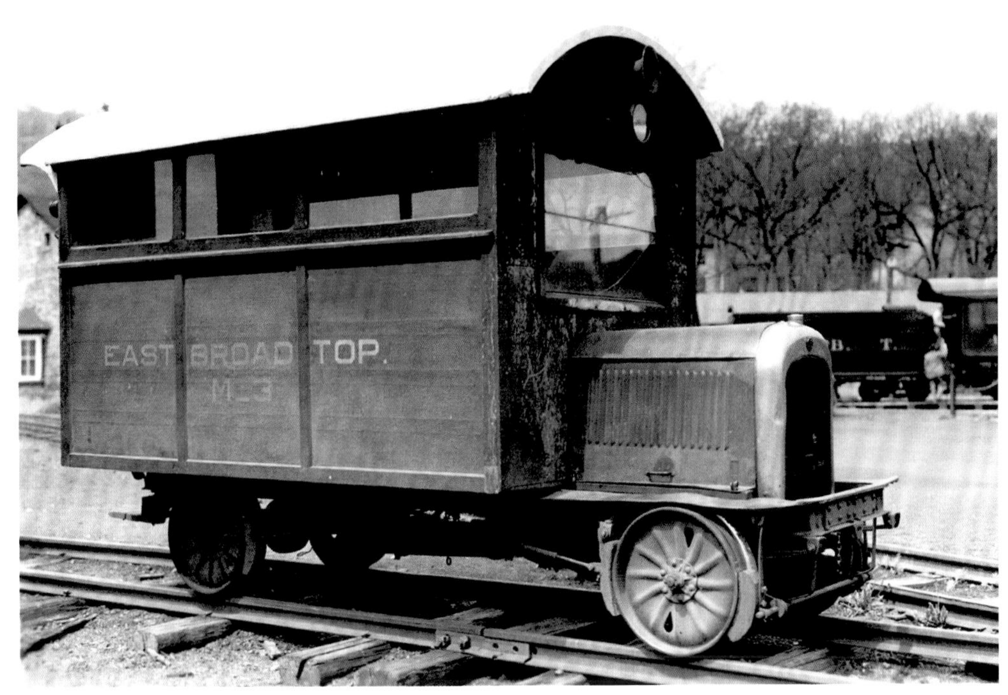

An East Broad Top rail motor in 1915.

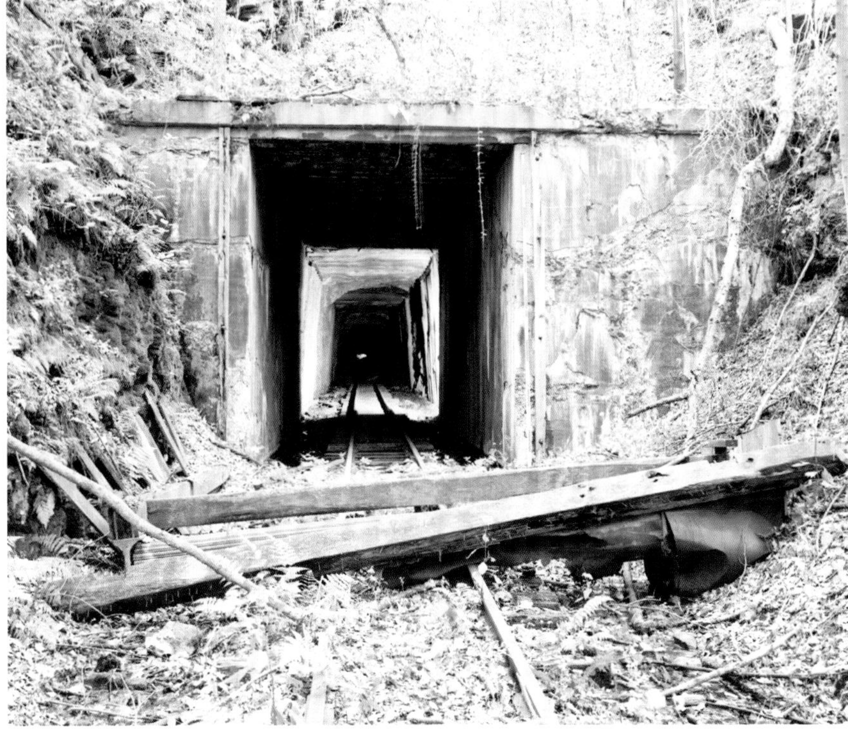

A photo of the abandoned Wray's Hill Tunnel in 1986.

The East Broad Top Railroad (EBT) served a valuable purpose in its day, transporting coal, iron ore, lumber, and ganister rock from south-central Pennsylvania's mountains for 80 years. Its "second life," however, might have given the railroad even greater acclaim. Thanks to restoration efforts and an interesting run as a historic tourist train, the EBT is one of the only places in North America where visitors can see an authentic narrow-gauge railroad, virtually intact, on its original tracks with original equipment.

One of the oldest railroads in America, the EBT was scheduled to get going in the mid-1800s. The Civil War delayed its debut, but its supporters began acquiring right-of-way in 1867, and the railway was built from 1872–74. Its first service was between the Pennsylvania towns of Mount Union and Orbisonia in 1873, with an extension to Robertsdale in 1874. In its heyday, the narrow-gauge railway possessed more than 60 miles of track.

The rugged area it spanned was an important region for mining, with coal needing to be transported from the east side of Broad Top Mountain to the Pennsylvania Railroad junction in Mount Union. That was job number one for the EBT, which also hauled concrete, rock, lumber, road tar, and agricultural goods. Coal traffic accounted for the vast majority of its revenue, especially after the iron industry declined in the area in the early twentieth century.

Unlike many railroads of its day, the EBT maintained steady profitability from its early days through the 1940s. That allowed it to modernize and make upgrades in equipment and infrastructure. The railroad soon acquired rolling stock from Philadelphia's Baldwin Locomotive Works. However, the 1950s brought a turn of events that led to the railroad's demise. As homes switched to oil and gas for heat, the demand for coal waned. Because coal made up the vast majority of the EBT's freight hauls, there was truly nothing the company could do to make up the gap. The railroad, and its sister coal company, were sold to Nick Kovalchick's salvage company in 1956.

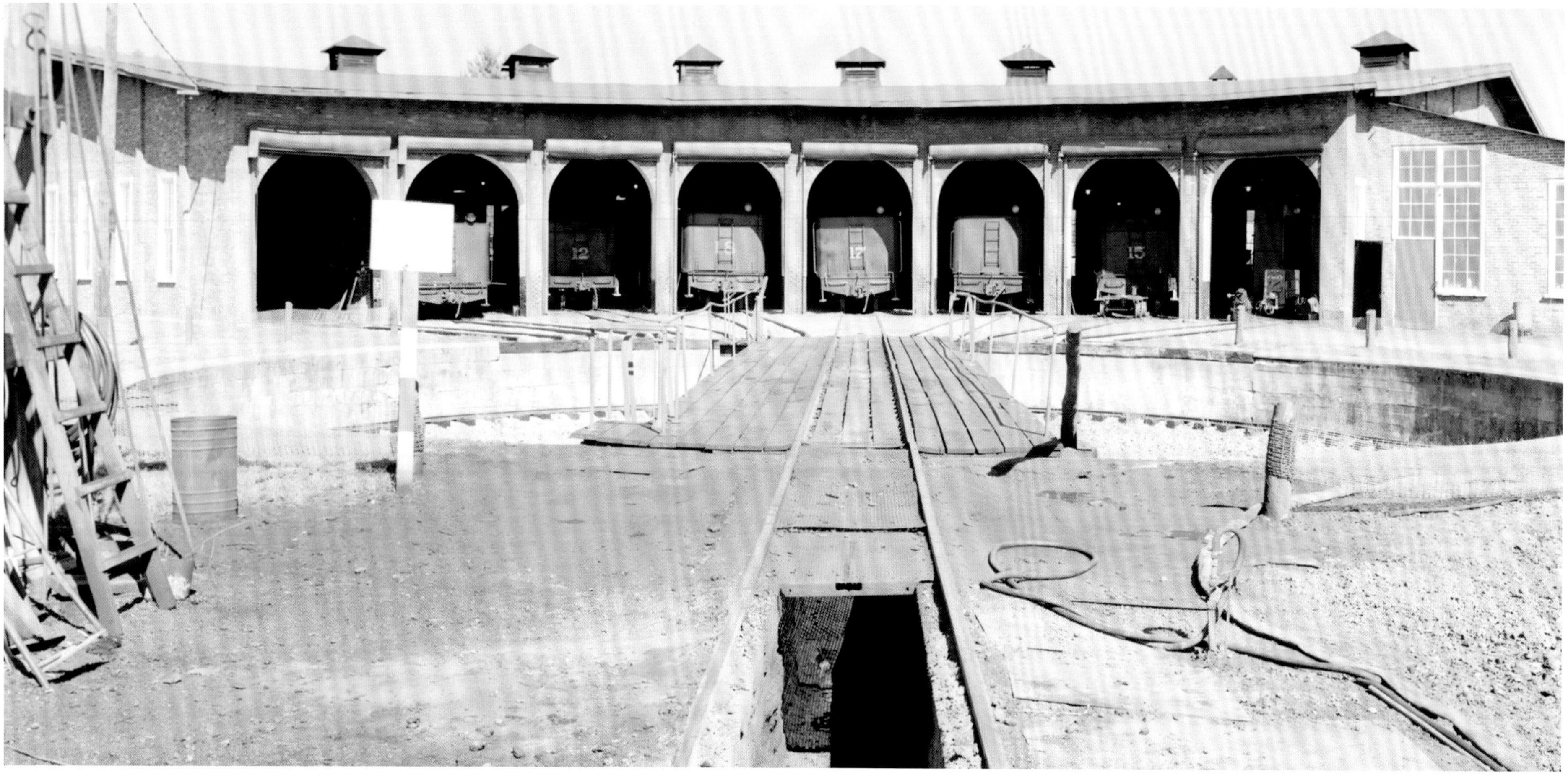

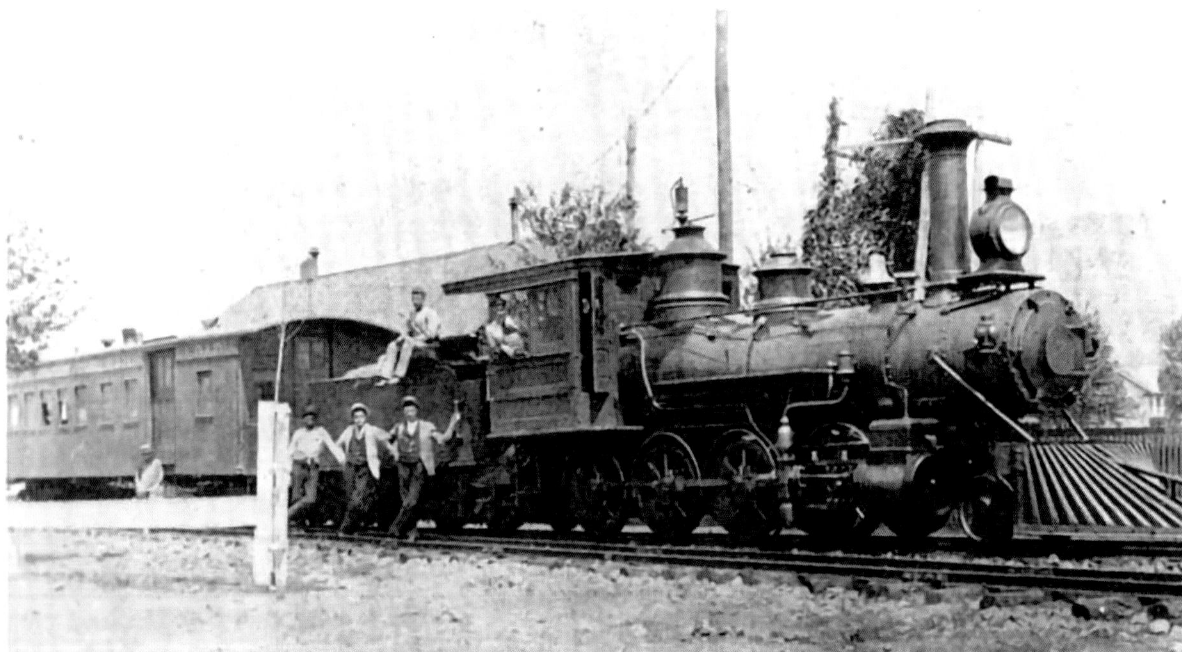

Top: The East Broad Top's roundhouse and turntable ash pit in Rockhill, Pennsylvania, photographed in 1986.

Bottom: A photo of the East Broad Top Railroad in 1889.

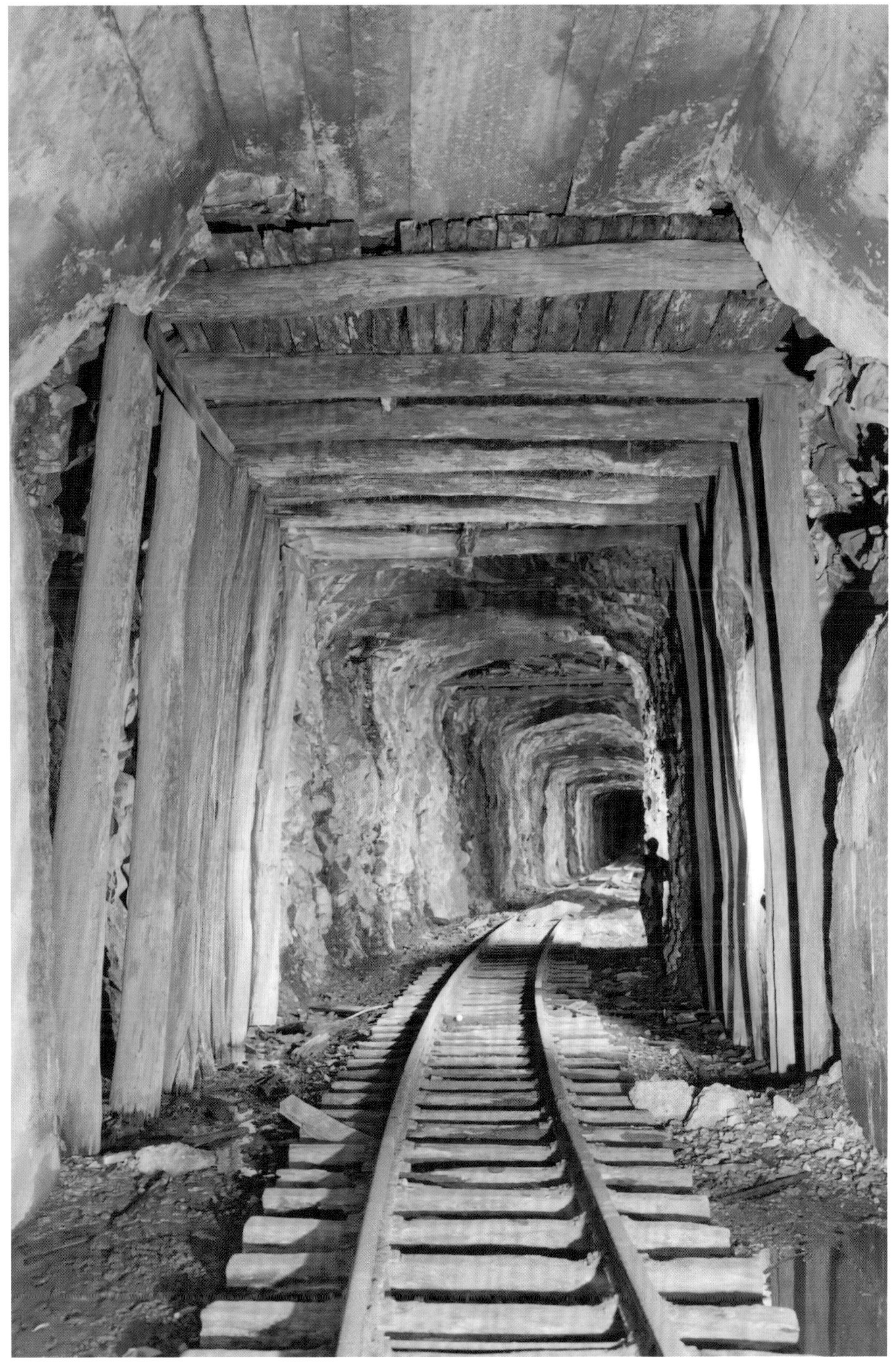

While other railroad stories end there, the EBT's story was just beginning. Kovalchick didn't scrap his purchase. Instead, after a few years while the remains of the old railroad sat dormant, he and his family decided to restore the railroad as a tourist attraction in time for the bicentennial celebrations of Orbisonia and Rockhill Furnace in 1960. Kovalchick refurbished two locomotives to run over four miles of track, and extended the line to five miles in 1961.

Along the way, the EBT earned several distinctions. In addition to being the last original narrow-gauge railroad operating east of the Rocky Mountains, it was designated as a National Historic Landmark in 1964 and earned a spot on the National Register of Historic Places two years later. It also made the National Trust for Historic Preservation's list of America's Most Endangered Places in 1996, but activists have tried to make sure the danger never becomes too real.

The trains ran for 50 years until operations were temporarily suspended in 2011. Train service resumed in 2021 after the EBT Foundation purchased the railroad in 2020 from the Kovalchicks. The EBT Foundation began a series of restoration projects with the help of the Friends of the EBT, a nonprofit group founded in 1983. Today visitors to the East Broad Top Railroad Heritage Park can enjoy a scenic ride on the oldest operating narrow-gauge railroad in the United States.

The abandoned Sideling Hill Tunnel photographed in 1986.

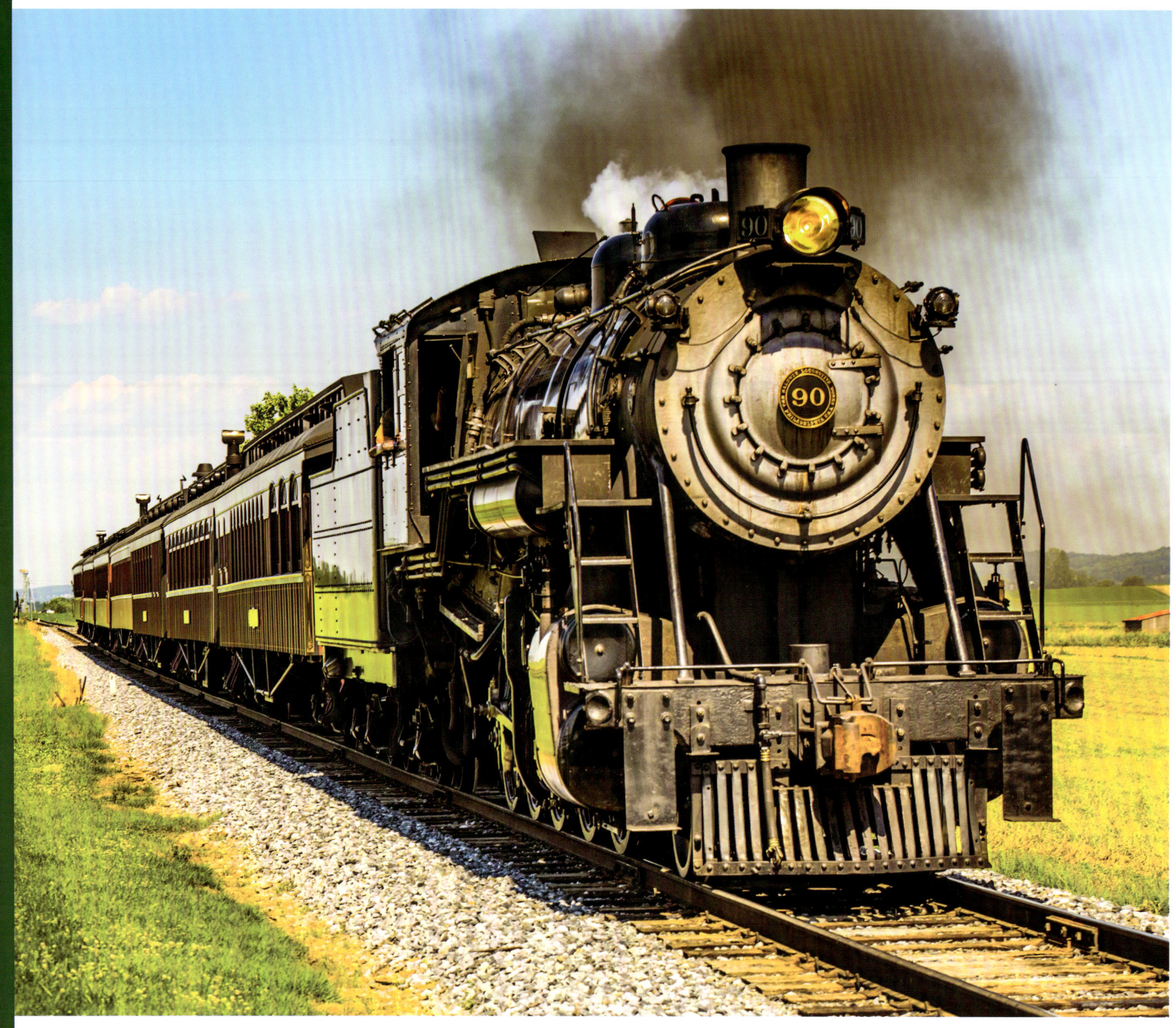

Built in 1924, the Strasburg Rail Road locomotive #90 is the railroad's largest and strongest engine. It was originally used for the Great Western Railway of Colorado.

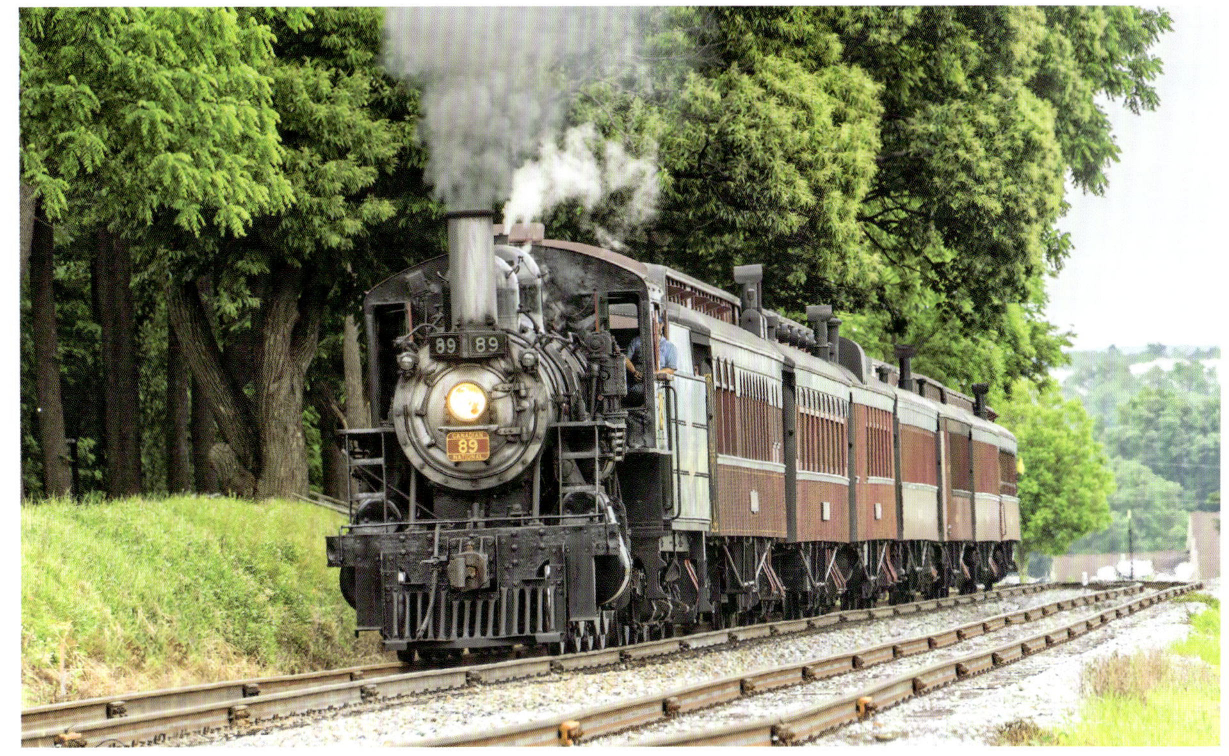

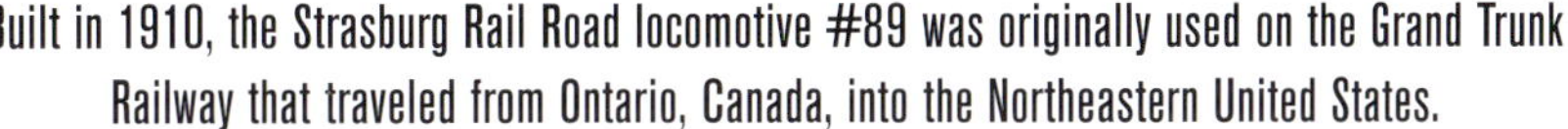

Built in 1910, the Strasburg Rail Road locomotive #89 was originally used on the Grand Trunk Railway that traveled from Ontario, Canada, into the Northeastern United States.

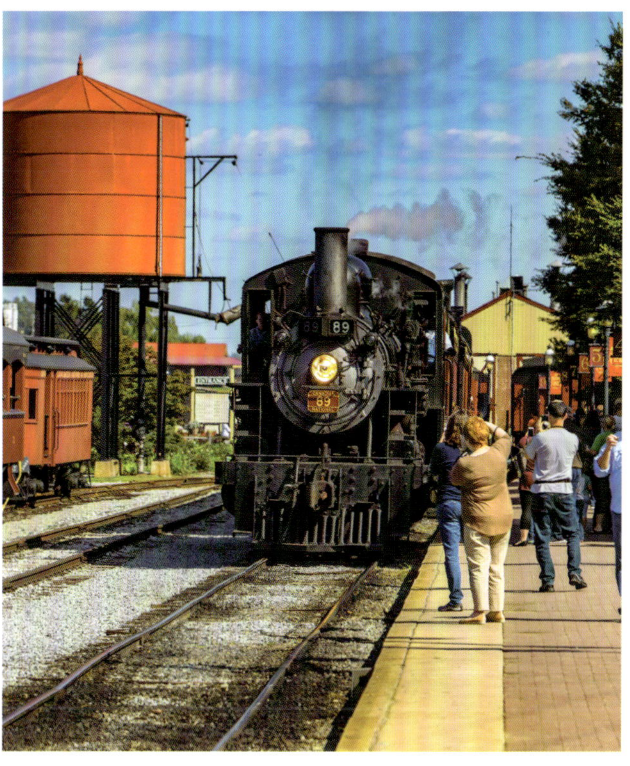

The Strasburg Rail Road is the only railway in America that will occasionally transport freight with a steam locomotive.

The Strasburg Rail Road was completed in 1837 as a freight service between Strasburg, Pennsylvania, and the junction at Leaman Place in order to save the economic future of Strasburg during the mid-nineteenth century's changing transportation mileu. In the early nineteenth century, transporting goods by canal was becoming more popular, leaving wagon roads and the Conestoga Wagon in the economic dust. With the opening of the Susquehanna Canal in southern Pennsylvania and the construction of the Philadelphia and Columbia Railroad, the goods that once passed through Strasburg on a wagon road to Philadelphia were no longer passing through town. Foreseeing a bleak outcome for Strasburg once its wagon route was obsolete, local businessmen petitioned the state government to build a railway that would connect Strasburg to the Philadelphia and Columbia Railroad. The governor granted the charter on June 9, 1832.

The early history of the railroad is not well known, but the grading of the roadbed began in 1835, with the tracks in operation two years later. The railroad operated with horse-drawn carriages until it was able to afford its first steam-engine locomotive. Throughout the nineteenth century, various owners controlled the operations of the railroad, passing from hand to hand. The railroad went through economic depressions, fire damage, and near abandonment, but it has survived to this day due to the passion of local enthusiasts. Declining freight business pushed the estate of former Pennsylvania State Senator John Homsher, who took control of the railway in 1918, to file a public utility abandonment with the state commission in 1958. But a group of local enthusiasts were able to raise enough funds to buy the railroad and keep it in operation as a heritage line.

Tourist excursions began on January 4, 1959, and they have continued to this day. The Strasburg Rail Road is the oldest continuously running railroad in the western hemisphere and also the oldest public utility in the state of Pennsylvania. For nearly five miles of historic Pennsylvania Dutch Country, passengers can travel on one of five historic steam engines along with the railway's large fleet of wooden passenger coaches. Today the railway no longer connects with the Philadelphia and Columbia Railroad but instead connects to the Amtrak Philadelphia–Harrisburg Main Line.

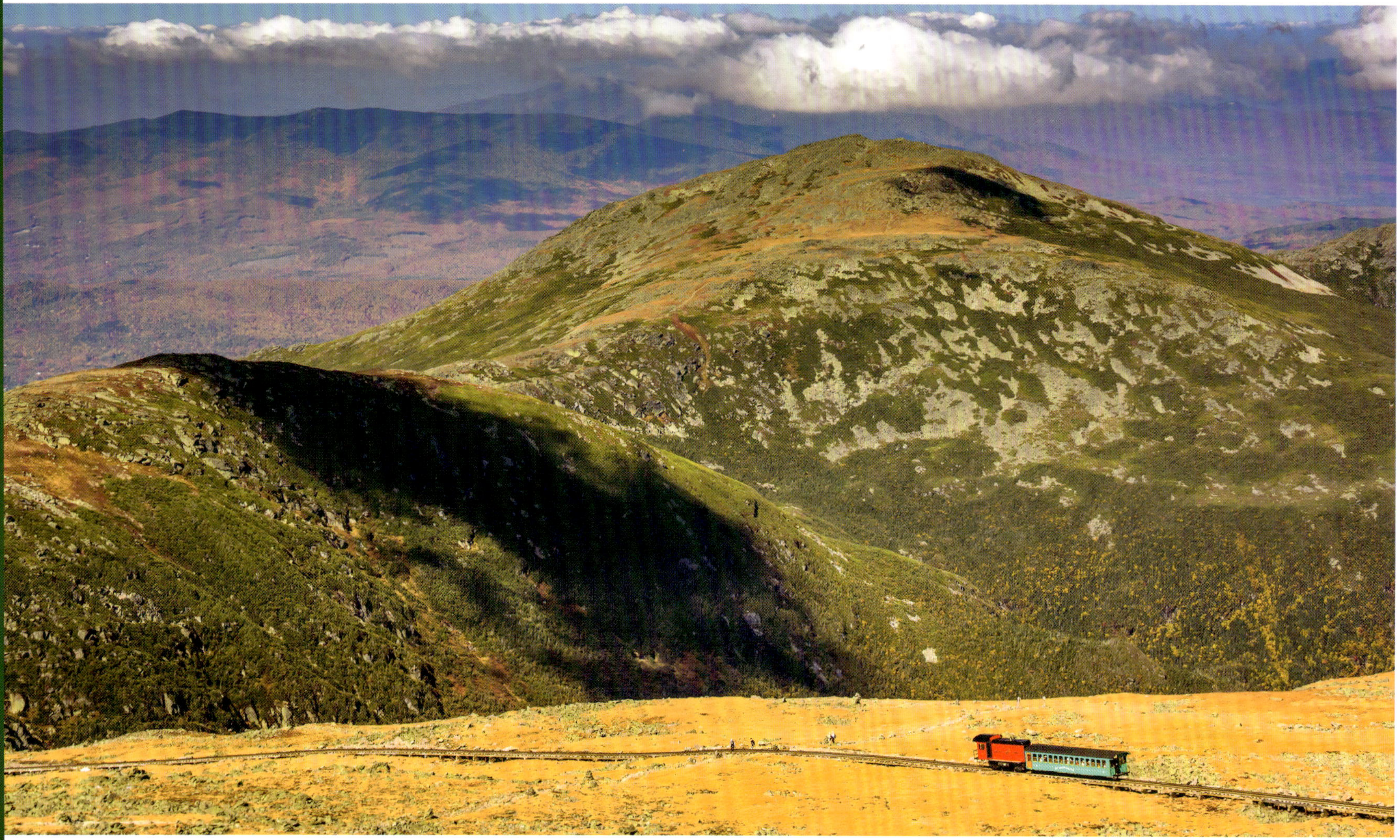

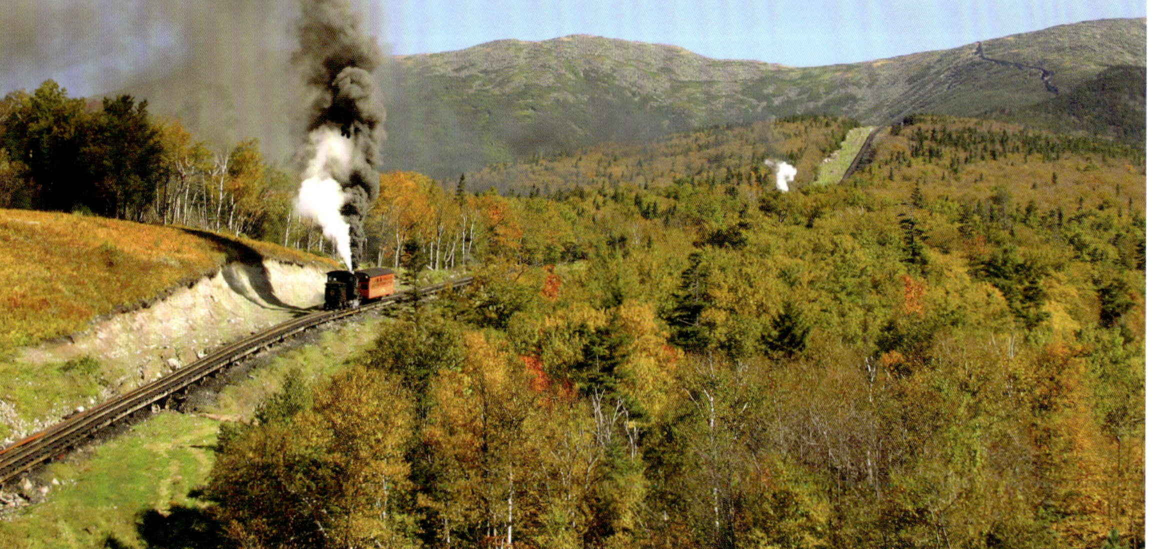

Top: The passenger carriages have a maximum capacity of seventy passengers.

Bottom: The environmental costs of the railway are very real. Each three-mile run of a steam locomotive on this railway requires 2,000 pounds of coal and 1,000 gallons of water.

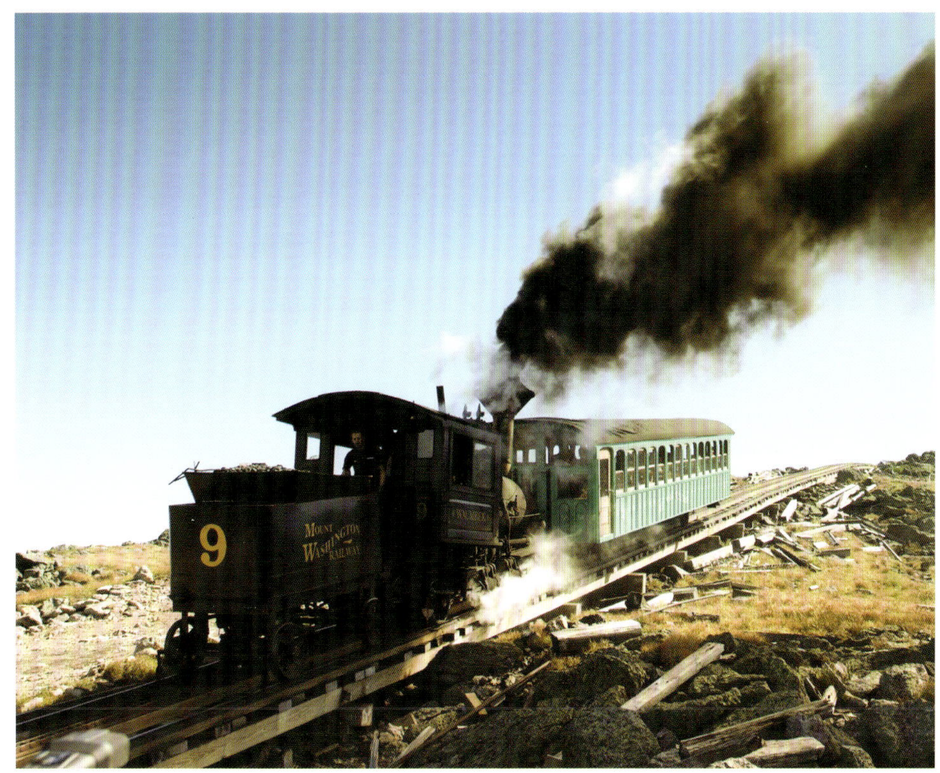

The *Waumbek* locomotive operating on the mountain. Built in 1908, this locomotive was converted for a short time to burn biodiesel but was reverted back to coal.

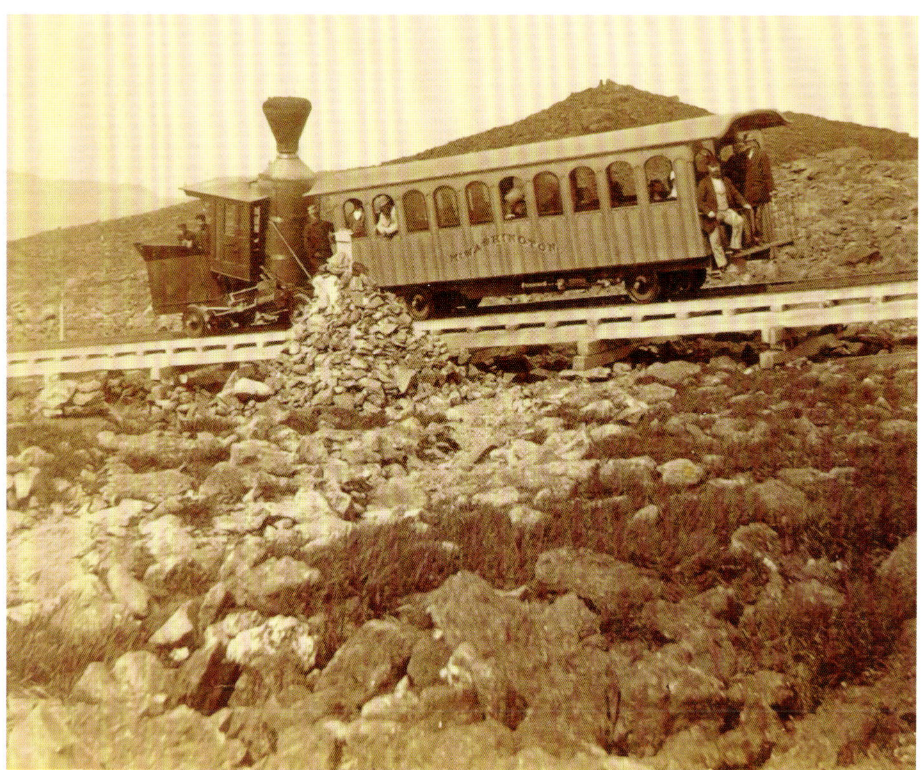

The train passing the cairn that memorializes the death of Lizzie Bourne, who died from cold exposure and exhaustion while climbing Mount Washington in 1855.

The first mountain-climbing and second steepest cog railway in the world, the Mount Washington Cog Railway of New Hampshire ascends the highest mountain in the Northeastern United States, Mount Washington. The railway may sound impressive by these stats, but the actual train ride might not be as exciting as you'd expect. Traveling at an average ascent speed of three miles per hour, it takes the Mount Washington Cog Railway 65 minutes to go the three miles up the mountain's western slope. The train travels at an average speed of five miles per hour on its descent, taking another 40 minutes to complete this slow and arduous trip.

Conceived of and built by Sylvester Marsh in the mid-1800s, the idea of the Mount Washington Cog Railway was often thought of as an impossible dream, but Sylvester Marsh thought otherwise. Marsh was given the charter to build the railway, despite the harm and risk of injury it presented to its future ridership, because the state legislature never thought that such a railway could ever be built. By 1858, Marsh attained the charter to build the road that would bring construction supplies to the track, but the plan was put on hold as the Civil War commenced. It wasn't until 1866 that construction began. Marsh built a working prototype and track of his cog railway—the first successful cog railway in America—and began to attract investors. Passengers started paying for rides in August of 1868, although the railway was not completely finished until July 1869. Even the president of the United States at the time, Ulysses S. Grant, took a ride to the top of Mount Washington in August of 1869.

There are only two stations along the Mount Washington Cog Railway, one at the base of Mount Washington and one at the summit. The train gains over 3,000 feet in elevation, starting at 2,700 feet above sea level and reaching 6,288 feet above sea level with an average gradient of 25 percent. The maximum gradient the train operates at is a jaw-dropping 37 percent on the cog system that Marsh himself invented and patented. The railway operated strictly with steam engines until 2008. Faced with rising criticism for environmental concerns and falling ridership, they introduced that year the first diesel locomotives on the railway in order to reduce reliance on steam engines and coal.

THE SOUTH

MEN
TRAIN ANNOUNCEMENT BOARD
SOUTHERN RAILWAY
NORTH BOUND
SOUTH BOUND

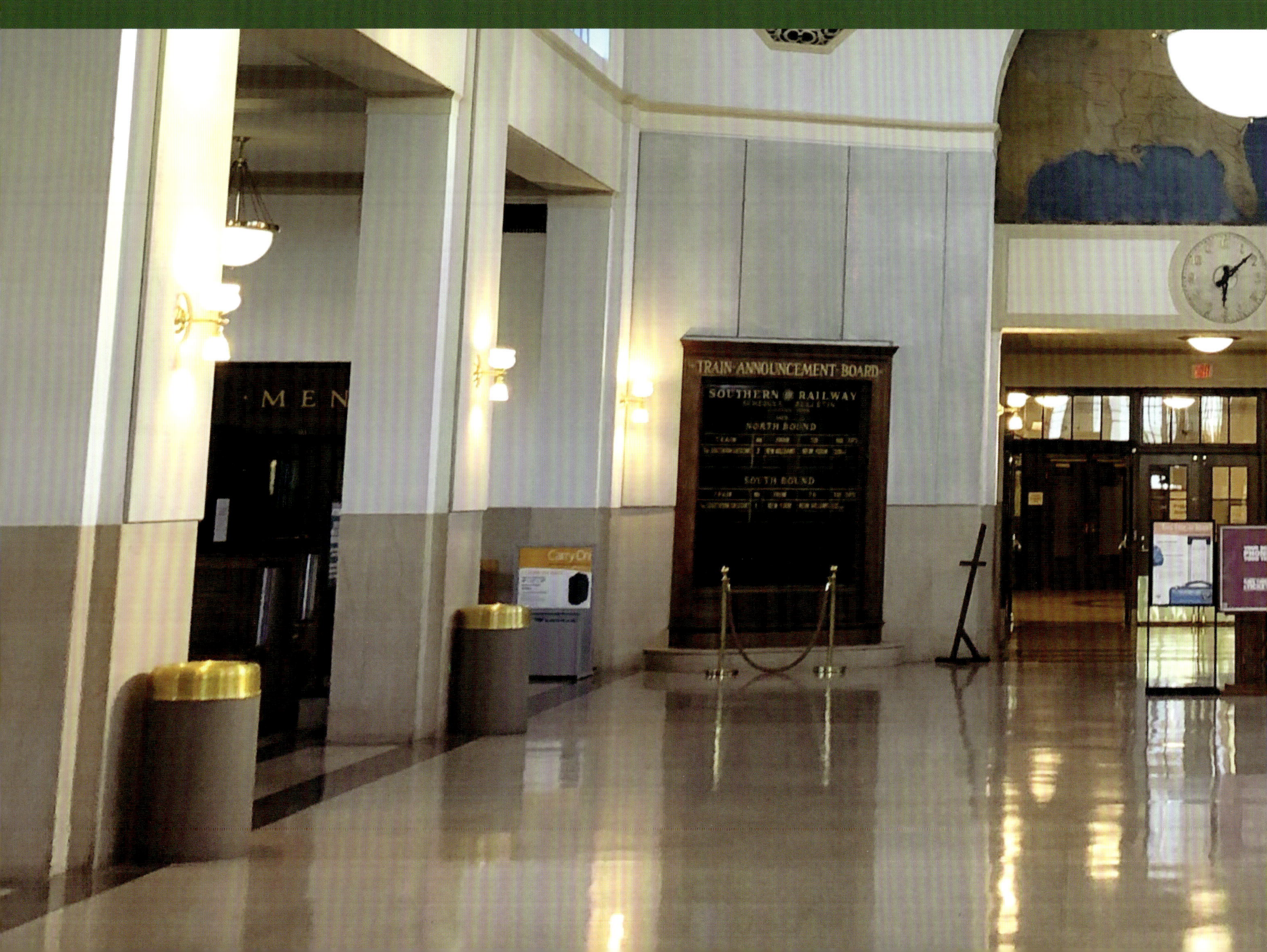

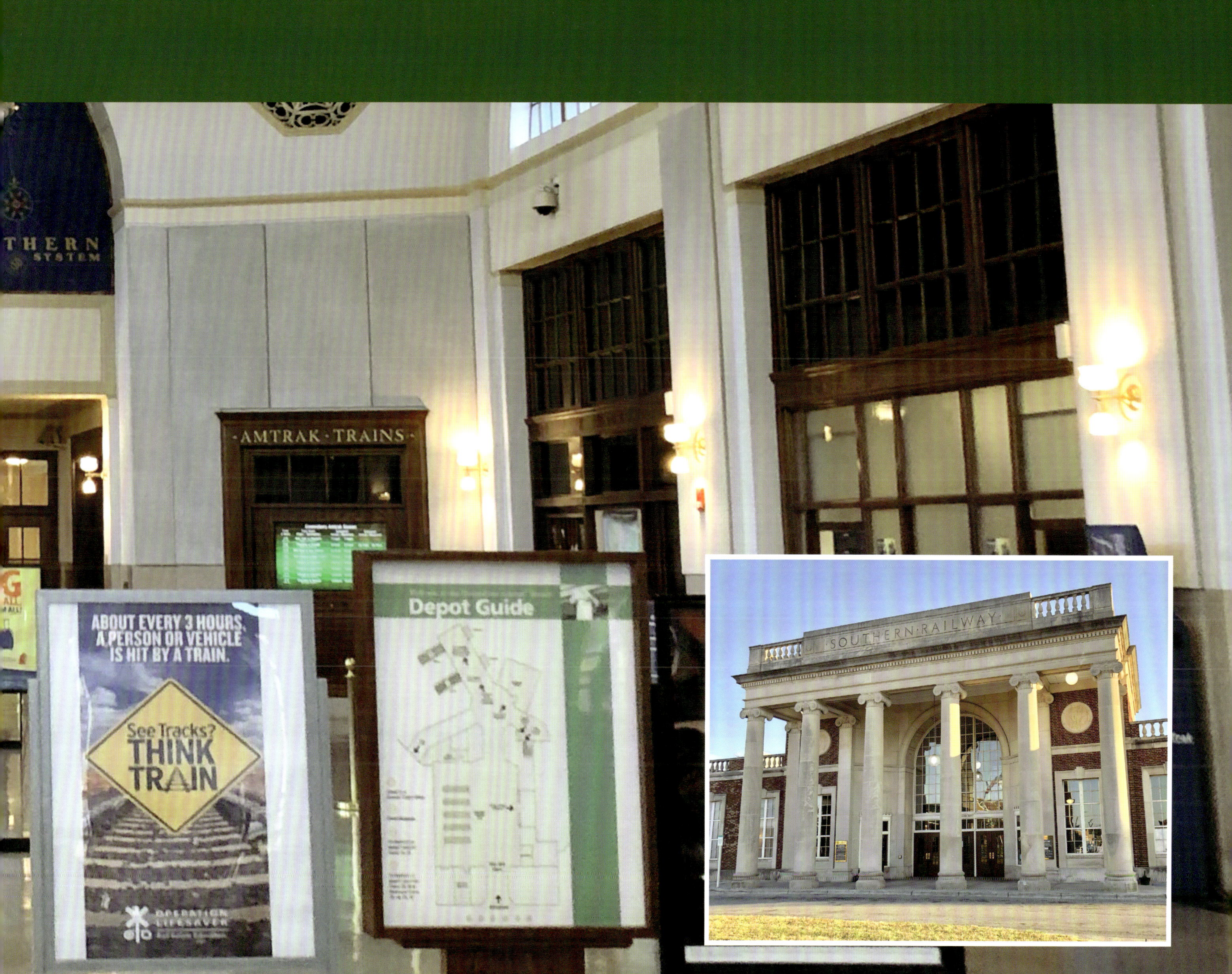
THERN
SYSTEM
AMTRAK · TRAINS
ABOUT EVERY 3 HOURS, A PERSON OR VEHICLE IS HIT BY A TRAIN.
See Tracks? THINK TRAIN
OPERATION LIFESAVER
Depot Guide
SOUTHERN RAILWAY

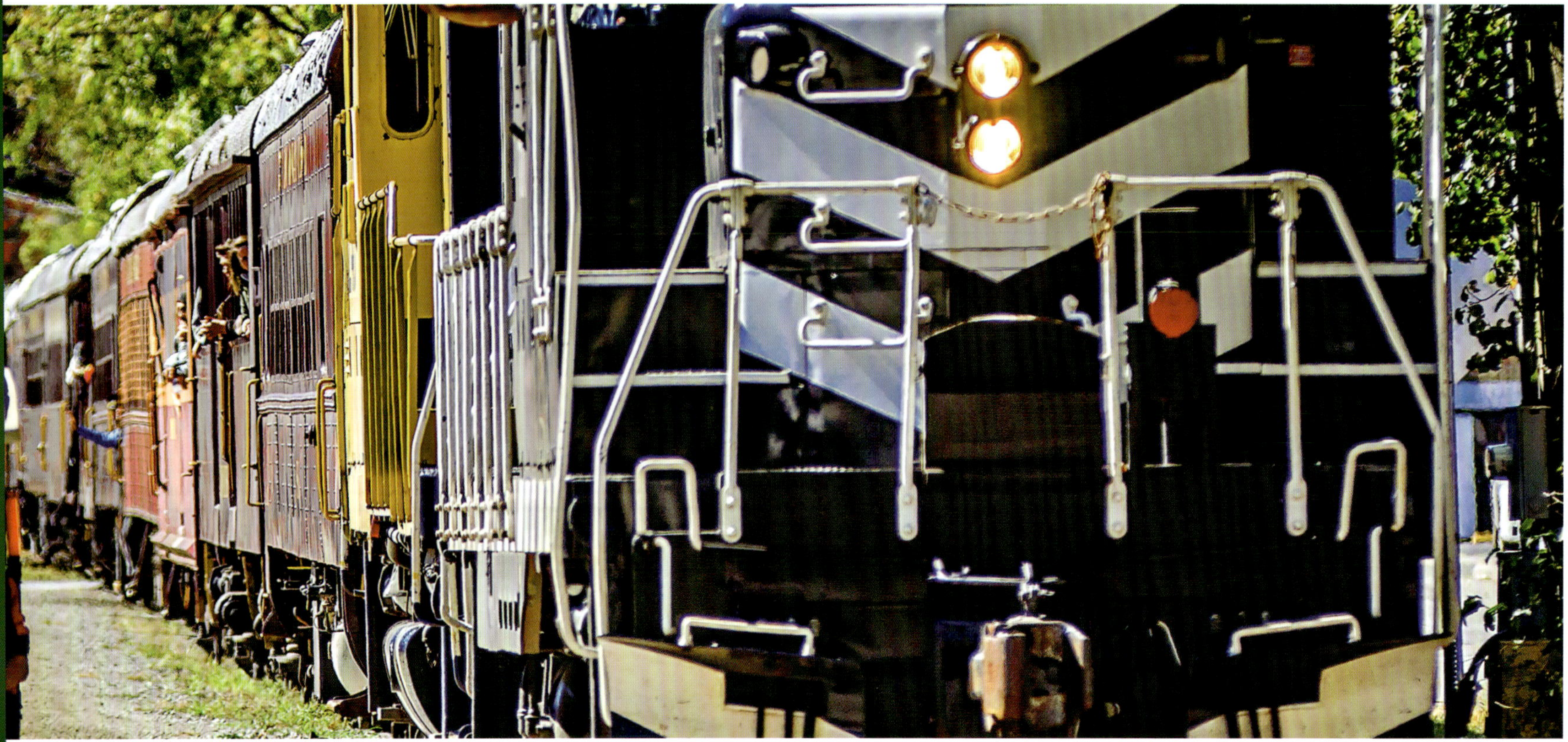

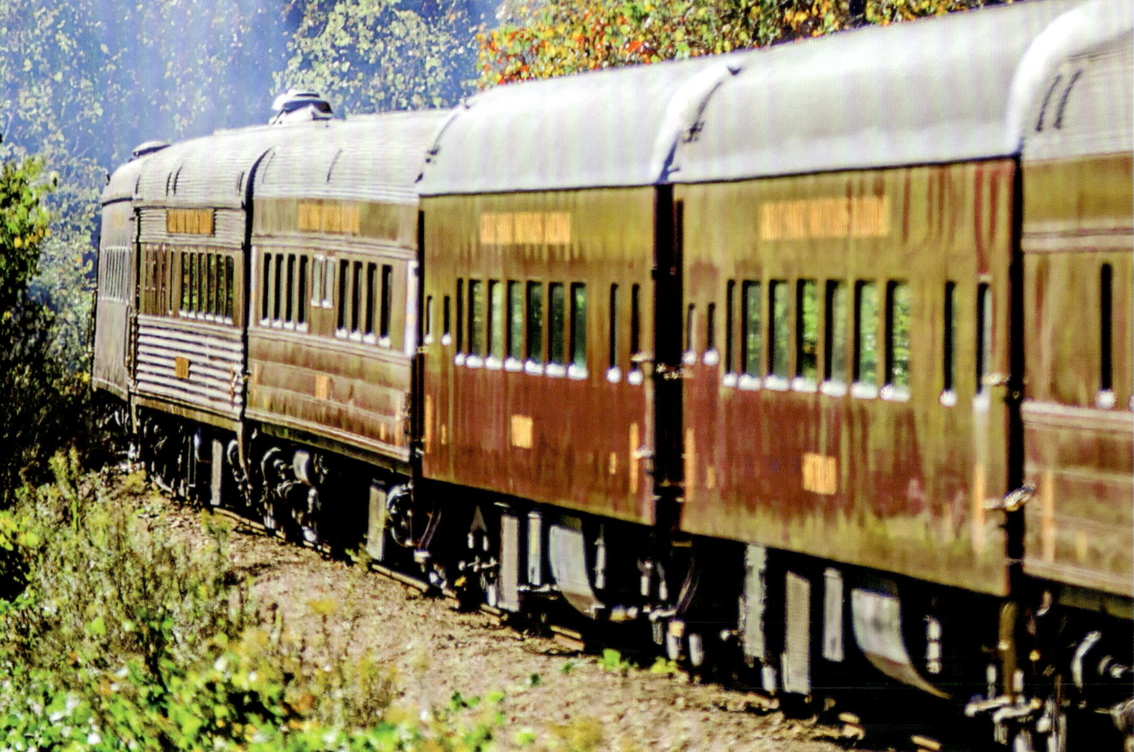

Top: The railroad owns seven diesel locomotives, two of which are no longer operational because they were wrecked during the filming of the movie *The Fugitive*.

Bottom: Gradients on the Great Smoky Mountains Railroad reach four percent in two areas.

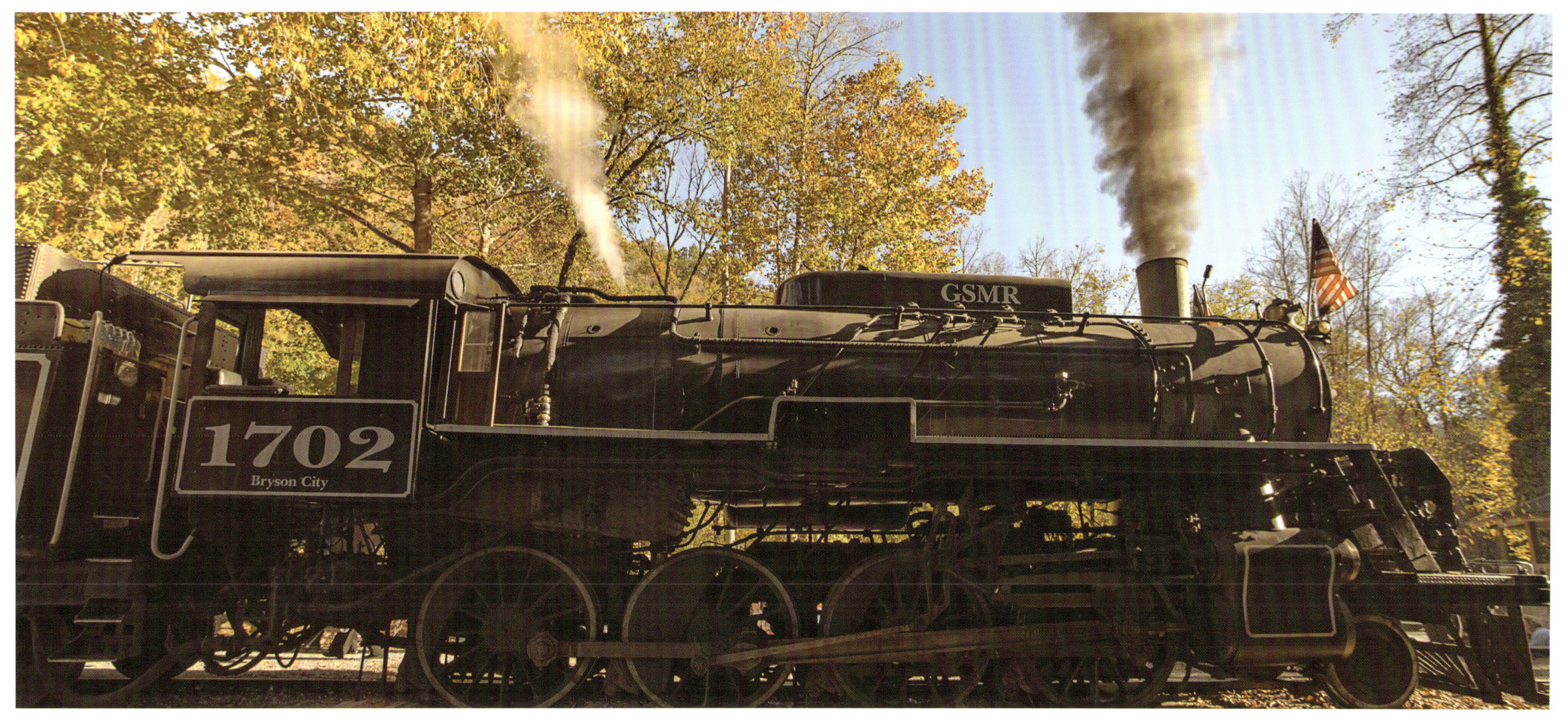

Steam locomotive #1702 was built in 1942 and was acquired by the Great Smoky Mountains Railroad in 1994. It was taken out of service in 2004, underwent a complete restoration, and returned to service in 2016.

Operating on 53 miles of the old Murphy Branch of western North Carolina, the Great Smoky Mountains Railroad travels through the scenic Nantahala National Forest between the mountain towns of Dillsboro and Andrews, North Carolina. Traveling through "fertile valleys, a tunnel, and across river gorges," according to the Great Smoky Mountains Railroad, the tourist train crosses some of Appalachia's finest terrain just south of the Great Smoky Mountains National Park.

The tracks were originally built as the Murphy Branch on what was then called the Western North Carolina Railroad. The Murphy Branch opened the mountainous and isolated areas west of Asheville, North Carolina, to the larger circles of commerce and culture that were connected to the Western North Carolina Railroad.

Built between 1881 and 1884 with convict labor, the Murphy Branch shipped timber and other products from the mountains for nearly a century. The Murphy Branch changed hands when the Western North Carolina Railroad was sold at foreclosure to the Southern Railway in 1894. In the 1980s the tracks laying west of Sylva, North Carolina, were closed due to declining freight traffic.

The North Carolina Department of Transportation (NCDOT) bought the portion of tracks that ran west from Sylva in 1988 and granted the right-of-way between Dillsboro and Andrews to the Great Smoky Mountains Railroad. In 1996, the Great Smoky Mountains Railroad bought the remaining section of track from the NCDOT.

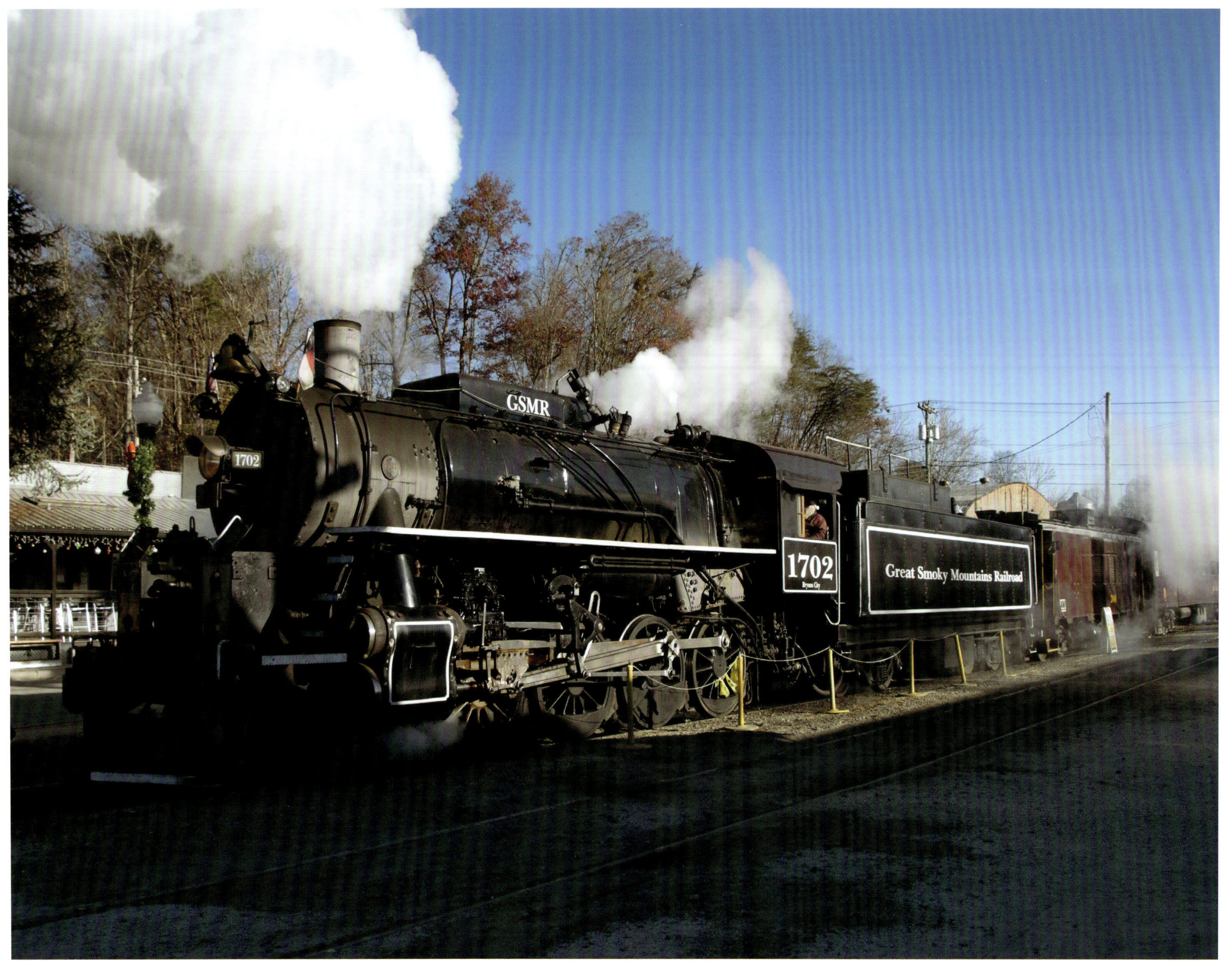

Winter is an excellent time to travel on the Great Smoky Mountains Railroad. It offers a ride inspired by *The Polar Express*, complete with hot cocoa, sweet treats, and a visit from Santa Claus.

Today the Great Smoky Mountains Railroad runs nearly 1,000 excursions a year on its tracks, traveling through the Great Smoky and Blue Ridge Mountains. It's one of the most popular tourist railroads in the country. Visitors can see over 125 varieties of trees, including hemlock, mountain ash, sugar maple, and umbrella magnolia. The area is also home to more than 100 types of fish and animals. This section of Appalachia is filled with waterfalls, rushing rivers, and picturesque streams that captivate the eye and imagination.

The railroad no longer operates all the way to Andrews due to damaged tracks. Service now runs between Bryson City and Nantahala or Dillsboro. The Nantahala Gorge Excursion is the longer trip of the two, traversing 44 miles from Bryson City to Nantahala and back again. The Tuckasegee River Excursion to Dillsboro clocks in at 32 miles round trip. Excursions take place on either a diesel or steam locomotive.

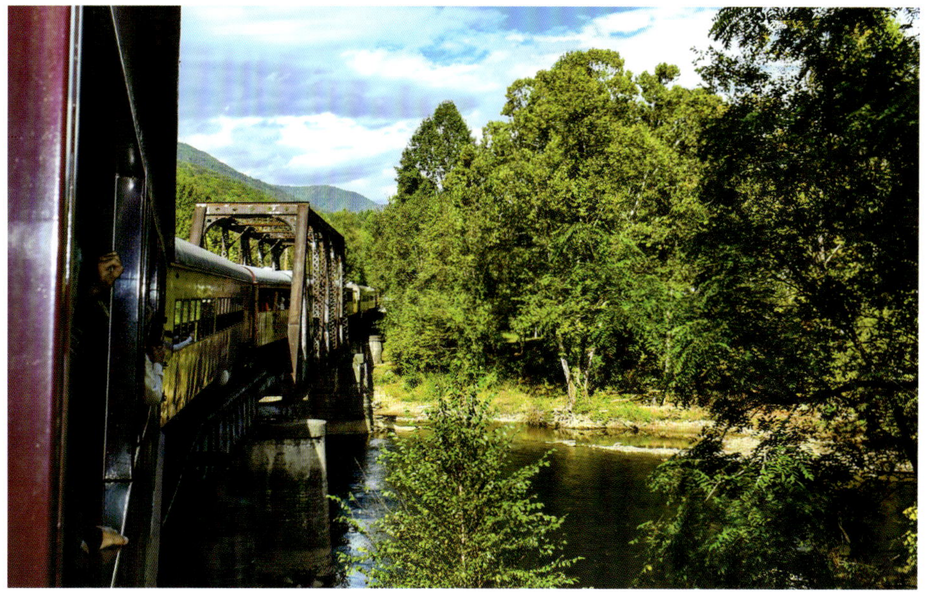

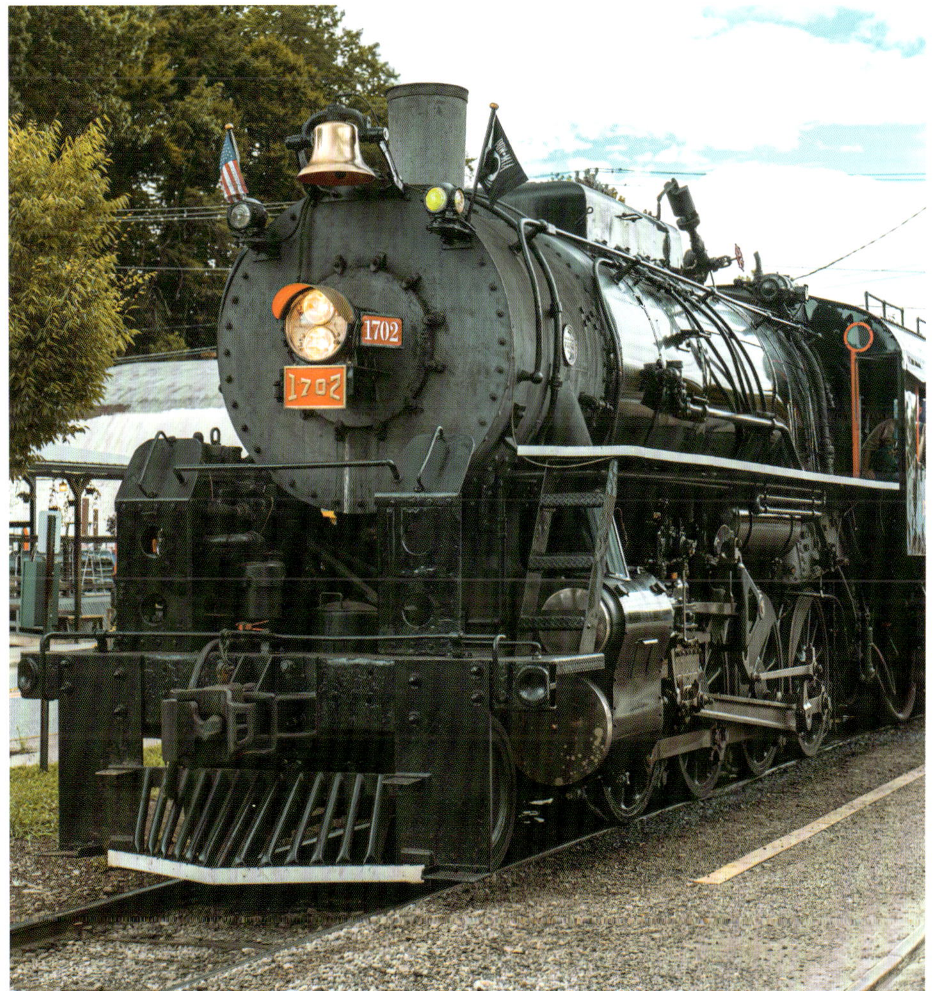

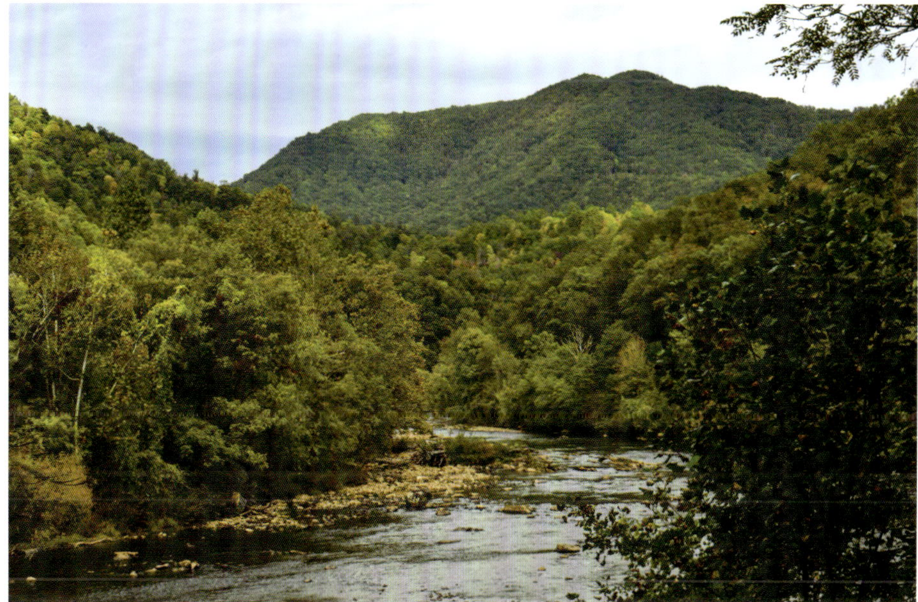

Left: The Great Smoky Mountains Railroad holds many special event excursions, featuring drinks and dining.

Right: Scenic views captured on the Tuckasegee River Excursion.

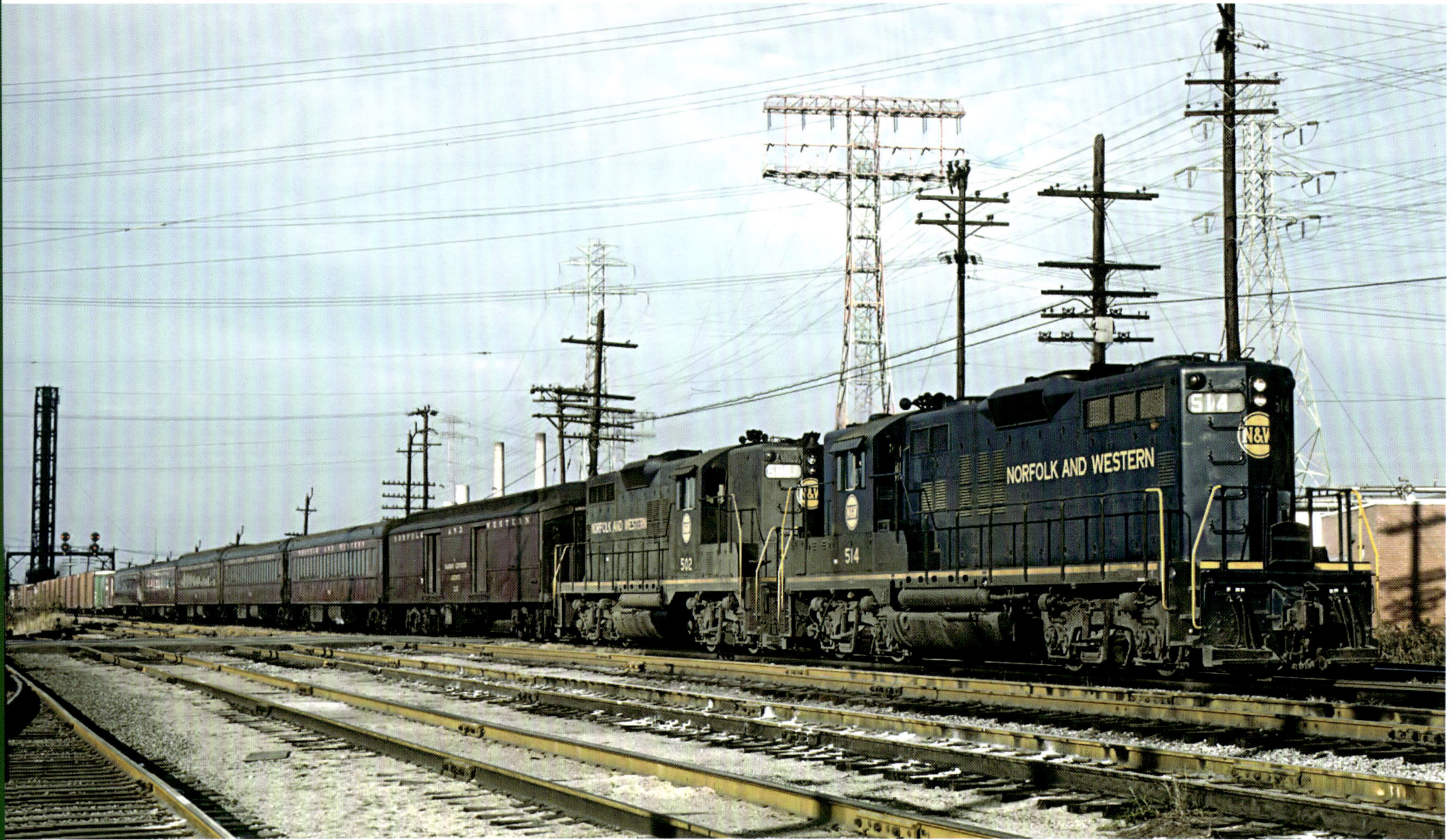

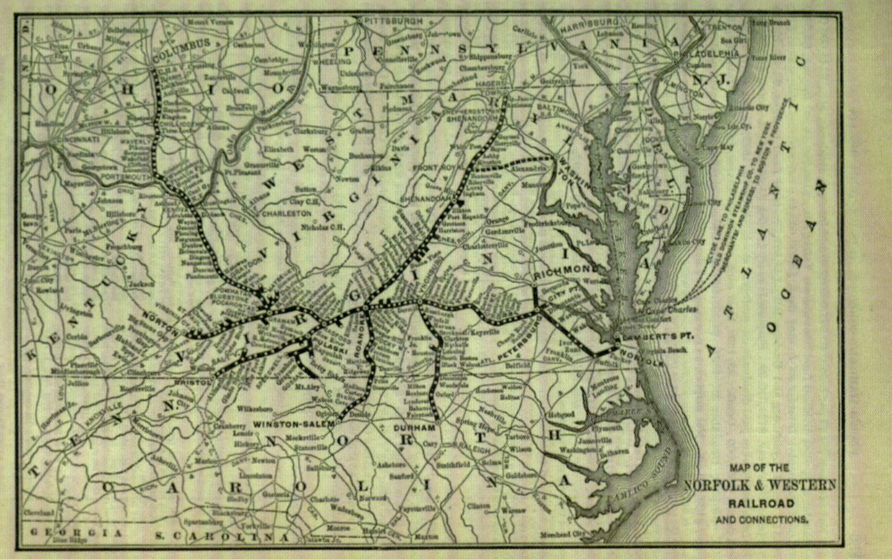

Top: The *Pocahontas* departing Norfolk, Virginia, in 1968.

Bottom: A map of the Norfolk and Western Railway from 1893.

The Norfolk and Western Railway (N&W) was originally chartered in 1838 as the City Point Railroad in Virginia. Over the following decades the name changed several times due to mergers and expansions. The railway profitably transported primarily agricultural products during that time, but it faced challenges during the Civil War and Panic of 1873. A foreclosure auction in 1881 led to its reorganization into the Norfolk and Western Railway. Frederick J. Kimball, a civil engineer, became vice president of N&W that year, and his interest in geology led the company to open the Pocahontas coalfields in Virginia and West Virginia, and transport coal along the railway.

During the late nineteenth and early twentieth centuries, N&W became renowned for its coal-hauling efficiency and powerful steam locomotives. The railroad built its own steam engines at the Roanoke Shops in Virginia, producing high-quality locomotives such as the Class A, Class J, and Y6. These locomotives were known for their durability and strength, allowing N&W to move heavy freight efficiently over mountainous terrain. It acquired the nickname "British Railway of America" because a railroad building its own steam locomotives was a rare practice outside of Britain.

In the mid-twentieth century, N&W expanded its lines and absorbed other railways, including the Virginian Railway, Nickel Plate Road, Wabash Railroad, and Pittsburgh and West Virginia Railway. These mergers diversified the company's freight base, adding routes into the Midwest and increasing its ability to transport goods beyond coal. While many other railroads transitioned to diesel power in the 1940s and early 1950s, N&W remained committed to steam well into the late 1950s, becoming one of the last major railroads in the U.S. to convert to diesel locomotives. By the 1970s, N&W had fully transitioned to diesel locomotives and adapted to modern railroading practices, maintaining strong profitability despite the struggles faced by other railroads during this period.

In 1982 N&W merged with the Southern Railway to form Norfolk Southern Railway, a move that created one of the largest and most powerful railroad systems in the United States. Today the former N&W mainlines remain critical corridors for freight transport, and its impacts on the industry and region are celebrated through preserved steam locomotives, historical societies, and museum exhibits, particularly in Roanoke, Virginia, the heart of N&W's steam locomotive legacy.

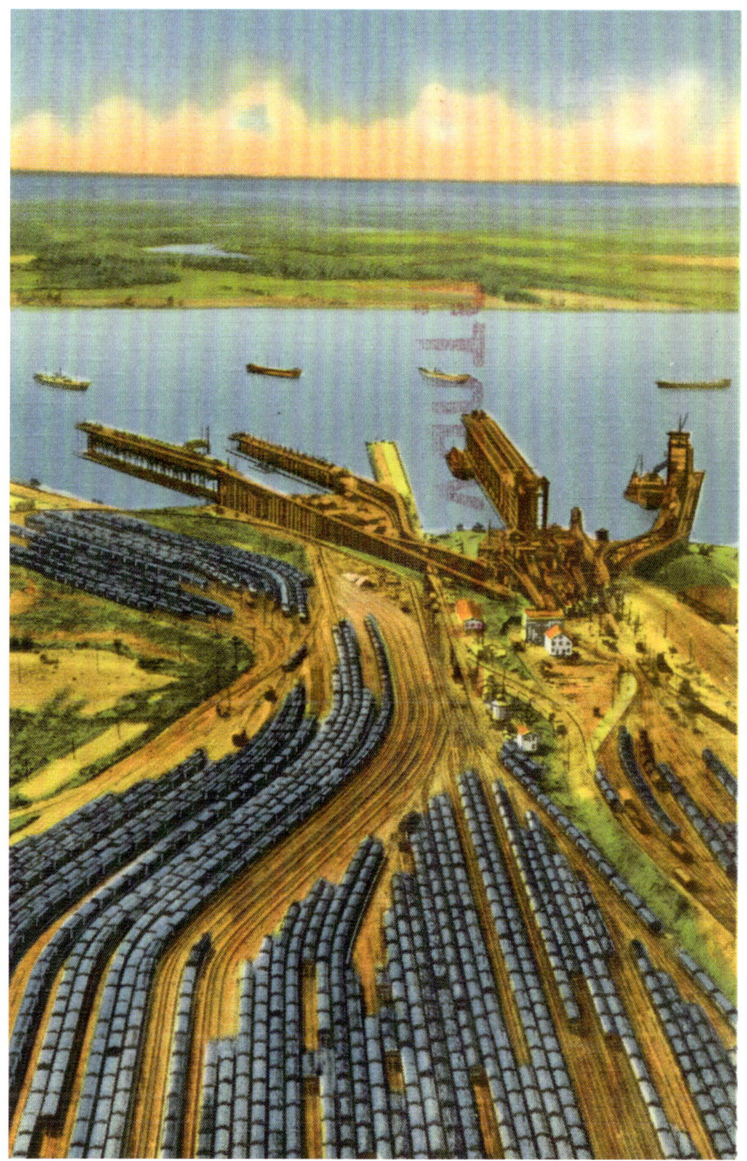

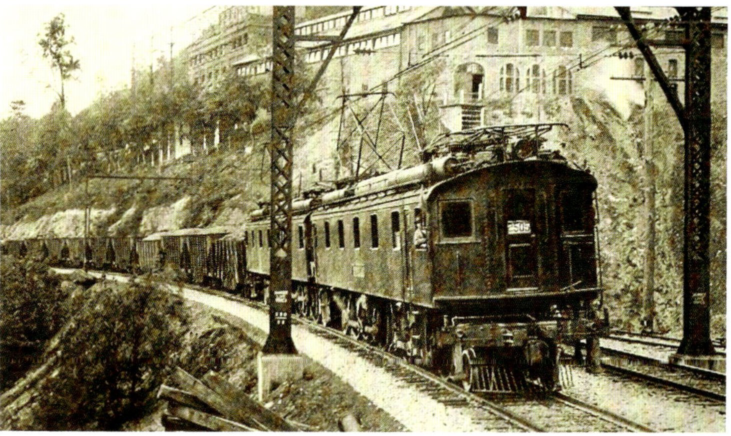

Top: The Norfolk and Western Railway was extended to the Lambert Point Coal Pier in Norfolk, Virginia, in 1886.

Bottom: An electric locomotive transporting coal on the Norfolk and Western Railway during World War I.

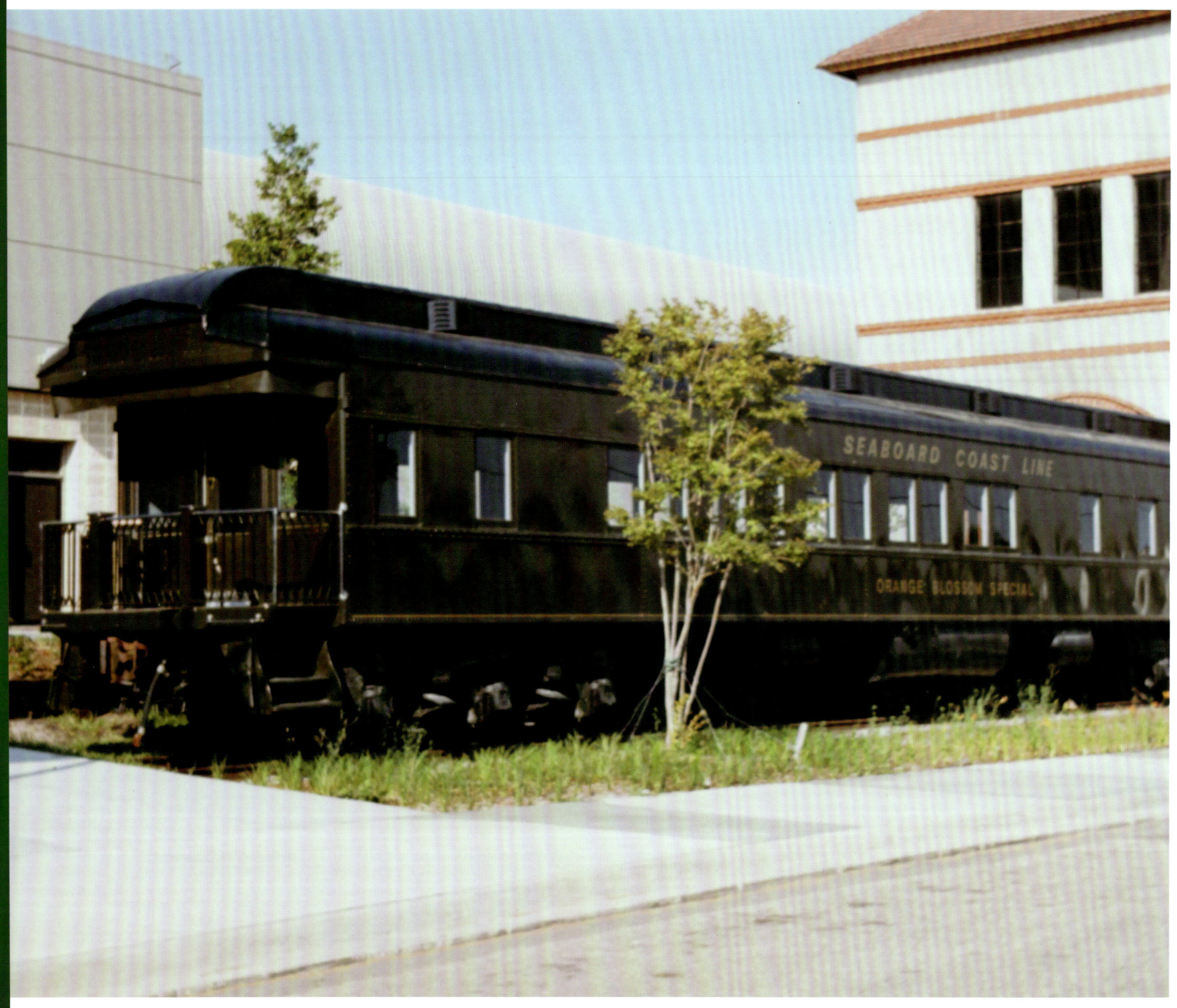

A Seaboard Coast Line Railroad coach, the *Orange Blossom Special*, in Jacksonville, Florida, in 1987.

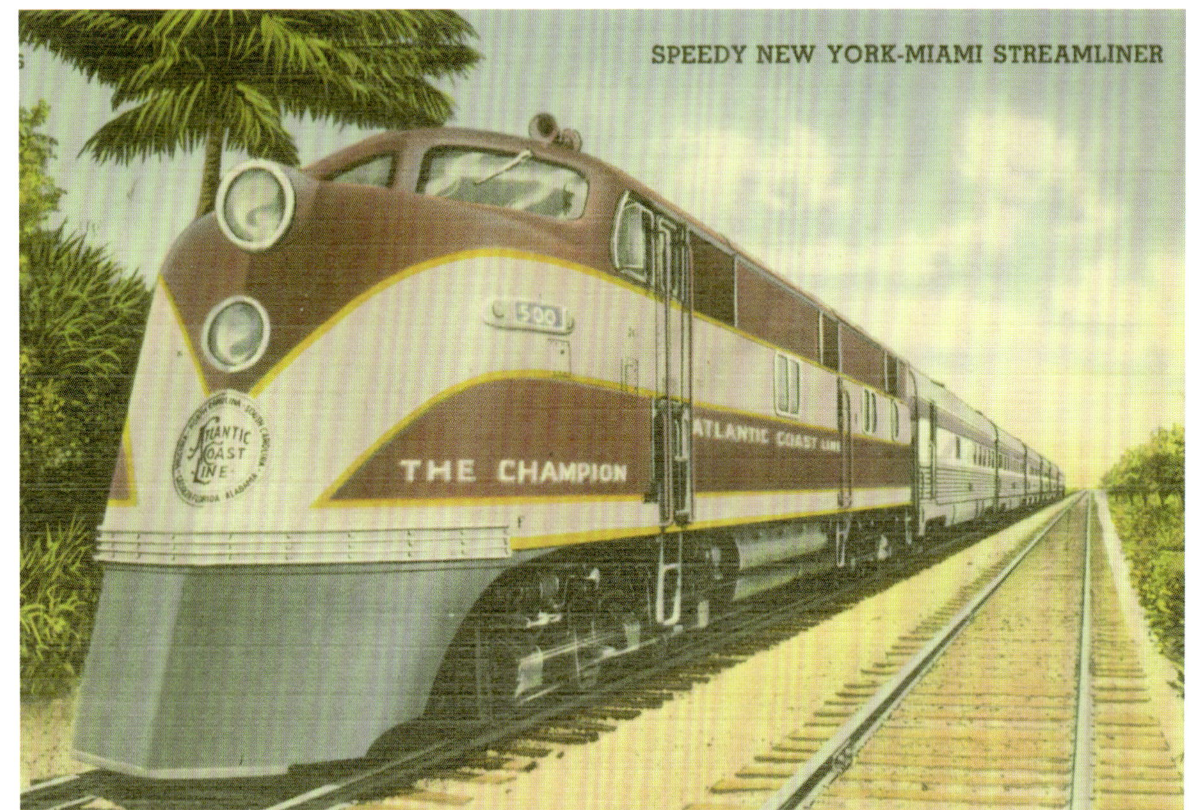

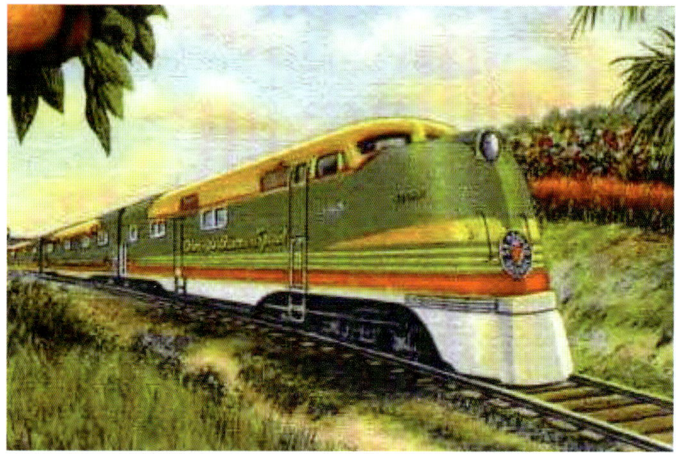

Left: An illustration of the *Champion*, the Atlantic Coast Line Railroad's foremost streamliner, from 1941.

Right: A postcard image of a Seaboard Air Line Railroad streamliner in Florida, circa 1940.

The Seaboard Coast Line Railroad (SCL) was formed in 1967 through the merger of two longtime competitors, the Atlantic Coast Line Railroad (ACL) and the Seaboard Air Line Railroad (SAL). This unification created a major rail system that stretched from Virginia to Florida and west into Alabama and Georgia, serving as a vital link for freight and passenger traffic in the Southeastern United States.

The ACL was one of the most financially stable and influential railroads in the South, with its roots dating back to the 1830s. By the early twentieth century, the ACL had developed a well-established mainline connecting Richmond, Virginia, to Florida, with key stops in cities including Charleston, Savannah, and Jacksonville. At its height, routes extended as far south as the Everglades and west to Birmingham, Alabama. It was known for its efficient freight operations that transported phosphate, coal, and manufactured products. The railroad's financial acumen made it resistant to transportation changes and economic downturns that plagued other railroads in the twentieth century.

In contrast the SAL was more innovative but often faced financial struggles. Despite its name, it was a traditional railroad, and the term "Air Line" referred to its direct, fast routes between major cities. The SAL's network, which also ran from Virginia to Florida, competed fiercely with the ACL. In 1939, the ACL launched an all-coach streamliner, the *Champion*, in response to the SAL's popular passenger streamliners, the *Silver Meteor* and *Silver Star*. The SAL was also ahead of its time in adopting diesel locomotives and lightweight trains, making it a leader in technological advancements despite its smaller financial footprint.

The merger of the ACL and SAL into Seaboard Coast Line Railroad in 1967 marked the end of their rivalry and the beginning of a more unified and competitive rail network in the Southeast. By combining their strengths, the SCL eliminated unnecessary duplicate routes, improved efficiency, and expanded freight services. Although the SCL retained its identity for a time, it became part of the Seaboard System Railroad in 1982 and later CSX Transportation in 1986, forming one of the largest rail systems in the United States. Much of its historic network remains in active use, continuing the legacy of its predecessors.

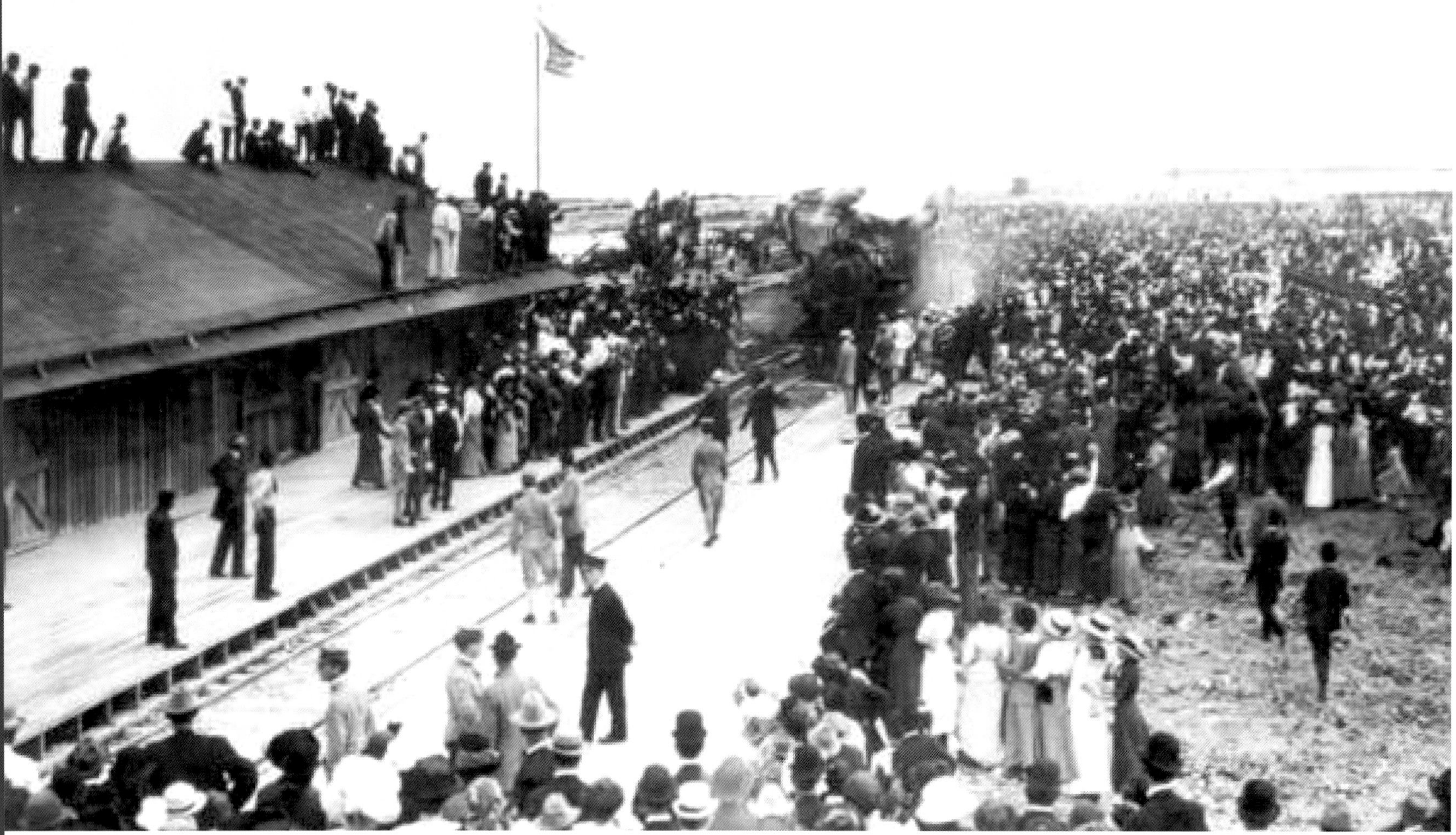

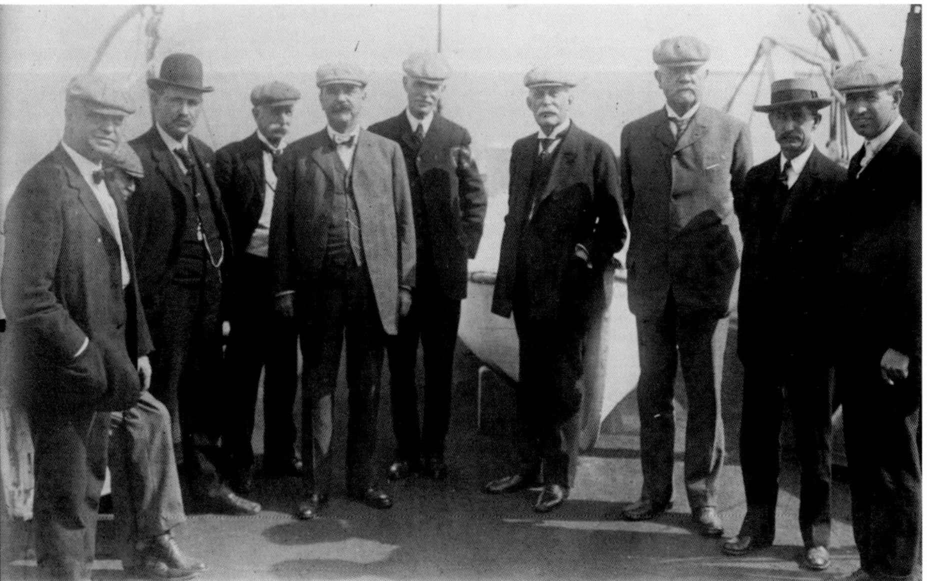

Top: The first FEC train arriving in Key West.

Bottom: Henry Flagler at the opening of the railway into the Florida Keys.

The Florida East Coast Railway (FEC) was influential in Florida's infrastructure development, largely due to the vision and ambition of its founder, Henry Flagler. Originally a co-founder of Standard Oil alongside John D. Rockefeller, Flagler turned his attention to Florida in the late nineteenth century, recognizing its potential as a tourism and agricultural hub.

Beginning in the 1880s, he upgraded a small Jacksonville–St. Augustine rail line, gradually extending it southward along Florida's east coast. During this expansion period, Flagler was also developing luxury hotels and resorts, transforming cities like Palm Beach and Miami into premier destinations. His railroad played a pivotal role in Miami's founding in 1896, as the city grew around the newly built the FEC tracks. When it was incorporated into a town with just a small population of 50 residents, the citizens wanted to name it Flagler, thanks to his role in its development, but he insisted it retain its Native American name, Miami.

One of the most remarkable feats in the FEC's history was the construction of the Overseas Railroad, an extension from Miami to Key West. Completed in 1912, this 128-mile stretch of rail crossed the open ocean, with massive concrete viaducts and bridges linking the Florida Keys. It was often called the "Eighth Wonder of the World" due to its engineering complexity. However, many workers perished during its construction, and the FEC was prosecuted for poor working conditions. The Labor Day Hurricane of 1935, one of the most powerful storms in U.S. history, destroyed much of the line. Unable to rebuild, the FEC sold the right-of-way to the state of Florida, which later converted it into the Overseas Highway, today's U.S. Route 1.

In the twentieth century, the FEC switched to diesel trains and embraced automation and intermodal shipping, which helped it survive while other Southeastern railroads struggled. The FEC's operations also depended heavily on rock trains, which transport limestone. Unlike many other major railroads that eventually merged into larger systems, the FEC remained an independent entity.

Today the FEC remains one of the most efficient and profitable regional railroads in the United States, specializing in freight transport between Florida's ports and the national rail network. It also played a key role in the development of Brightline in 2018, a privately operated intercity passenger train service. Brightline uses the FEC's historic corridor to connect Miami, Fort Lauderdale, West Palm Beach, and Orlando, continuing Flagler's vision of an interconnected Florida.

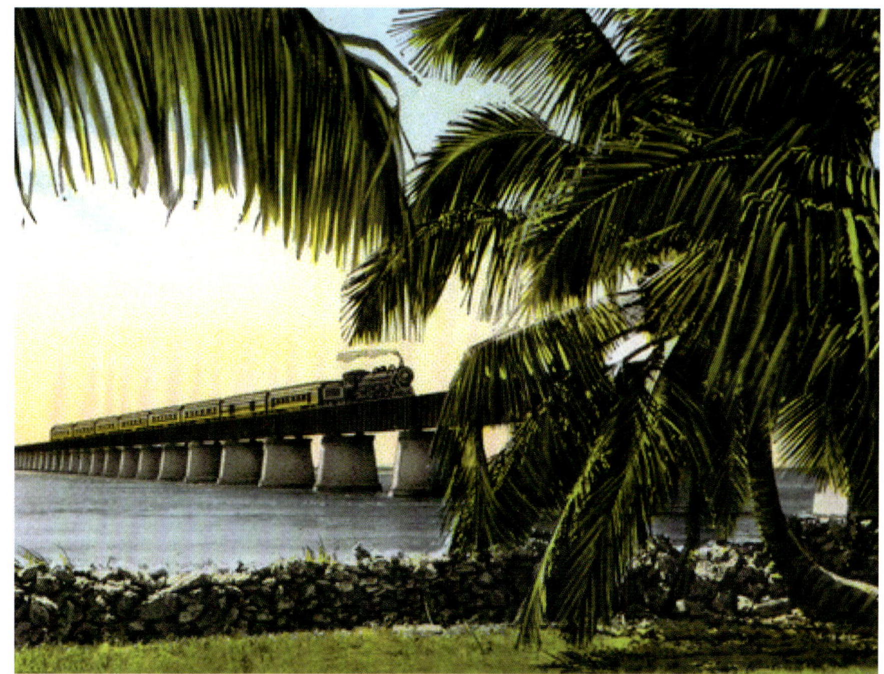

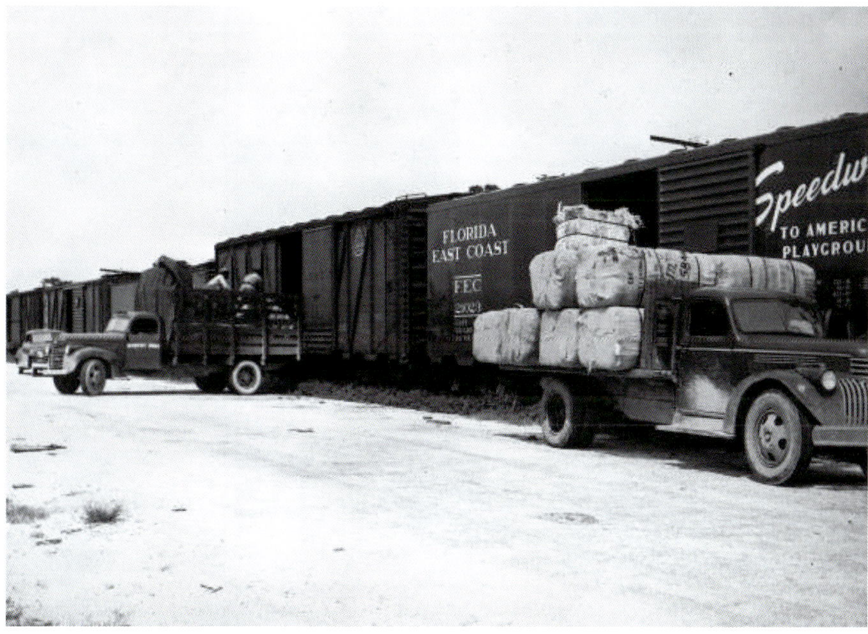

Top: A depiction of the Overseas Railroad crossing the Moser Channel into Key West, in 1920.

Bottom: Workers loading produce into refrigerated FEC cars circa 1950.

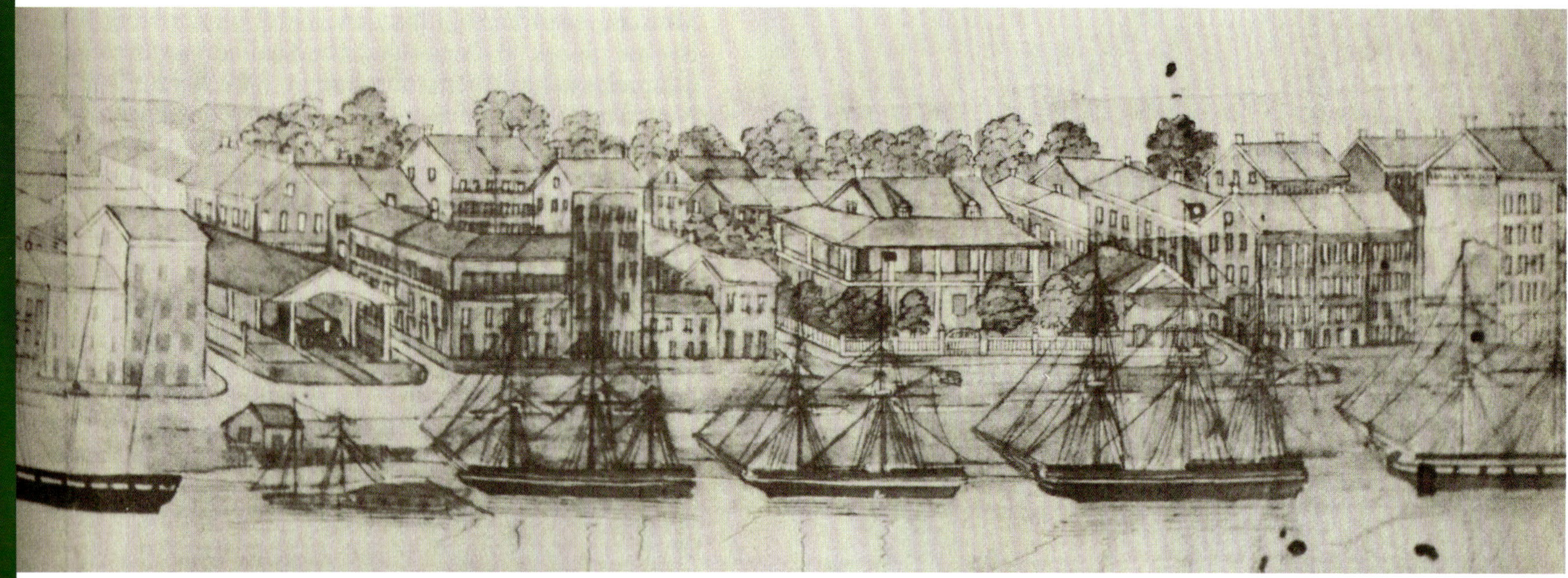

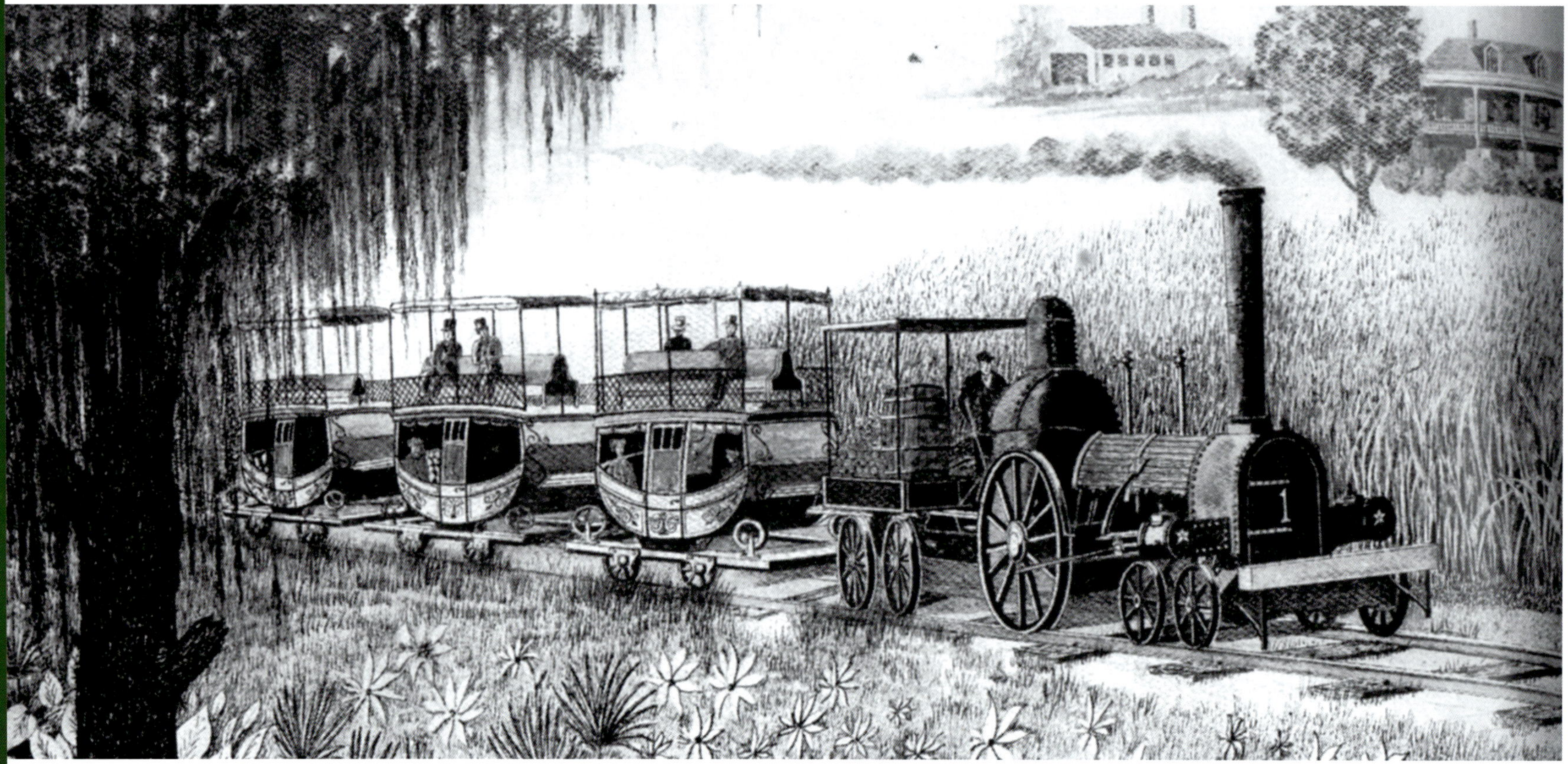

Top: An 1836 sketch of the Mississippi riverfront in New Orleans by G. W. Sully. On the left is a Pontchartrain Railroad station.

Bottom: An early depiction of a locomotive and carriages on the Pontchartrain Railroad near Elysian Fields Avenue and the shore of Lake Pontchartrain.

The Pontchartrain Railroad was among the earliest railroads constructed in the United States. It was chartered in 1830 to connect the Mississippi River to Lake Pontchartrain in New Orleans. It was also the third common carrier railroad, meaning its services were available to the general public and did not require contracts, to open in the U.S.

The railway facilitated trade and transportation between the city of New Orleans and ships arriving at Lake Pontchartrain, where lakeside ports served as an important alternative to the Mississippi River. During the railroad's construction, swampland was filled in along parts of the route to create a bed for the tracks. The five-mile-long line opened in 1831 and ran along the iconic Elysian Fields Avenue from the city's riverfront Faubourg Marigny neighborhood to the lakefront Milneburg neighborhood.

When the Pontchartrain Railroad opened, horses pulled the trains and carriages, but by 1832, steam locomotives were introduced. The line quickly became a key transportation route for many goods, helping to fuel the economic growth of New Orleans. The Pontchartrain Railroad also played a role in early tourism, as New Orleanians used the train to reach Milneburg, where they could enjoy lakefront resorts and other attractions.

Despite its early success, the Pontchartrain Railroad struggled to remain relevant in the twentieth century as freight transport outgrew the line. When Milneburg resorts closed due to a land reclamation project in the late 1920s, passenger service also declined sharply. The last passenger service occurred on March 15, 1932, and the city filled the transit gap with bus services. The line changed ownership multiple times before eventually being abandoned in 1935, marking the end of over a century of service.

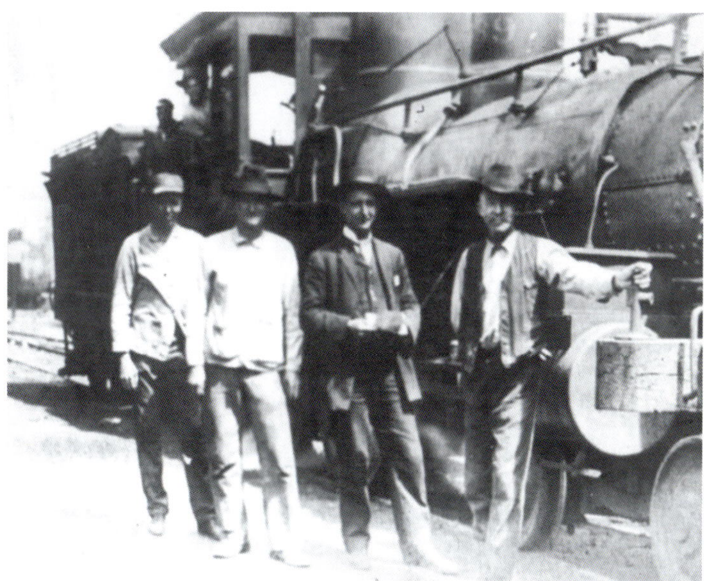

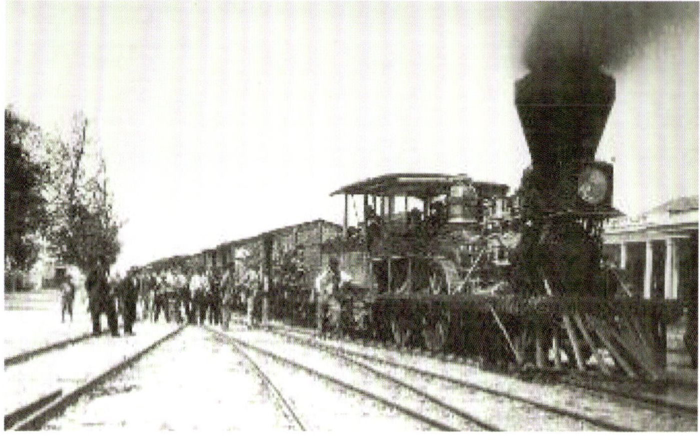

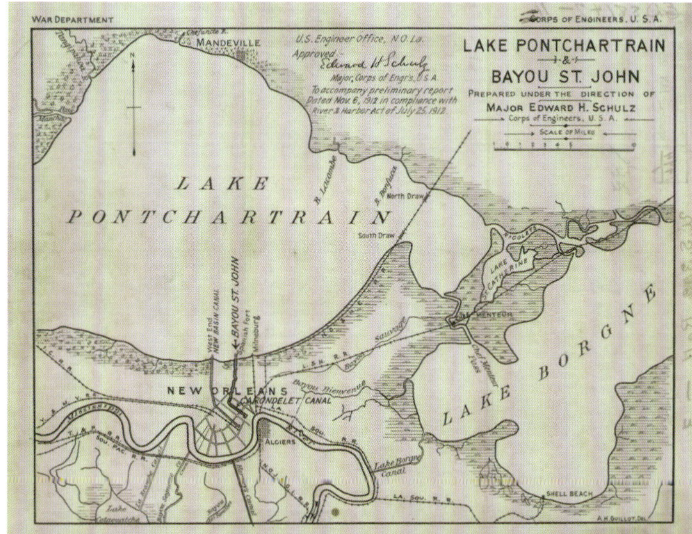

Top: Engineers, conductors, and other rail workers posing by the *Smoky Mary* locomotive in 1904.

Middle: A photo of a Pontchartrain Railroad train in Milneburg, New Orleans, taken sometime in the 1860s.

Bottom: A map from 1912 showing the bayou and Lake Pontchartrain region. Ocean-faring ships often exchanged goods at Lake Pontchartrain, and the railroad would transport cargo to and from the Mississippi River for riverboat transport in the interior United States.

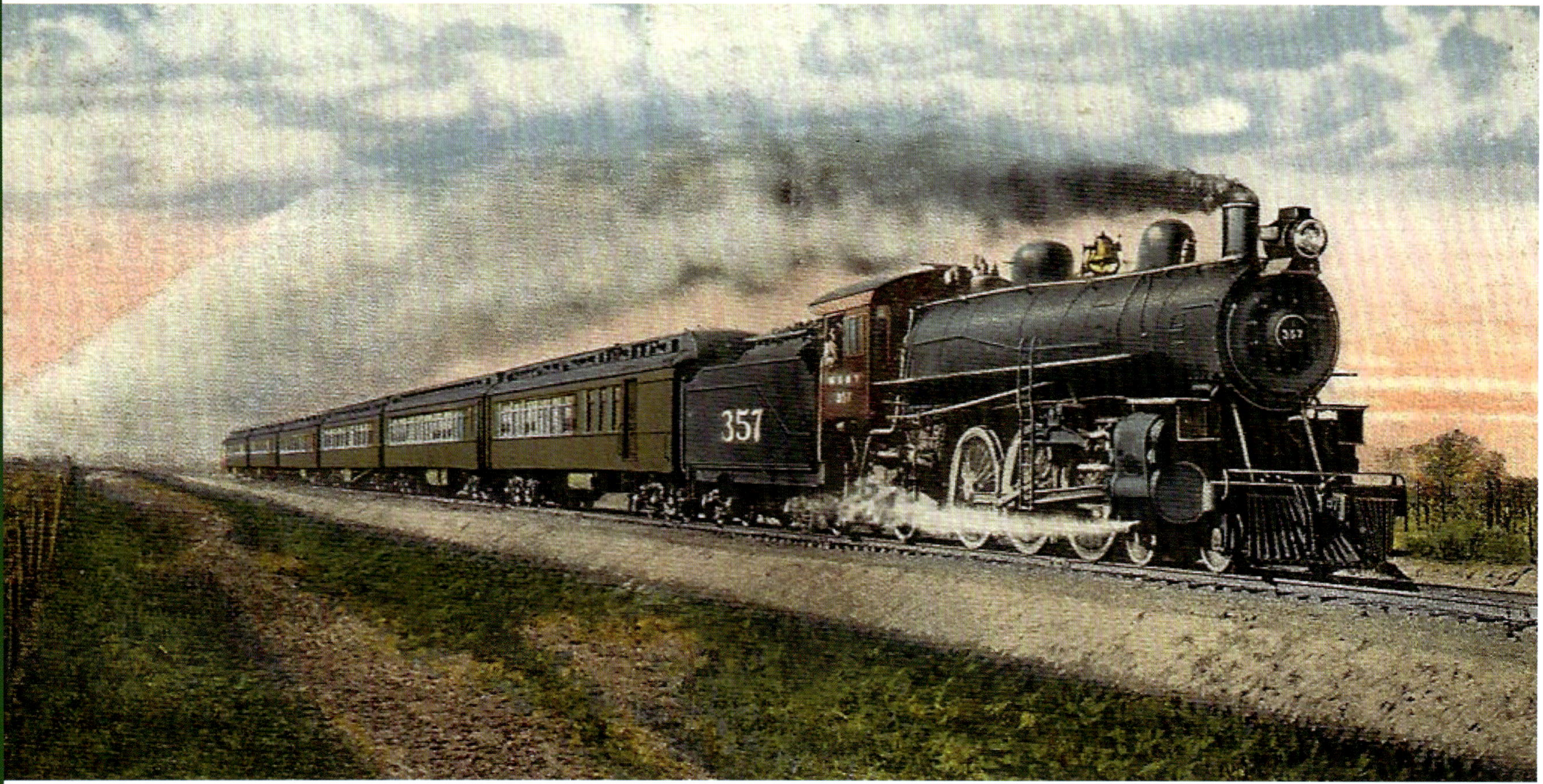

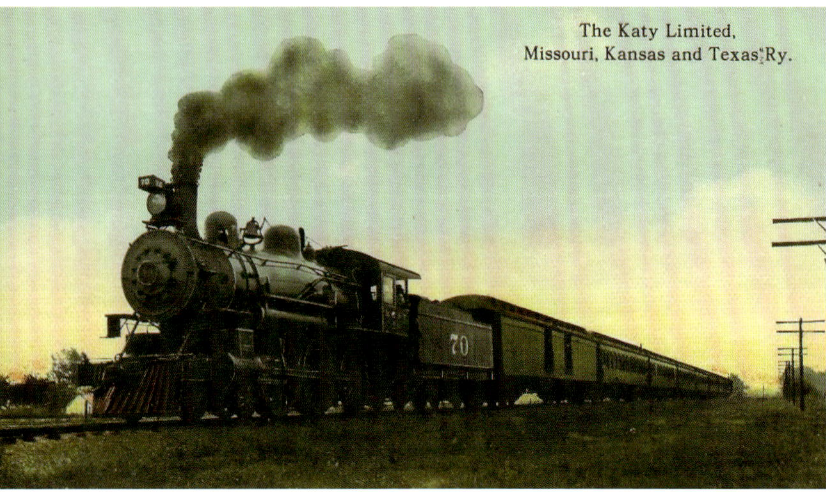

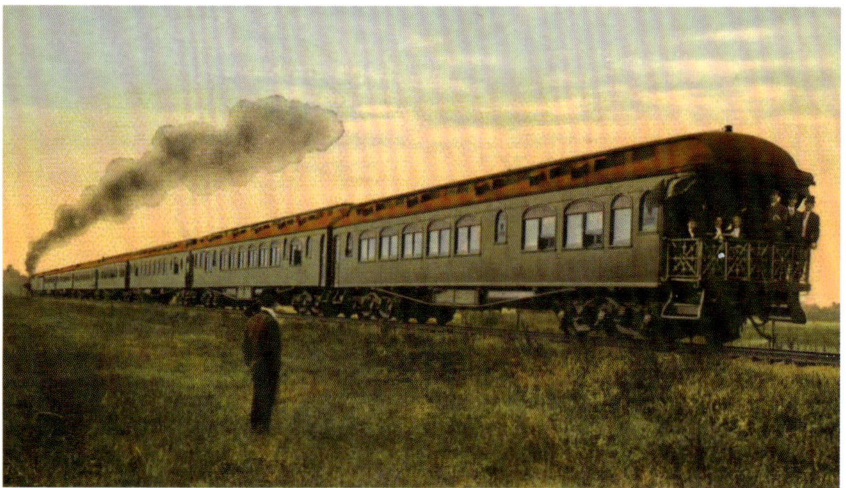

Top: A postcard of the *Texas Special* circa 1919.

Bottom Left: Based on the paint scheme, this postcard of the *Katy Limited* dates to between 1900 and 1923.

Bottom Right: A postcard showcasing the *Katy Flyer* passenger train, which would run between San Antonio and St. Louis or Kansas City.

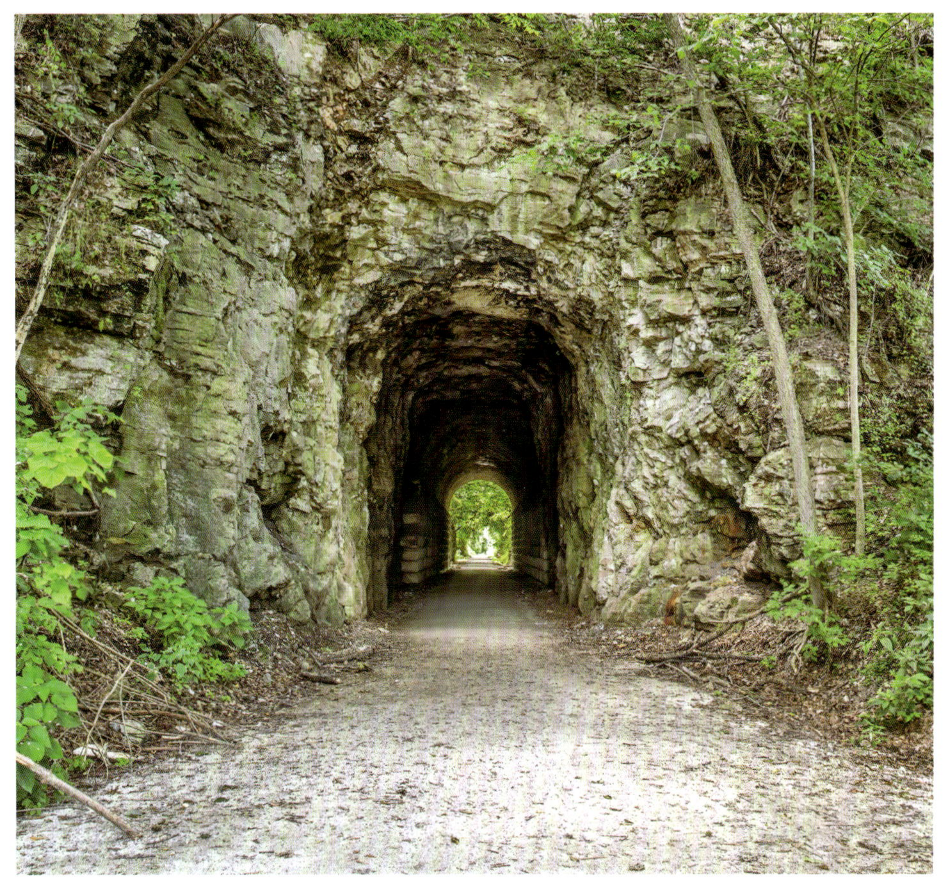

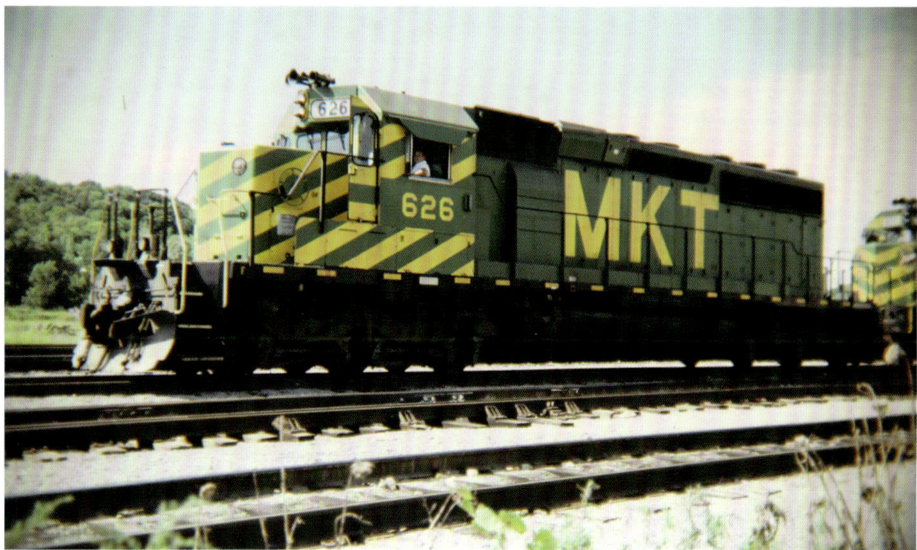

Left: A path passing through an old MKT tunnel in Katy Trail State Park just outside of Rocheport, Missouri.

Right: An MKT locomotive passing through Kansas City, in 1986.

The Missouri–Kansas–Texas Railroad (MKT), commonly known as "the Katy" due to its timetable and stock exchange abbreviation, was incorporated in 1870 with the goal of building a rail line through Kansas and into Indian Territory in present-day Oklahoma. Congress had promised land grants to the first railroad to reach the southern Kansas border, but courts overturned the grants because the land belonged to Native American tribes.

During its first year the railroad did receive other government land grants to connect military posts in Kansas, Oklahoma, and eventually Texas, which included Fort Riley, Fort Scott, and Fort Worth. Despite the other failed land grants promise, the MKT succeeded in reaching the southern Kansas border ahead of its competitors—with the help of 182 miles of track it purchased from the Union Pacific Railroad, Southern Branch during the MKT's incorporation. The railroad continued to extend its tracks southward from Junction City, Kansas, through the Indian Territory, reaching the Red River and the Texas border in 1872, where it established the town of Denison as its first Texas hub. Expansion continued throughout the late nineteenth and early twentieth centuries, with the MKT extending its network to key Texas cities, including Dallas, Houston, Waco, Austin, and San Antonio. The railroad faced competition from other major lines but maintained a strong presence in the region. The MKT also became known for its passenger service, introducing popular trains such as the *Texas Special*, the *Katy Limited*, and the *Katy Flyer*, which provided connections between Texas and Midwestern cities.

The railroad eventually faced financial struggles and declining passenger service in the mid-twentieth century. In 1988 the Missouri Pacific Railroad and the Union Pacific Corporation purchased the MKT, returning it to the company from which it acquired its original tracks more than one hundred years earlier. A portion of the MKT's tracks now form the country's longest continuous rail trail at 240 miles in Missouri's Katy Trail State Park.

THE MIDWEST

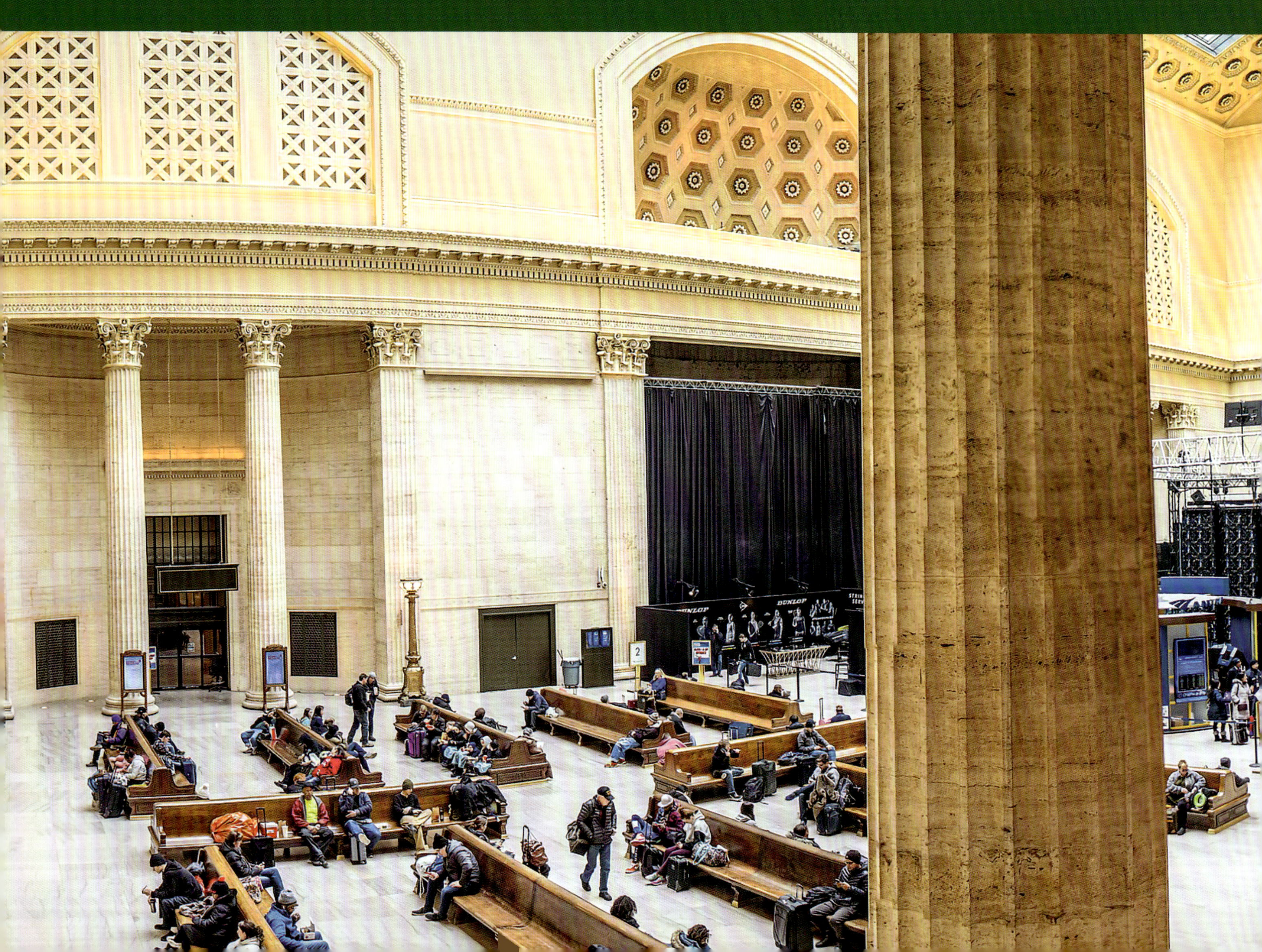

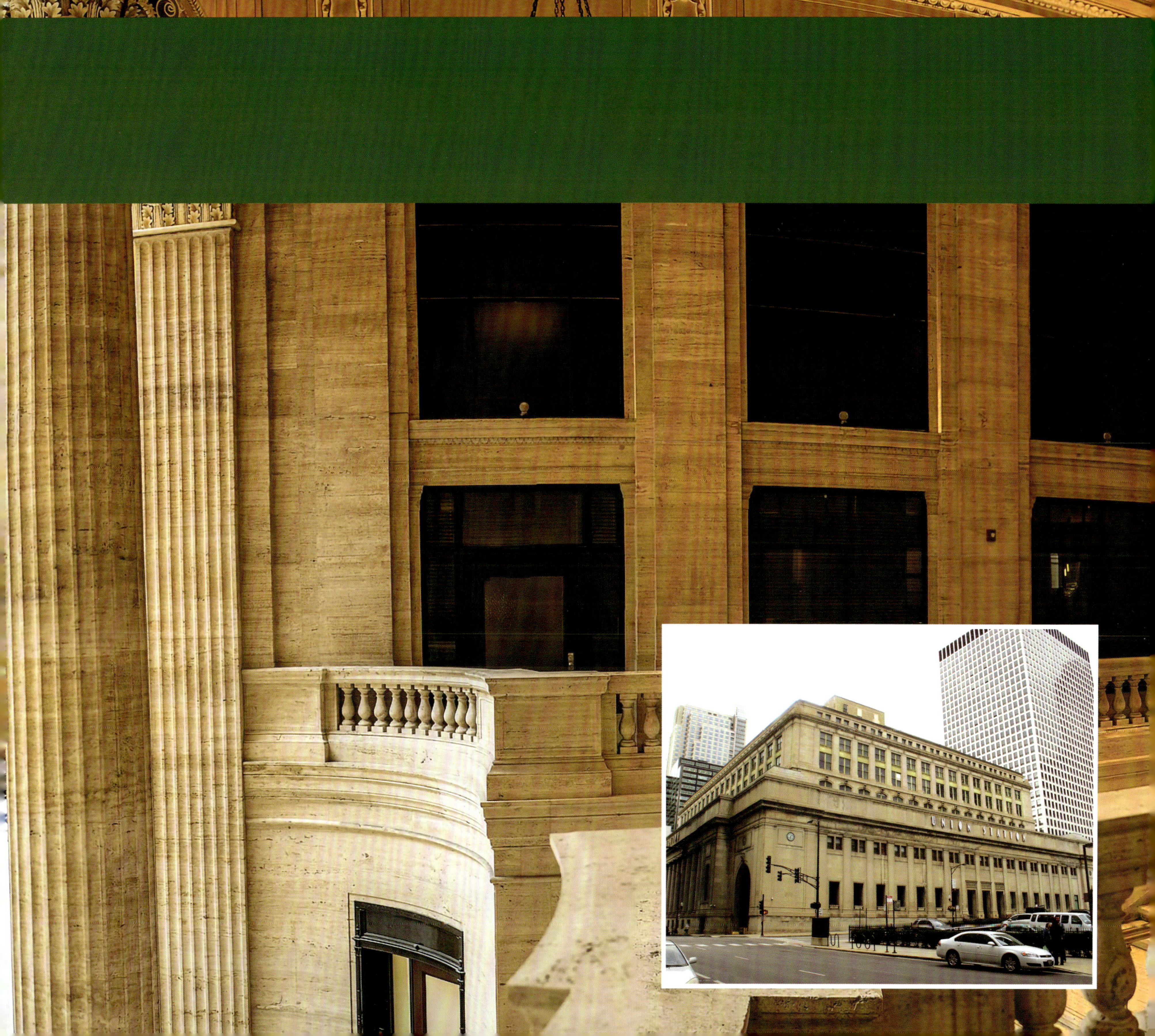
UNION STATION

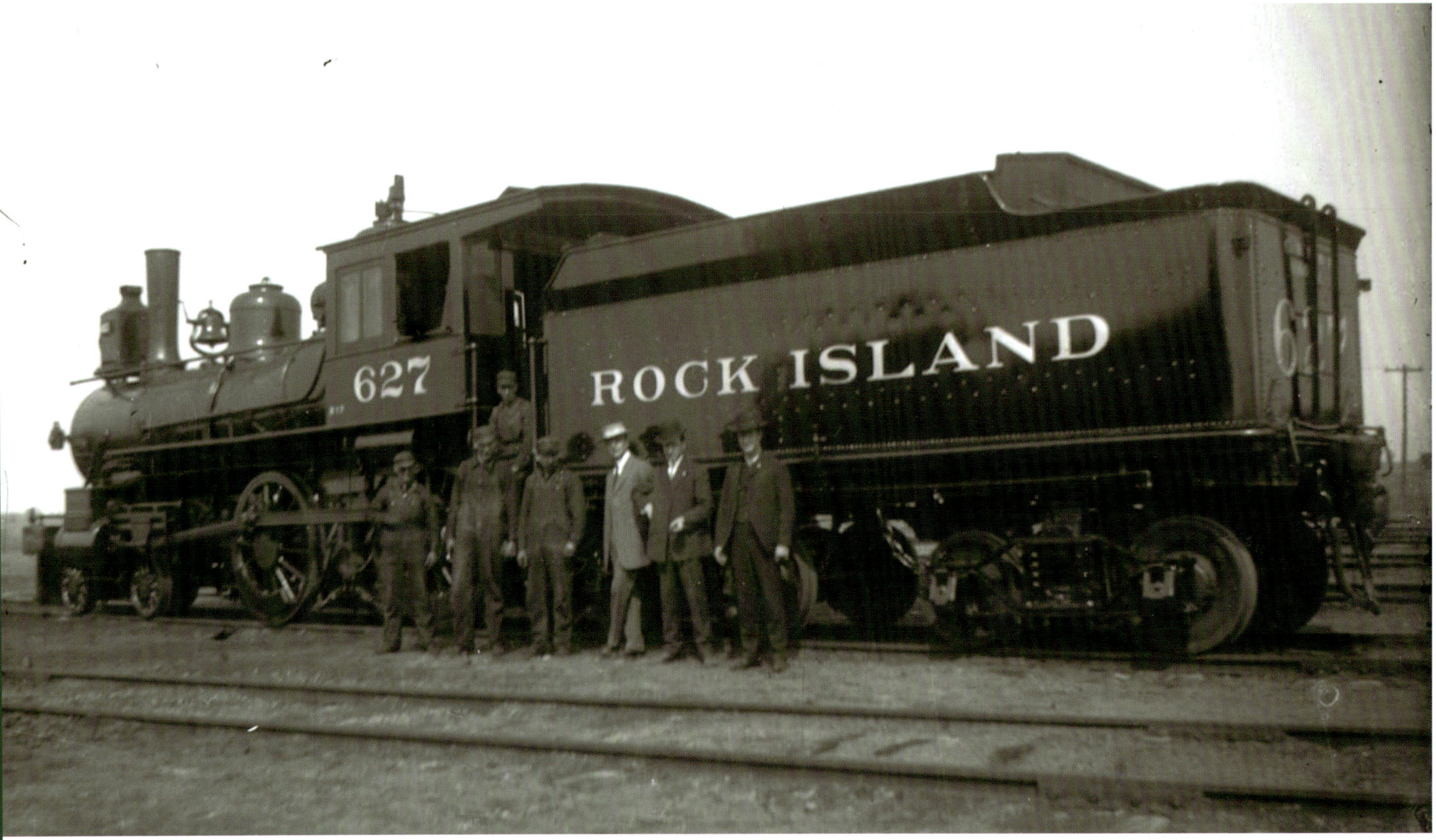

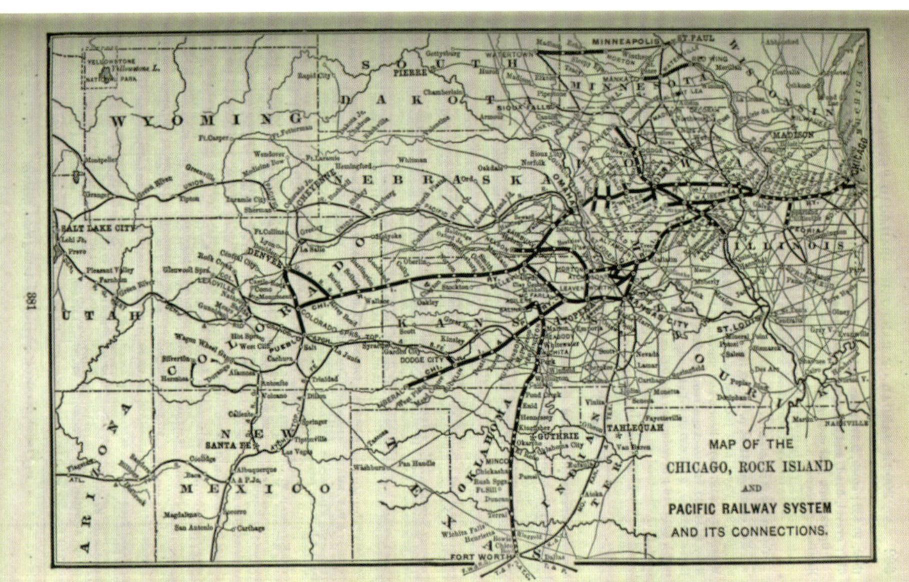

Top: A Rock Island locomotive circa 1910.

Bottom: A map of the expansive Rock Island Line from 1897.

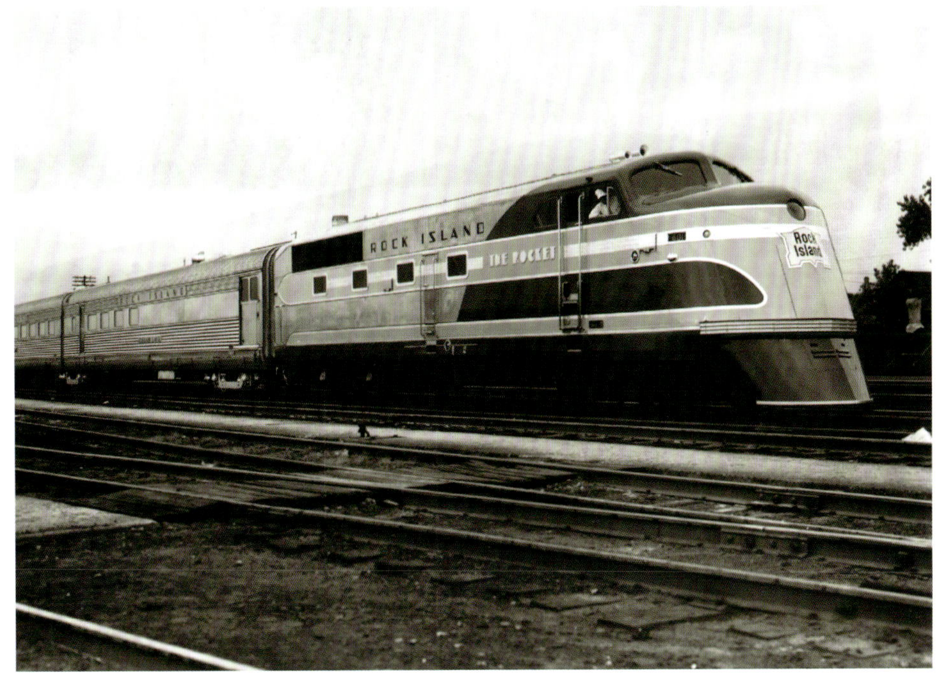

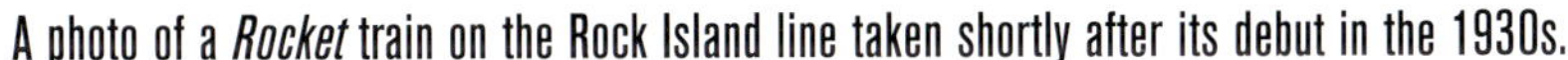
A photo of a *Rocket* train on the Rock Island line taken shortly after its debut in the 1930s.

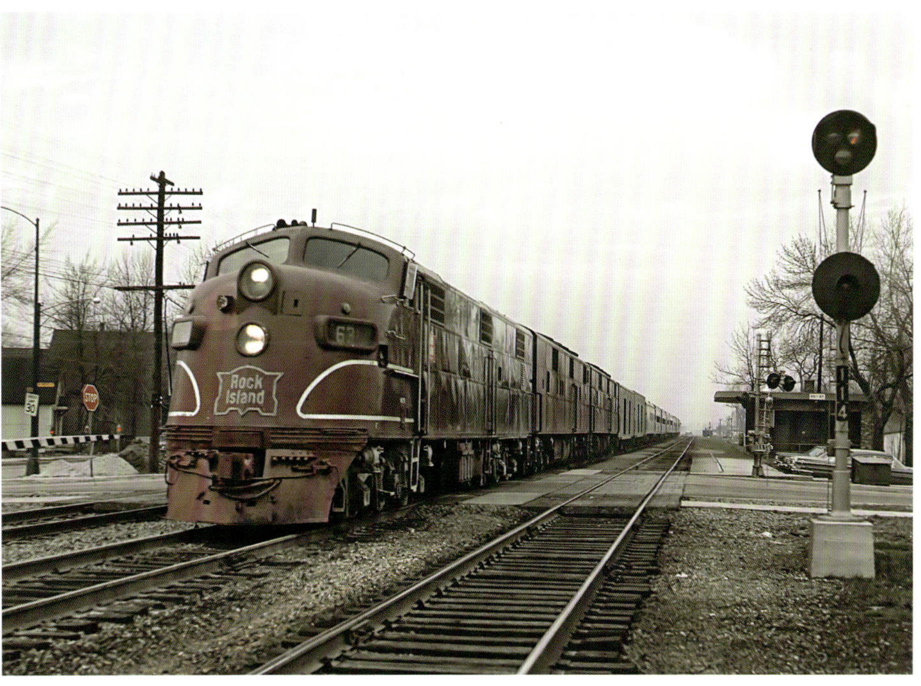

A *Golden State* passenger train in Washington Heights, Illinois, in 1965.

The Chicago, Rock Island and Pacific Railroad, commonly known as the Rock Island Line, was a major Midwestern railway that operated from the mid-nineteenth century until its bankruptcy in 1980. Founded in 1852, it initially connected Chicago with Joliet, Illinois. In 1854, it completed a route to Rock Island, Illinois, and became the first railroad to connect Chicago to the Mississippi River. Over time, the railroad expanded west and south, building an extensive network that reached across Iowa, Missouri, Kansas, Nebraska, and South Dakota and into Tennessee, Texas, Colorado, and New Mexico.

The Rock Island Line played a crucial role in the development of the Midwest by transporting agricultural products, livestock, and manufactured goods via its freight services. In the mid-twentieth century, it was also known for its fast and modern intercity passenger trains. The railroad introduced its famous diesel-powered *Rocket* streamliners in the 1930s, competing with larger railroads for long-distance passengers. The *Golden State Limited*, a joint service with the Southern Pacific Railroad, provided a connection for passengers traveling between Chicago and Los Angeles, with stops in Kansas City and El Paso. At its inauguration in 1902, the *Golden State Limited* had the longest route in the United States, at 2,762 miles.

Despite these expansive routes, the rise of automobile travel and airline competition in the mid-twentieth century eroded the Rock Island Line's passenger revenues. By the 1960s and 1970s, the company was in financial distress and employed cost-cutting measures, which led to poor train and track maintenance. A proposed merger with the Union Pacific Railroad also fell apart after a decade of hearings and regulatory delays.

In the 1980s the Rock Island Line was officially liquidated, marking one of the largest railroad shutdowns in U.S. history. The liquidation raised $500 million, enough to pay off the railroad's creditors. Portions of its network were acquired by other railroads, such as the Chicago and North Western and the Oklahoma, Kansas and Texas Railroad. Today remnants of the Rock Island Line can still be found in intercity routes, rail trails, and modern freight networks.

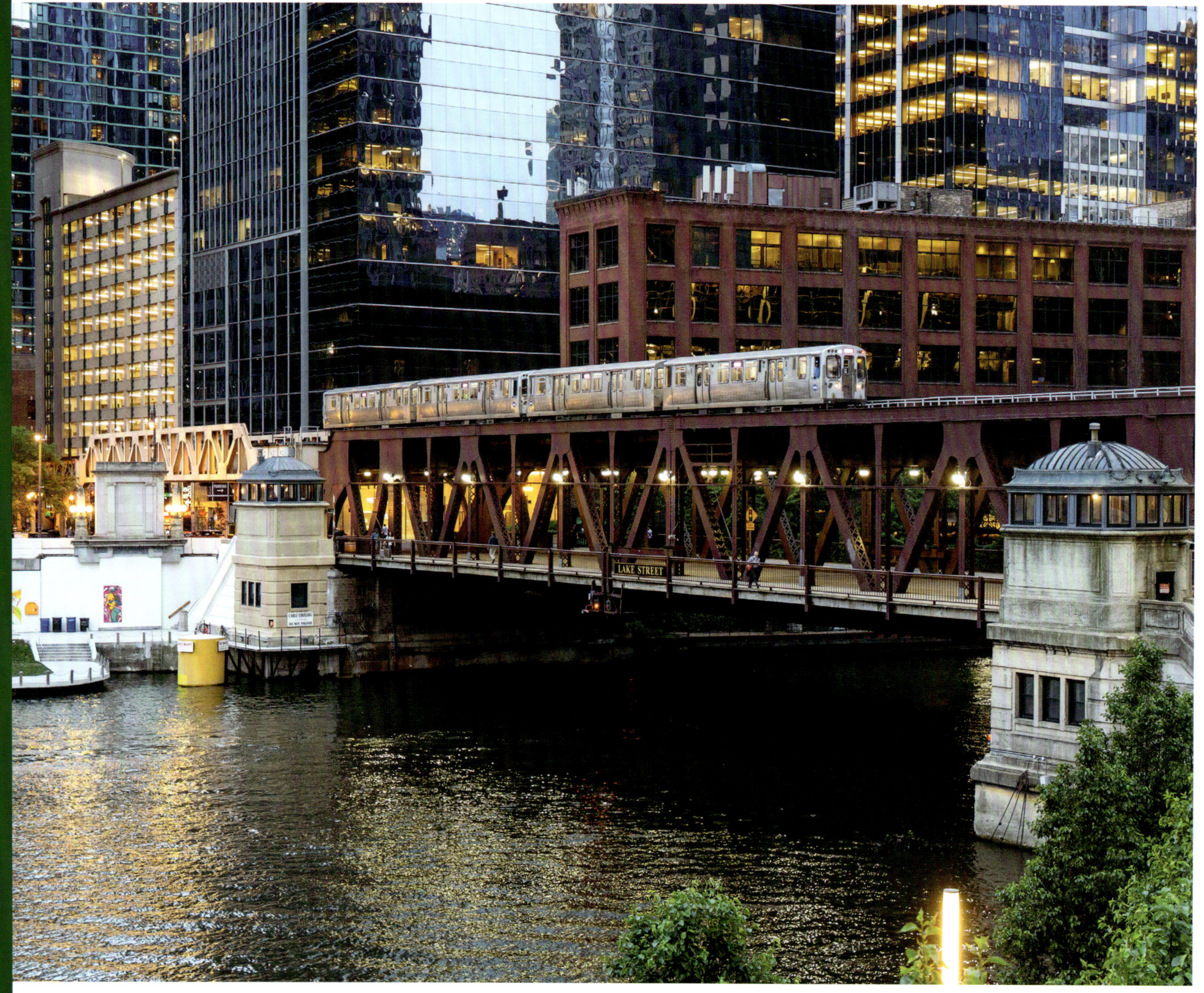

A Green Line train crossing the Chicago River over the Lake Street drawbridge.

While not as long or as busy in getting commuters around as its Red Line sister, Chicago's Green Line holds a few important distinctions among the Chicago Transit Authority (CTA) rail system. It's the only line in the CTA system that stays elevated above street level from start to finish, putting the "L" in elevated rail. It also uses the oldest segments in the system, including a segment dating back to 1892.

Of course, older is not always better when it comes to train technology. For a while, the Green Line operated in a state of bad disrepair that had Chicagoans concerned about safety. In 1994, the Green Line closed for the largest reconstruction project in Chicago public transportation history. For more than two years, commuters accustomed to riding the Green Line from the western suburbs of Oak Park and Forest Park to and from downtown were out of luck due to the closing of the line. Six stations were closed permanently upon completion of the project—a move that did not sit well with commuters near those stops. Another uproar followed two years later in 1998 with the cutting of 24-hour service to the Green Line due to budget constraints.

All in all, however, the CTA earned its share of praise for the project that brought the Green Line up to the quality, reliability, and safety standards of other commuter rails. It has undergone more changes and upgrades than any other CTA line since its construction. The line runs from the Harlem/Lake station to the west, through the downtown Loop, and then south to branches for either King Drive/Cottage Grove or Halsted/Ashland/63rd. The Green Line covers 31 stops over 20 elevated miles, with an average of 22,000 commuters boarding the trains every weekday. The most recent addition to the line opened in 2024 at Damen Avenue and Lake Street.

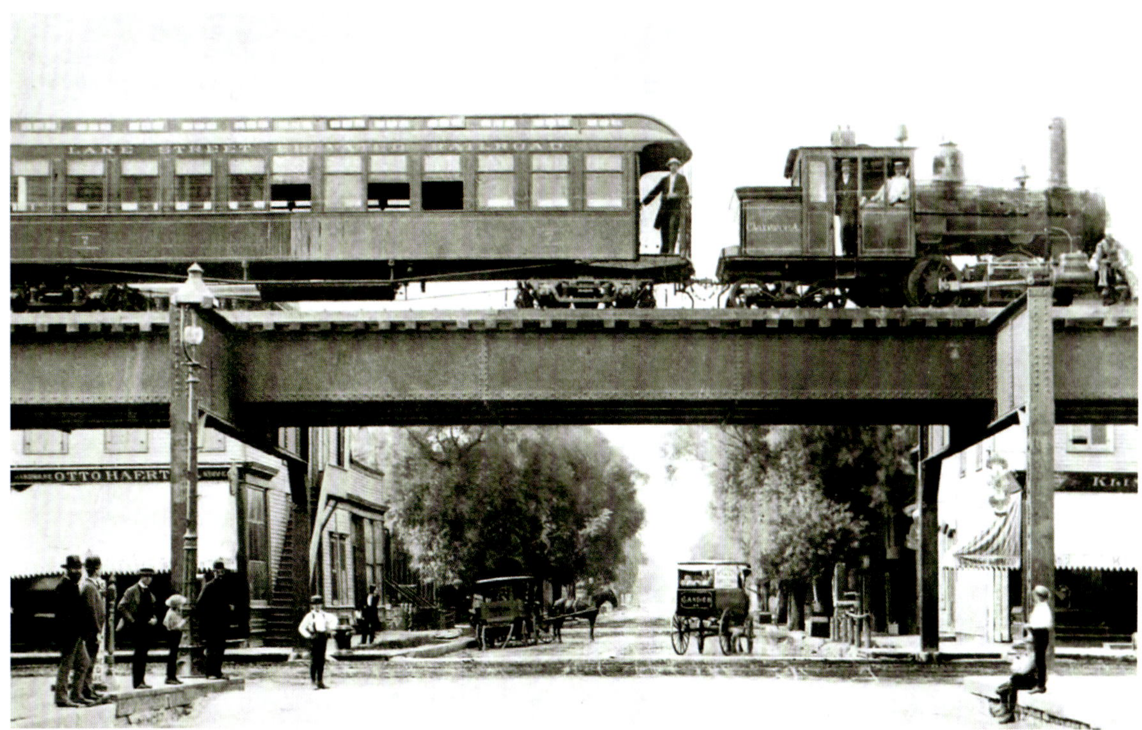

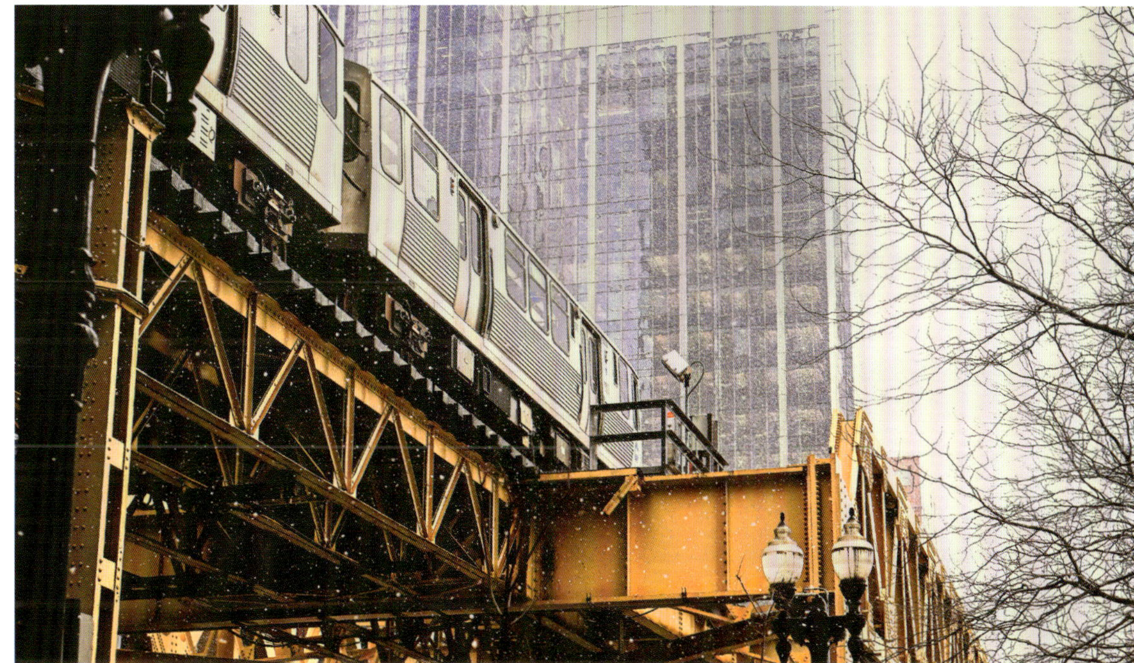

Top: A photo of the Lake Street Elevated Railroad from 1893. The Lake Street Elevated Railroad's original section of track ran from Canal Street west to California Avenue, eventually running all the way to Austin Avenue in the western suburbs by 1899.

Bottom: The Green Line running in downtown Chicago.

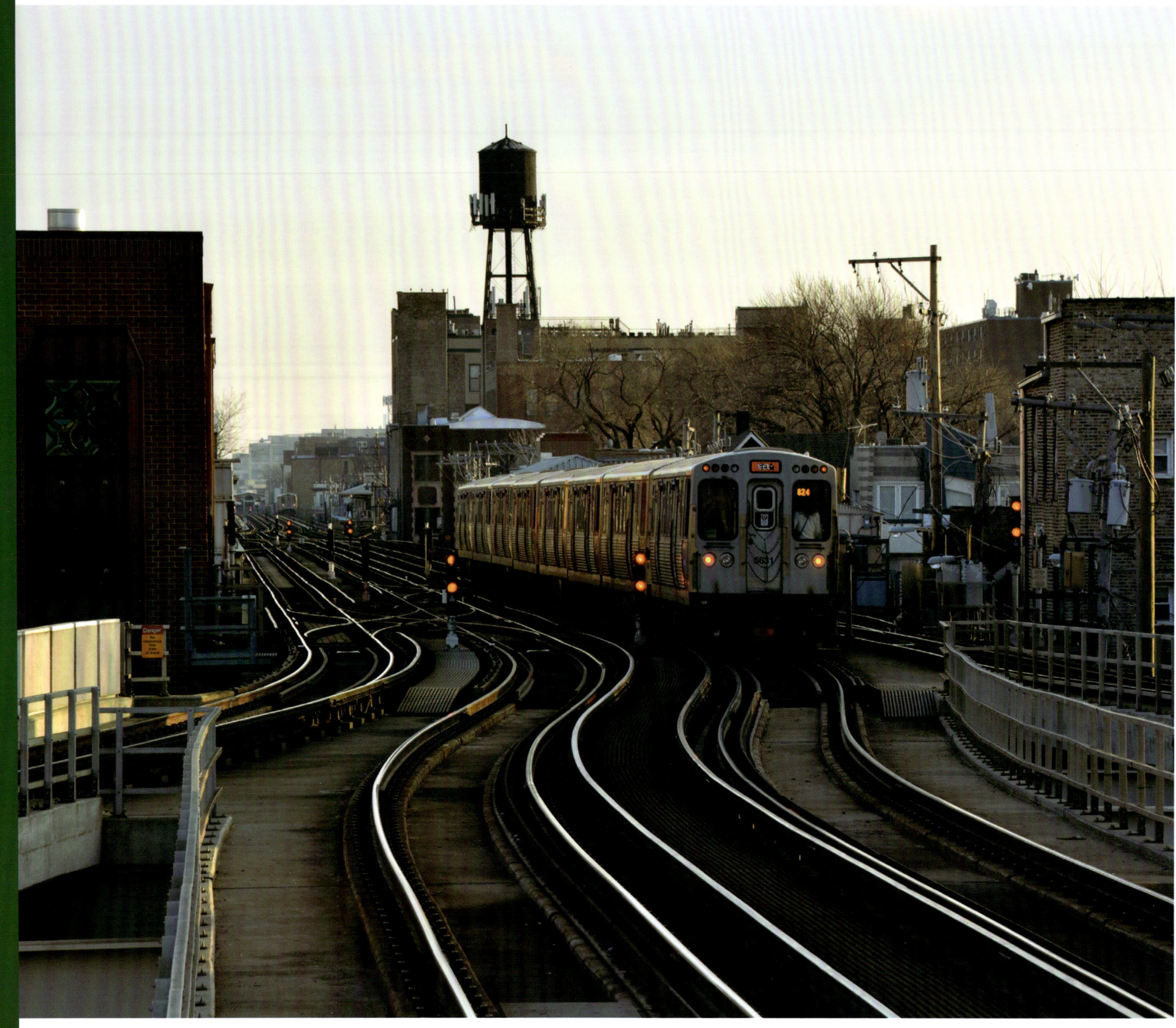

A photo of the Red Line leaving the Belmont station headed toward the Loop. This is the oldest section of the Red Line's track, which was built by the Northwestern Elevated Railroad in 1900. Today three different routes use this section of track, including the Red, Brown, and Purple Lines.

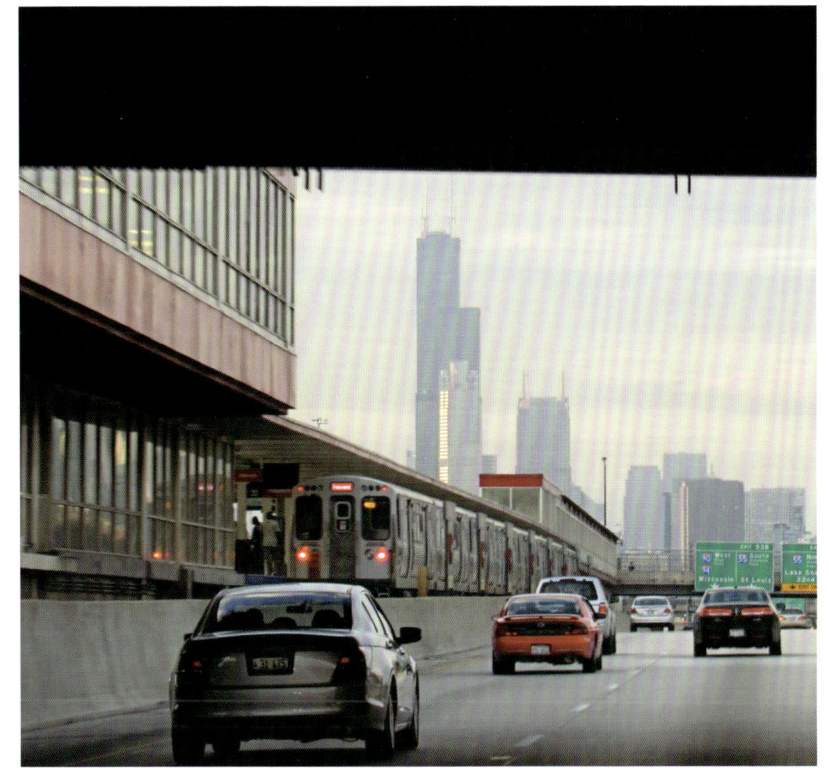

A Red Line train headed northbound toward the Loop on the Red Line's Dan Ryan branch.

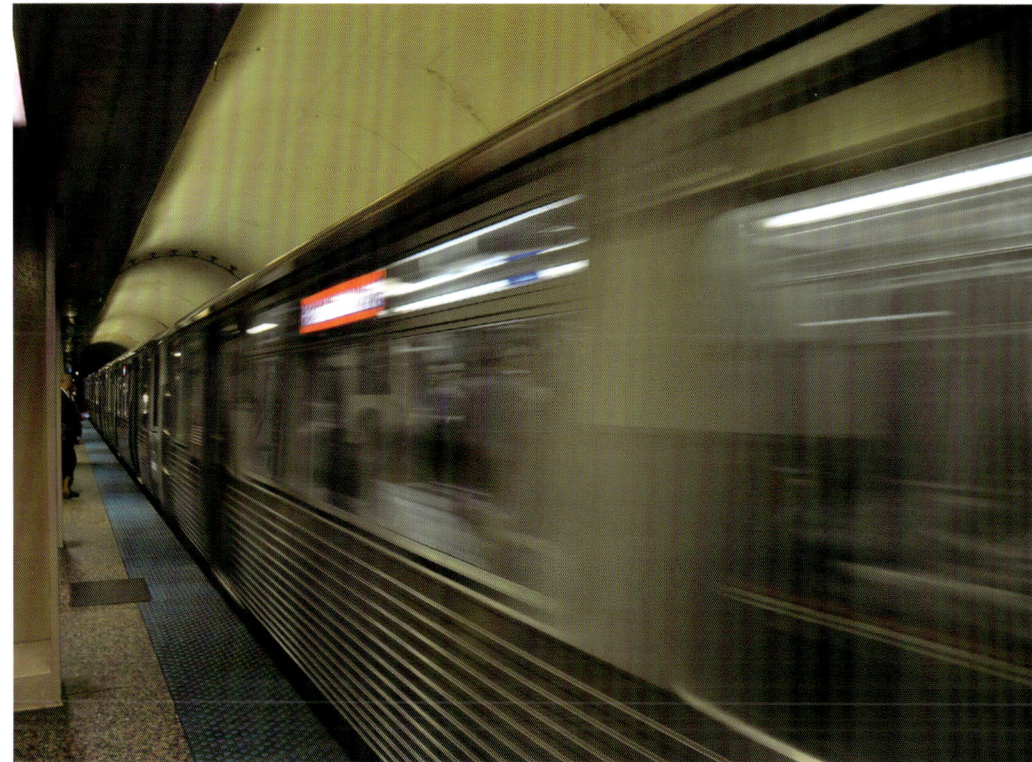

The Red Line in the State Street Subway under the Chicago Loop.

The busiest line on Chicago's famed "L" (elevated rail) system, the Chicago Transit Authority (CTA) Red Line runs north and south through the city. Sometimes called the "Howard–Dan Ryan" line, it moves over 400,000 people on weekdays, helping Chicagoans stay away from busy highways and roads on their way to and from work.

The CTA operates as an independent government agency created by Illinois State lawmakers. Though it did not come about until 1947, the Red Line's roots date way back to 1900. That's when the Northwestern Elevated Railroad opened a section of track from the Loop (downtown Chicago) to Wilson, which remains the oldest section of track on the Red Line.

That early route was extended north to Evanston over the next few years, and in 1913 the North Side "L" was routed through the Loop and connected to the South Side "L." For more than 100 years, Chicagoans have been able to ride from one side of town to the other.

Over time, and at the urging of the CTA once that entity took the reins, the North–South route was adjusted to provide better efficiency, reduce sharp curves, and enhance the experience for its growing number of passengers. It came to be known as the Red Line when the CTA went to a color-coded system in 1993, pairing the Howard branch of rail with the Dan Ryan branch. The Lake Street branch merged with the Englewood and Jackson Park branches to form the Green Line, and an Orange Line was added to get travelers to and from Midway Airport.

The Red Line, however, stands as the king of the CTA system in terms of stations (33), miles (21-plus), and number of passengers. It begins at Howard Station near Rogers Park on the city's north side, where passengers can transfer to and from the suburban Yellow and Purple Lines. One of the most popular north side stops is Addison, where Chicago Cubs baseball fans flock to the "friendly confines" of Wrigley Field from April to (with any luck) October.

The State Street Subway segment of the Red Line starts near the Chicago History Museum and takes folks downtown and then south past Museum Campus—the Field Museum, Shedd Aquarium, and Adler Planetarium—near Soldier Field. Continuing south, the Red Line's Dan Ryan branch begins at Chinatown. It continues south to stop at Rate Field for White Sox fans and then culminates, for now, at 95th/Dan Ryan.

Though Chicago commuters know it as a tried and true option, the Red Line is on its way to even bigger and better days. The CTA announced in 2016 plans to extend the route south to 130th as part of the "Red Ahead" improvement project for the line. Construction is slated to begin at the end of 2025, finishing as early as 2029. The plan calls for the addition of four stations near 103rd, 111th, Michigan Avenue, and 130th.

The extension is one of four planned improvements under the "Red Ahead" program. It would mark the first extension of the "L" system since the 1993 opening of the Orange Line. "It is one of the biggest, boldest equity investments in the history of this great city," CTA President Dorval Carter said of the project, which is projected to cost more than $5 billion.

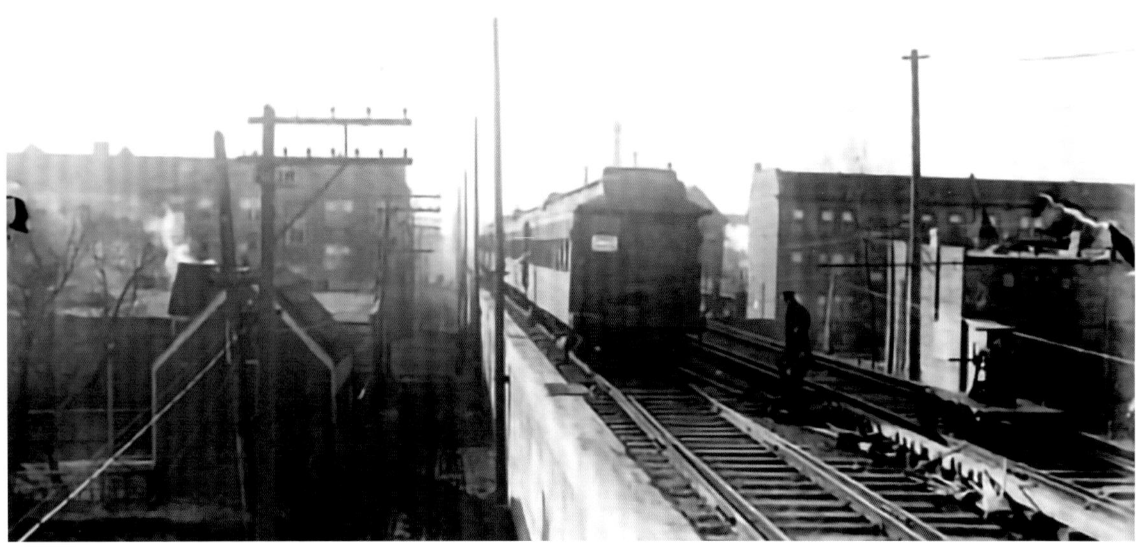

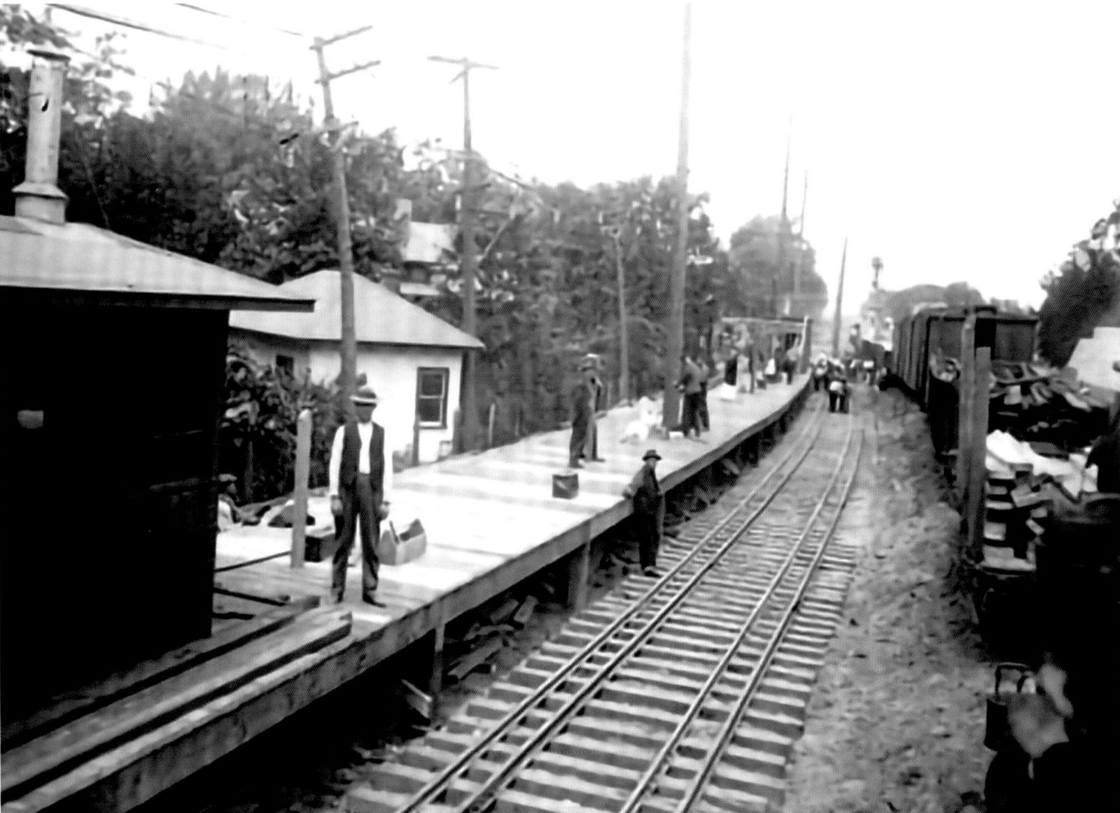

Top: The Red Line, then the Northwestern Elevated Railroad, near the Argyle station in Chicago's historic Uptown neighborhood in 1916.

Bottom: Red Line tracks outside of the Jarvis station in the far northside neighborhood of Rogers Park. This section of track was added onto the original Northwestern Elevated Railroad in 1908.

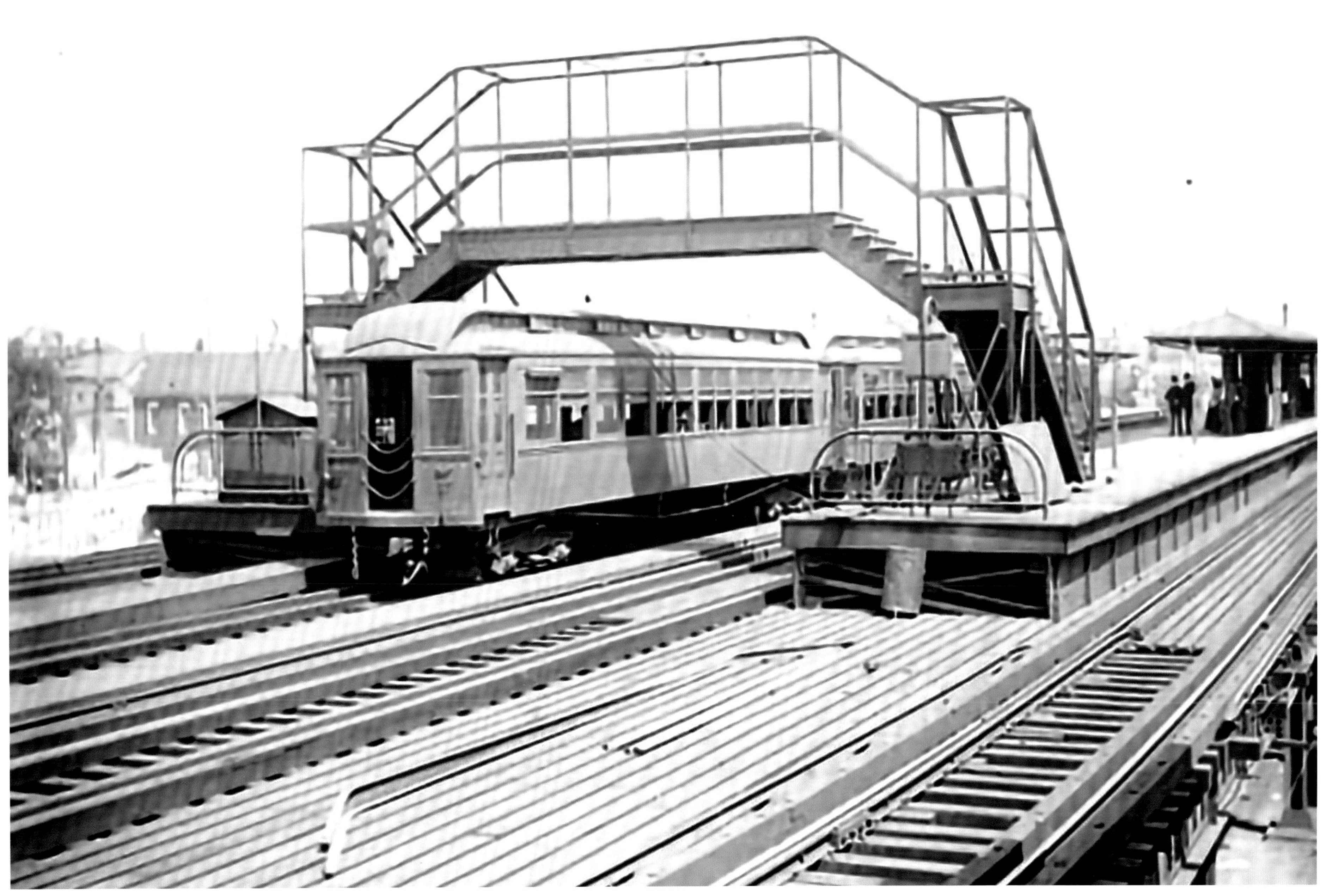

A train car on the Northwestern Elevated Railroad in 1907 at the Belmont station. Overhead is the passenger transfer bridge that allowed passengers to cross between the northbound and southbound platforms.

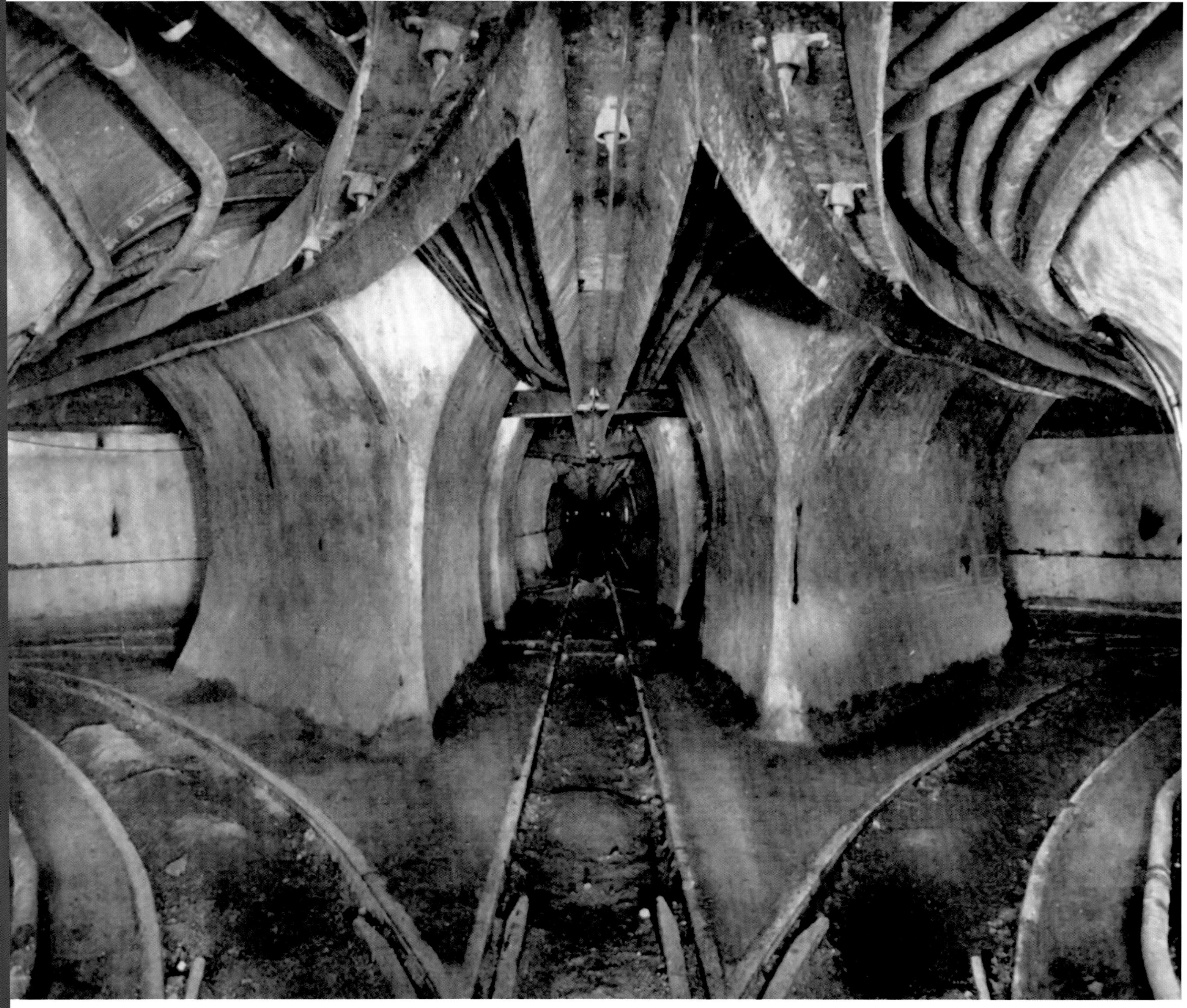

A grand union junction of two tracks under the Chicago Loop. A grand union allows for a train car on a track to turn left or right onto the intersecting track or to continue forward on the track that it is on.

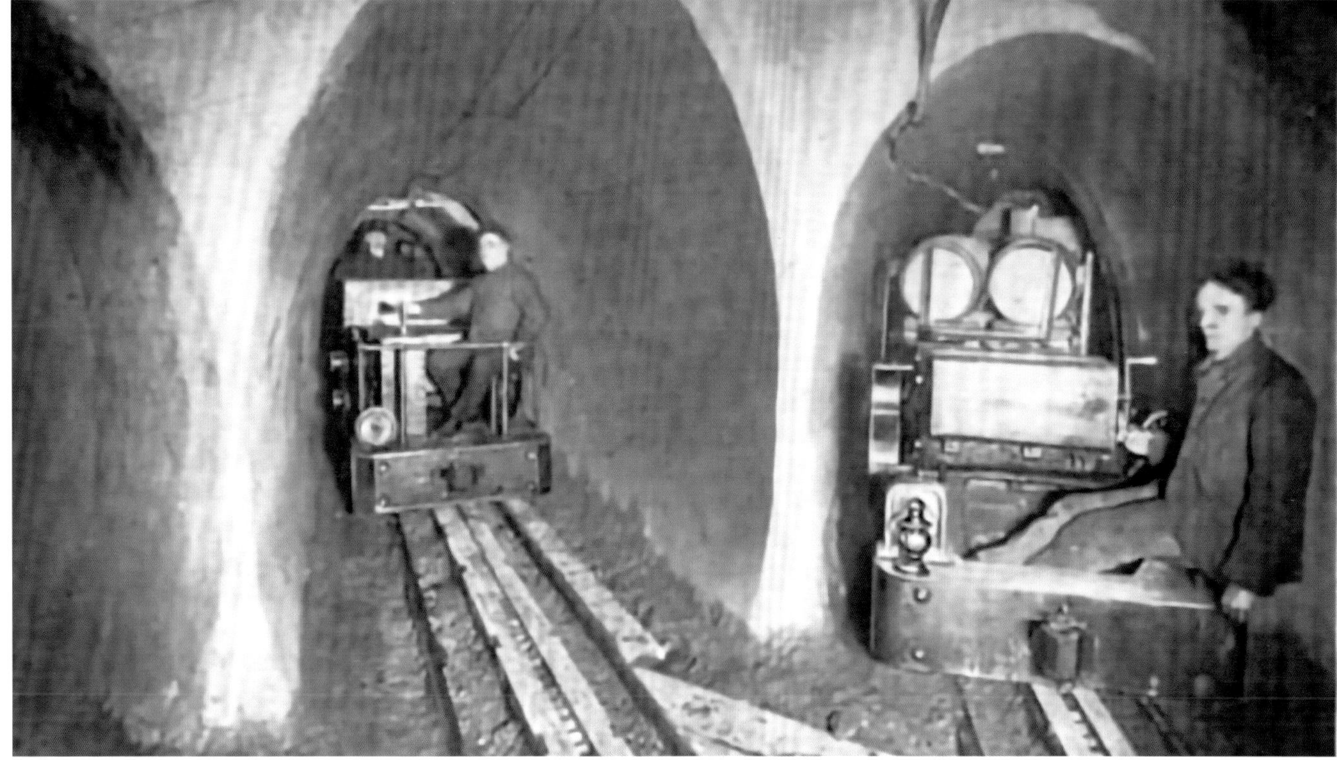

A publicity photo for the Chicago Tunnel Company from 1904. The photo was taken on the tracks beneath the intersection of State and Randolph Streets.

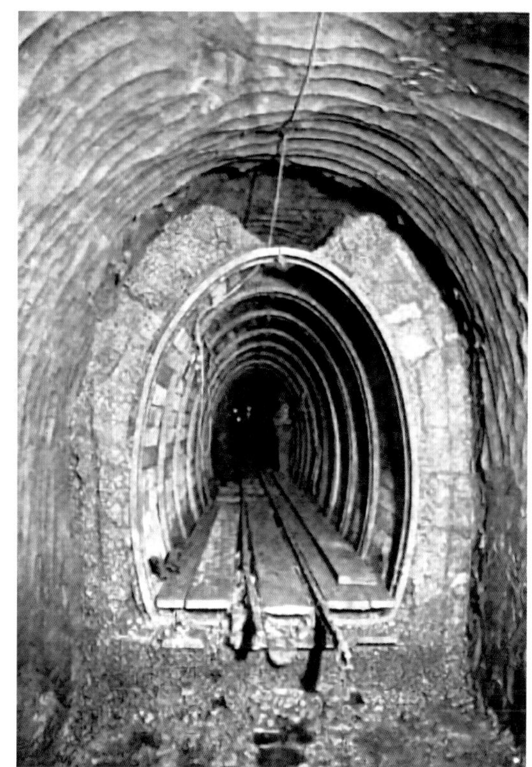

A tunnel under construction in 1902.

Though most Chicagoans will never see them, a series of underground tunnels winds under their downtown—reminders of an aggressive and ultimately failed business plan from years gone by. In 1899 the Illinois Telephone and Telegraph Company began hauling blue-clay soil out from under the city, planning to plant its network of telephone cables there.

When the city refused to grant permission for manholes through which the cable could be fed underground, the company came up with the idea to run underground rails that could move the cable spools. By 1903, the company had even bigger ideas. It changed its franchise with the city, deciding the rail system could also accommodate coal, mail, and merchandise deliveries.

Building 60 miles of tunnels was a bold plan for the Chicago Tunnel Company (CTC), and the project began without a single client committed to using the rail system. Those clients eventually came. Tunnels were dug under the likes of the Board of Trade, City Hall, the Federal Reserve Bank, Chicago Tribune, Civic Opera House, and Field Museum, among others. The CTC even had success selling "tunnel air" to some of the downtown clients, pumping cool, 55-degree underground air into buildings that were sweltering in the summer heat.

Small electric trains ran for decades under the city, delivering merchandise and materials away from the traffic congestion above. Street names were painted on the walls for train operators to follow. Business began in the 1910s and was booming in the 1940s and 1950s. Tough times followed, though, as the trucking industry hit its heyday above ground and a series of investments became difficult for the company to pay.

The CTC went out of business in 1959, selling its fleet of small trains as scrap for $64,000. The *Chicago Tribune* continued to use the tunnels until 1981 to transport newsprint from a warehouse to Tribune Tower. In 1992, employees of a dock company accidentally made a hole in a tunnel roof, sending millions of gallons of water flooding into the tunnel system at a damage cost of about $2 billion. The tunnels have been sealed ever since.

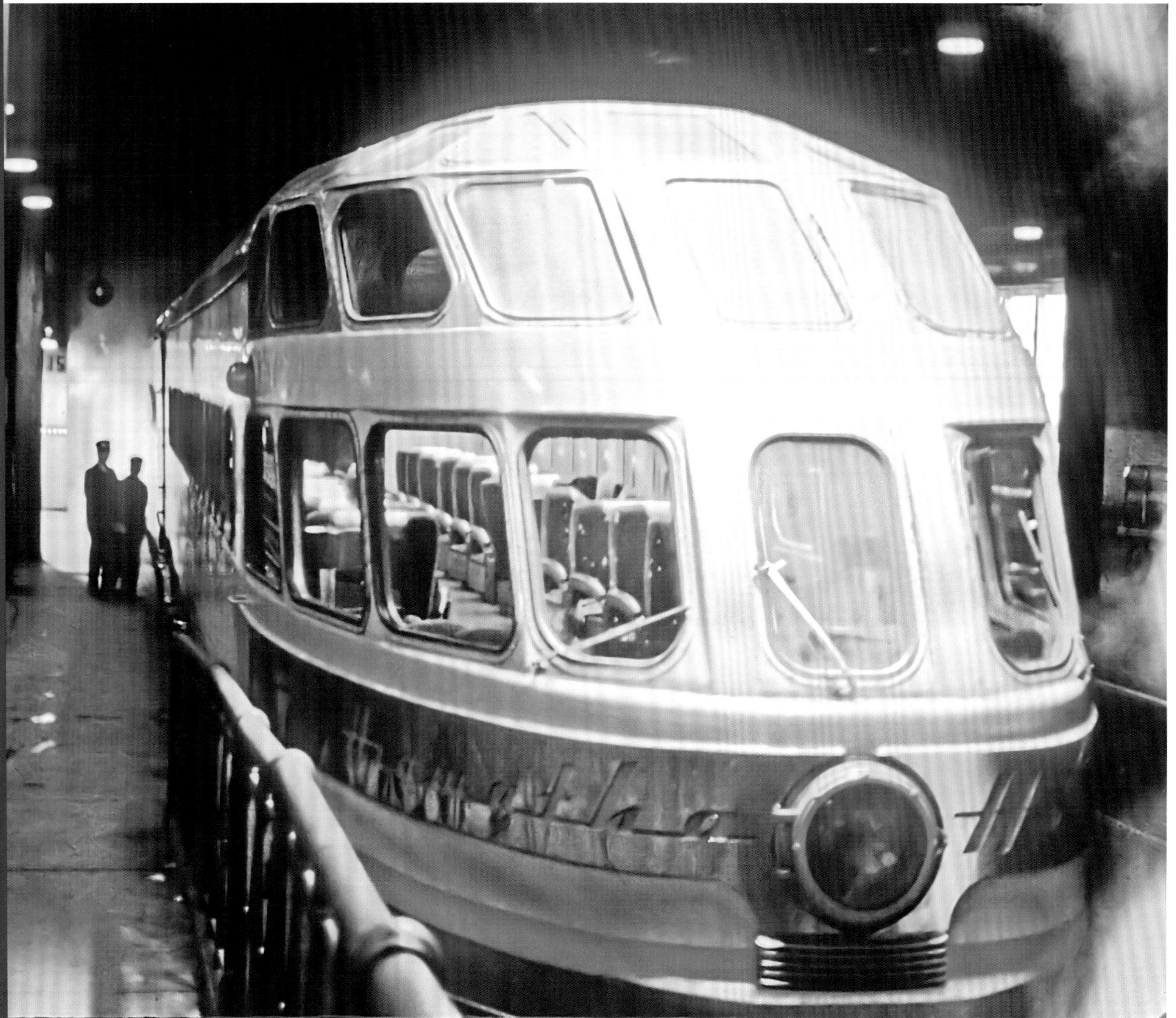

A skytop *Hiawatha* train at Chicago's Union Station.

Mentions of the Chicago, Milwaukee, St. Paul and Pacific Railroad (CMStP&P), better known as the Milwaukee Road, still tug at the heartstrings of railroad buffs some four decades after it went out of business. A major railway running through the Midwest and Northwest from 1847 to 1980, it cut through mountain ranges, drew trackside spectators, and remains honored in museums and private collections across the country.

It began as the Milwaukee and Waukesha Railroad in Wisconsin in 1847, when it connected Milwaukee to the Mississippi River for shipping. It became the Milwaukee and Mississippi in 1850, the Milwaukee and St. Paul in 1867, and the Chicago, Milwaukee and St. Paul in 1874 before absorbing the Chicago and Pacific Railroad Company in 1879.

Whatever one calls it, the Milwaukee Road was fearless. The independent railway competed with the big boys for both Midwestern agricultural business and the growing desire to reach the West, taking on Pacific Northwest and its transcontinental mission during the early 1900s. At one time, the Milwaukee Road was the longest end-to-end railroad in the country, stretching from Louisville to Puget Sound, thanks to some incredible engineering feats.

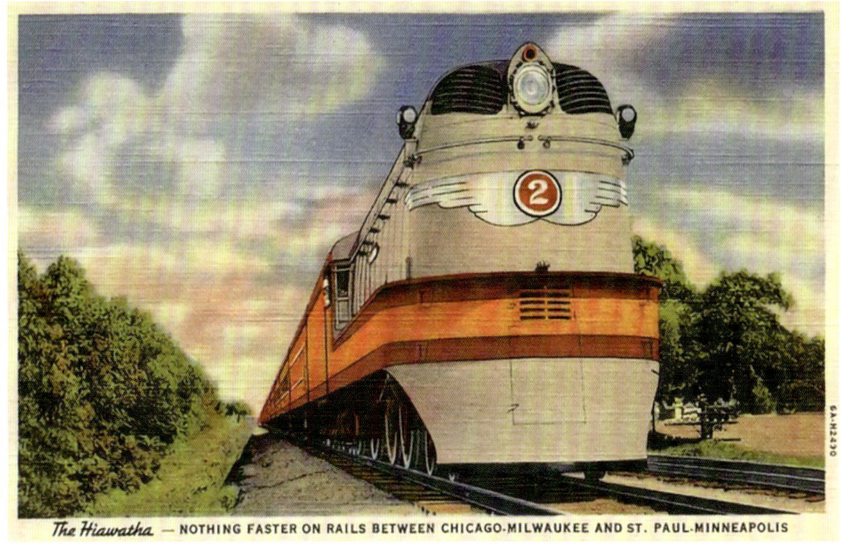

Top: A postcard for the Milwaukee Road promoting their new locomotives in 1935.

Bottom: A train leaving Chicago's Union Station on the Milwaukee Road in October of 1955.

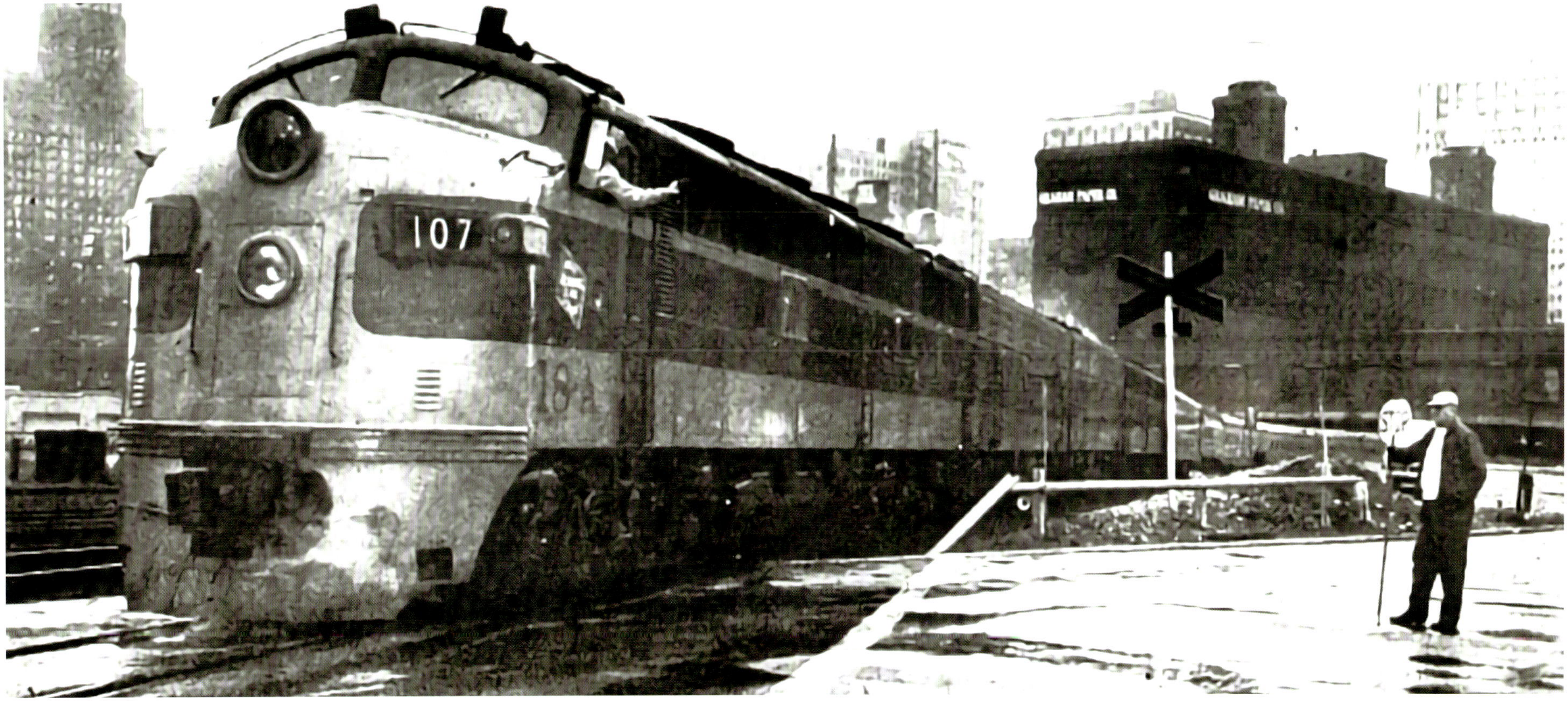

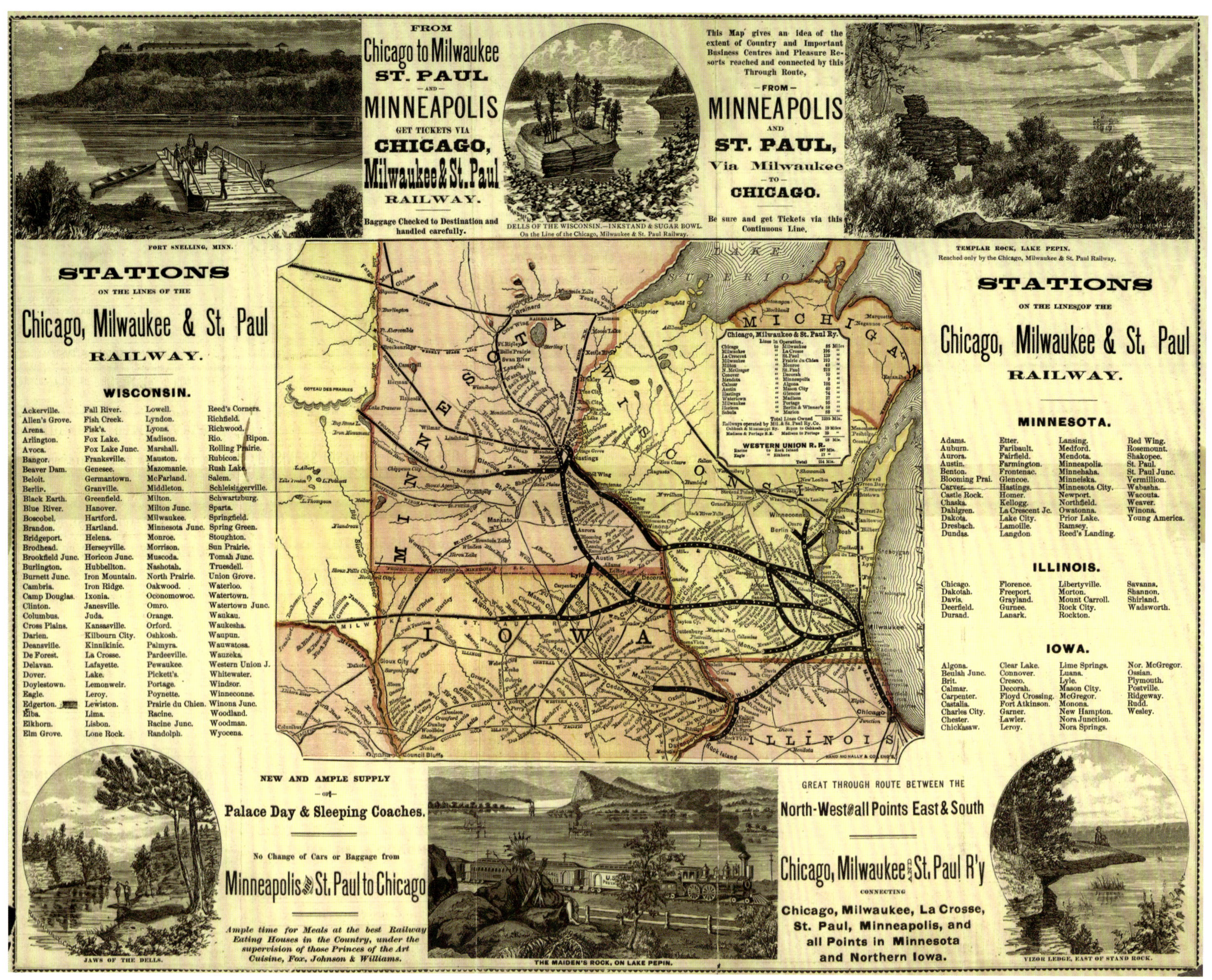

A Milwaukee Road brochure featuring stations, illustrations, and a map circa 1874. The reverse side shares timetables with a title boasting "The Very Best Line!"

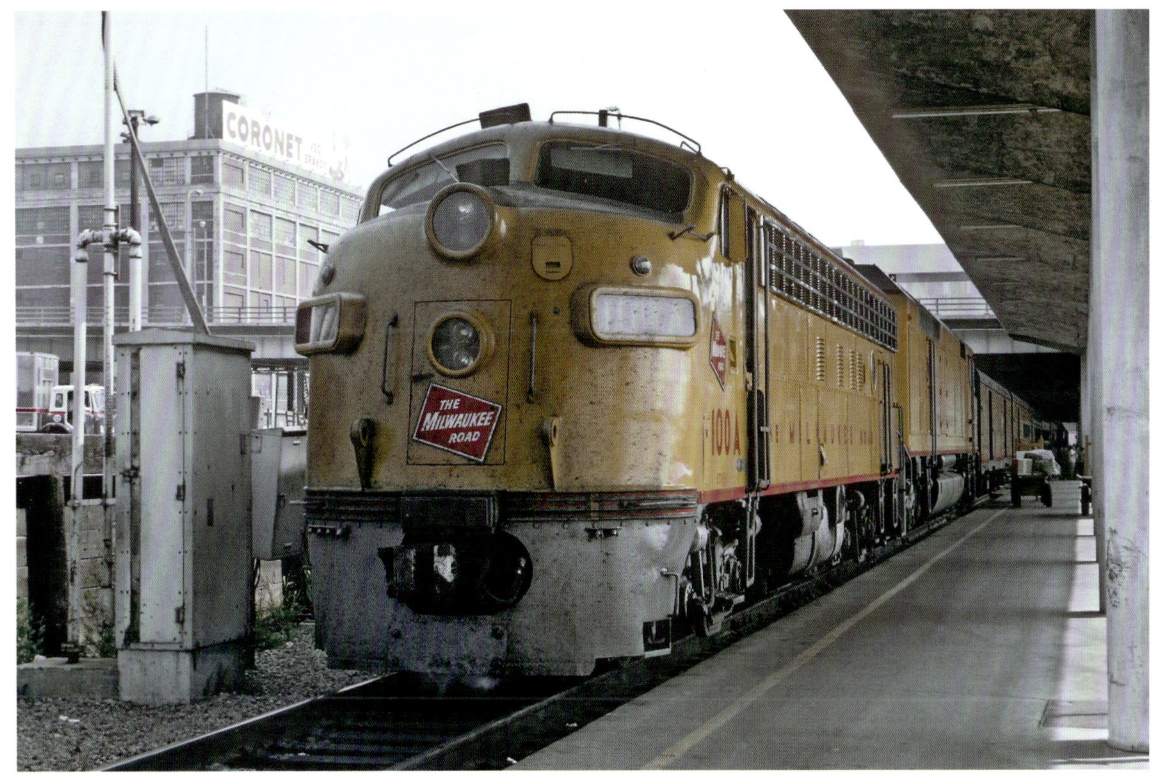

One of the Milwaukee Road's lasting legacies is the *Hiawatha*, a fleet of unique steamers that captured the attention of Midwesterners and train fans nationwide. Patrons gathered around the tracks to cheer on the maiden run of the *Hiawatha* on May 29, 1935. The trains were built to connect Chicago with several other Midwestern cities, and tickets were usually sold out, with some 16,000 passengers riding in the first six weeks.

Without the support of government agencies or some of the nation's wealthy train moguls, it's not surprising that the Milwaukee Road ran into serious financial troubles like many other railroads in the early twentieth century. The company entered bankruptcy in the 1920s and 1930s, went through multiple reorganizations during the following decades, and finally ran out of steam in the 1970s. Though some of its lines remain in use by other companies, the final Milwaukee Road service departed Tacoma on March 15, 1980.

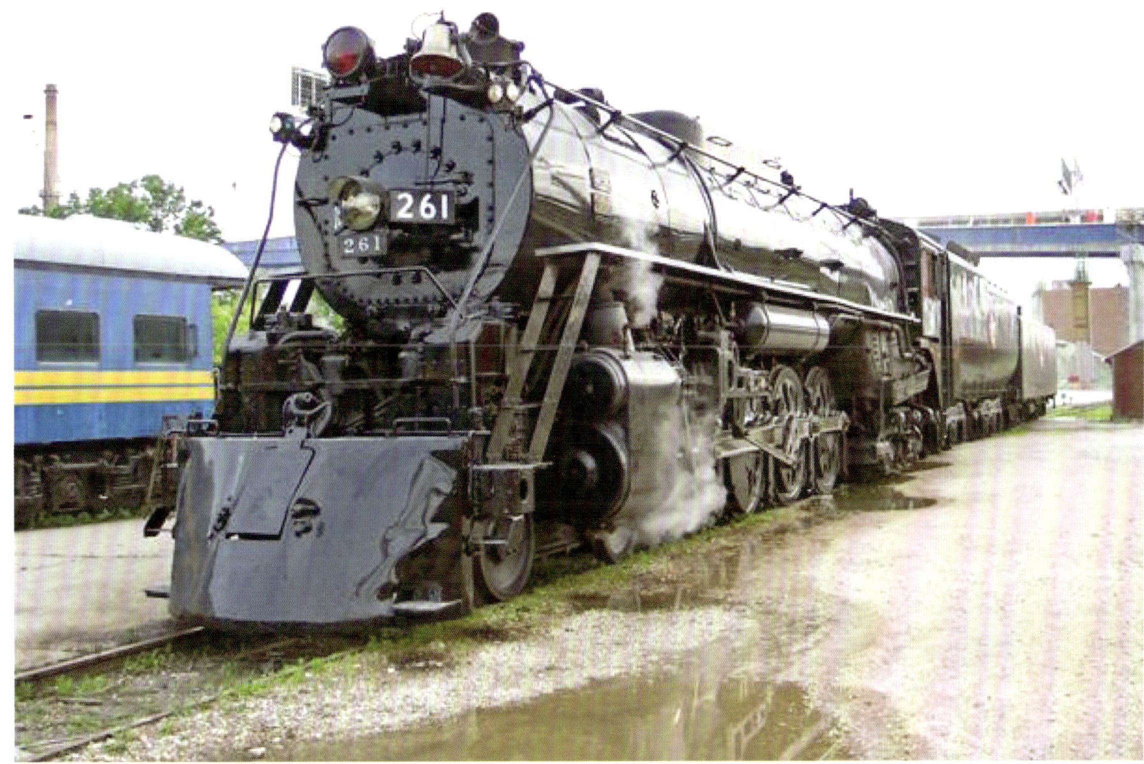

Top: The *Morning Hiawatha* stopping in Milwaukee in 1969 as part of the Chicago-Milwaukee-Twin Cities mainline route.

Bottom: The Milwaukee Road 261 steam locomotive still runs occasional excursions thanks to the nonprofit organization Friends of the 261.

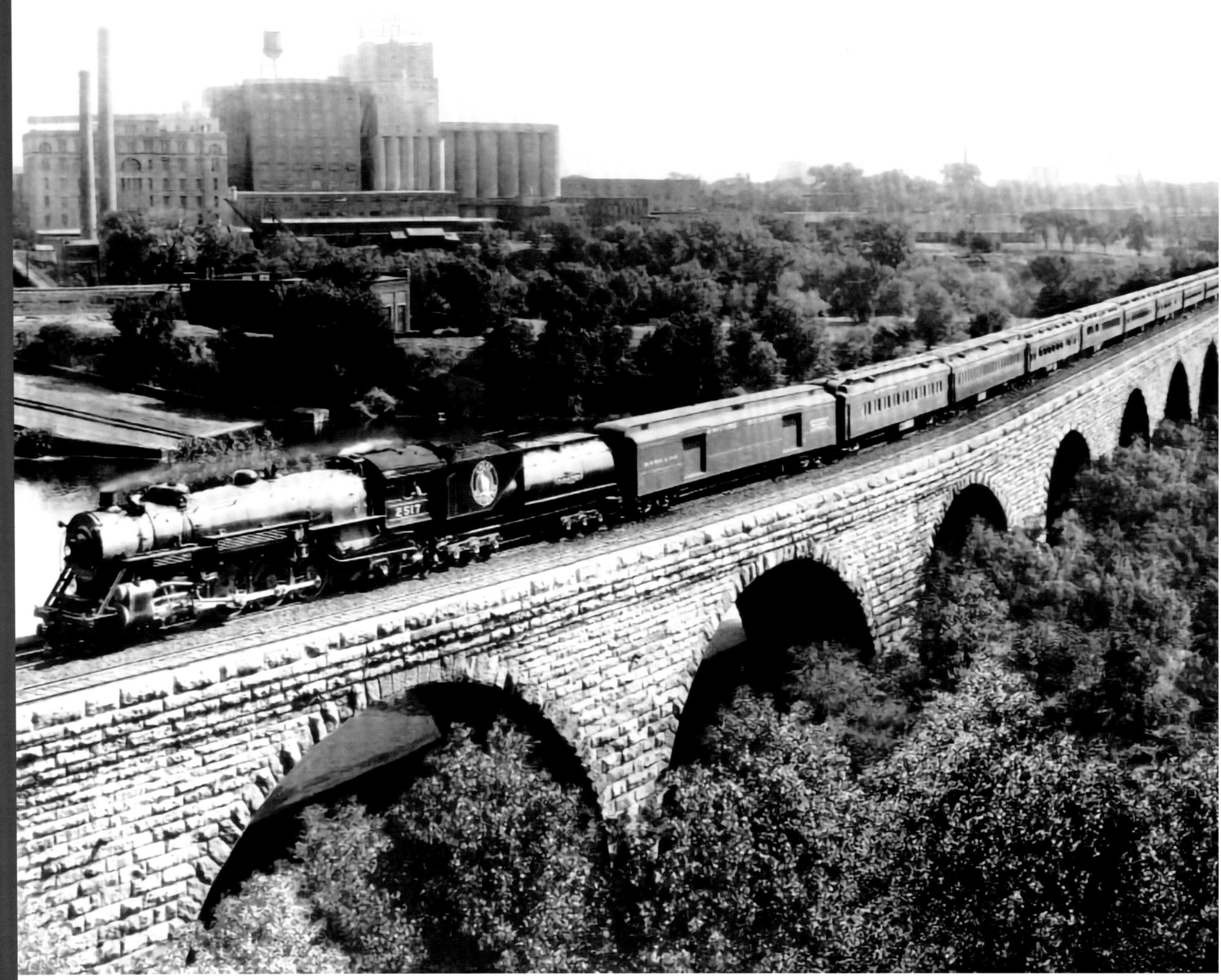

The GN's *Empire Builder*, which ran from Chicago to Seattle, pictured in front of downtown Minneapolis. The stone arch bridge it is crossing is used today as a pedway.

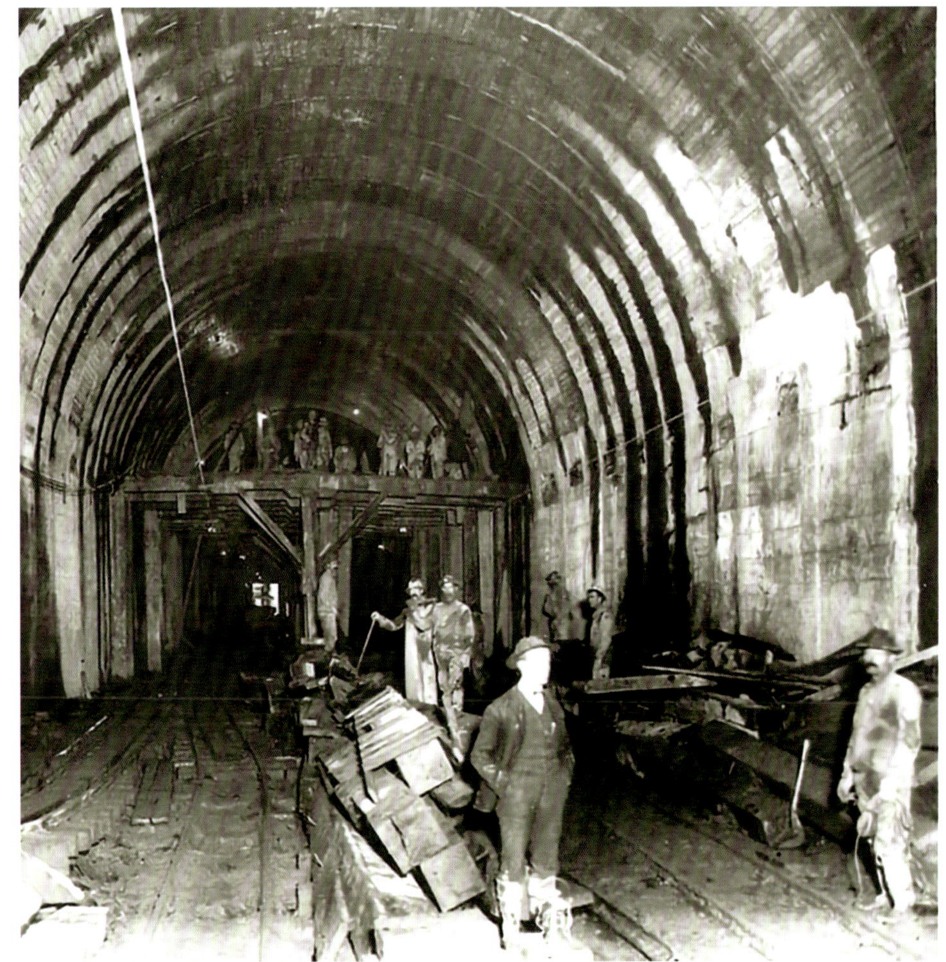

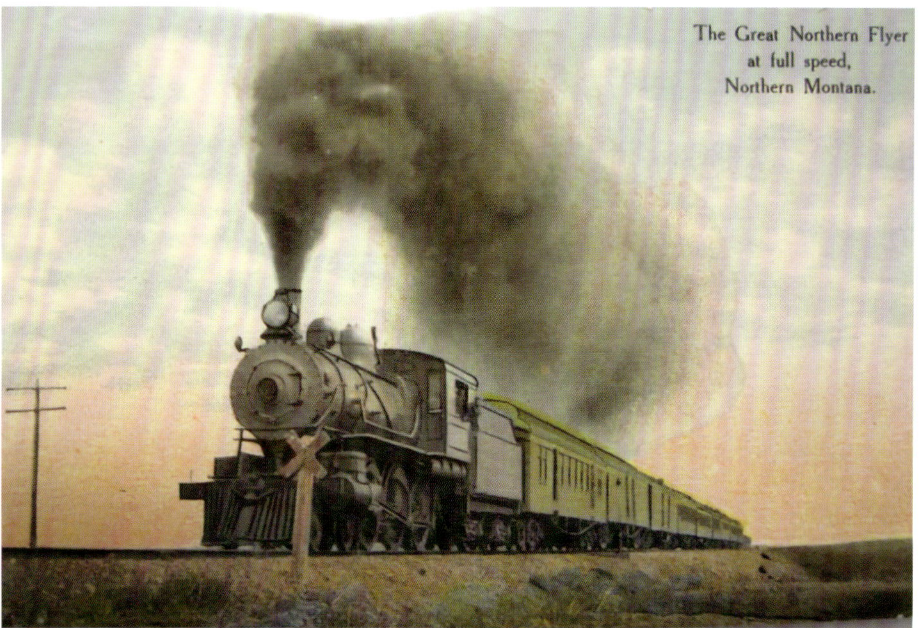

Left: The GN built the Great Northern Tunnel in 1905. The Great Northern Tunnel was bored under Seattle's downtown and was the tallest and widest railway tunnel in the nation at the time.

Right: A postcard of the *Great Northern Flyer* circa 1900.

The Great Northern Railway (GN) holds two major distinctions among transcontinental railways in United States history. First, it was the northernmost transcontinental, running from the shores of Lake Superior in Duluth, Minnesota, all the way to Seattle, Washington. And second, it was the only one among the transcontinental railways that was built entirely with private funds—not a cent of federal money was put into the project.

The GN was the brainchild of Ontario-born James J. Hill, who became known as the "Empire Builder" for his success with it. As an 18-year-old in 1856, Hill was trying to get from St. Paul, Minnesota, to visit a friend in Manitoba, but the last ox-cart caravan had already left for the season. Grounded in Minnesota, he not only saw first-hand the need for a railway to transport passengers across the upper Midwest, but he also landed a shipping clerk job at a Mississippi River steamboat company—his initiation into the transportation business.

Hill's train debut came with the St. Paul and Pacific Railroad in 1866. A decade later, he and two partners purchased the line, which became the St. Paul, Minneapolis and Manitoba Railway Company in 1879 and the Great Northern Railroad in 1889. By the end of the following year, the GN was operating on 3,260 miles of track across the upper Midwest, and it was eyeing the Rocky Mountains as its next prize.

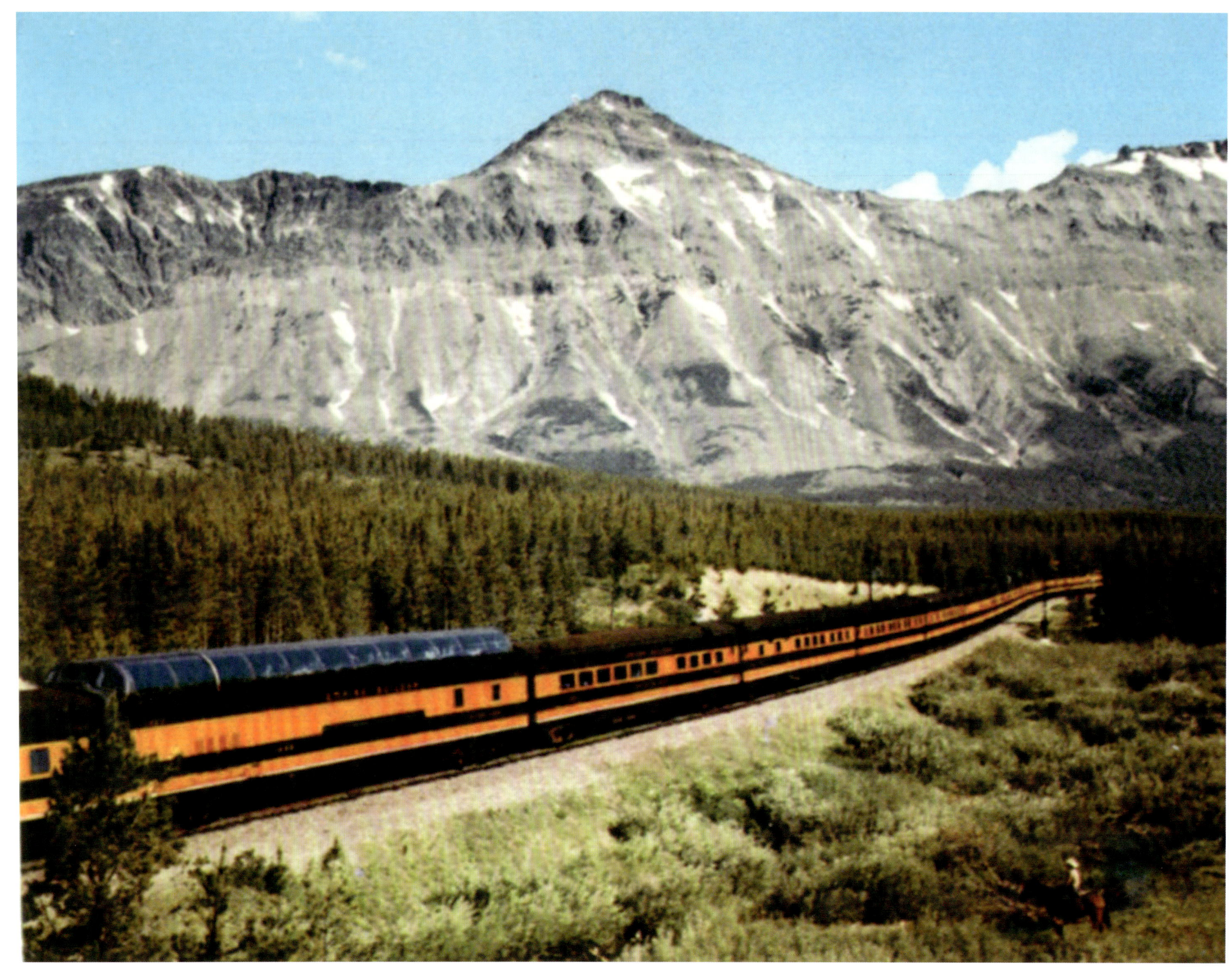

A postcard of the *Empire Builder* at Marias Pass in Montana.

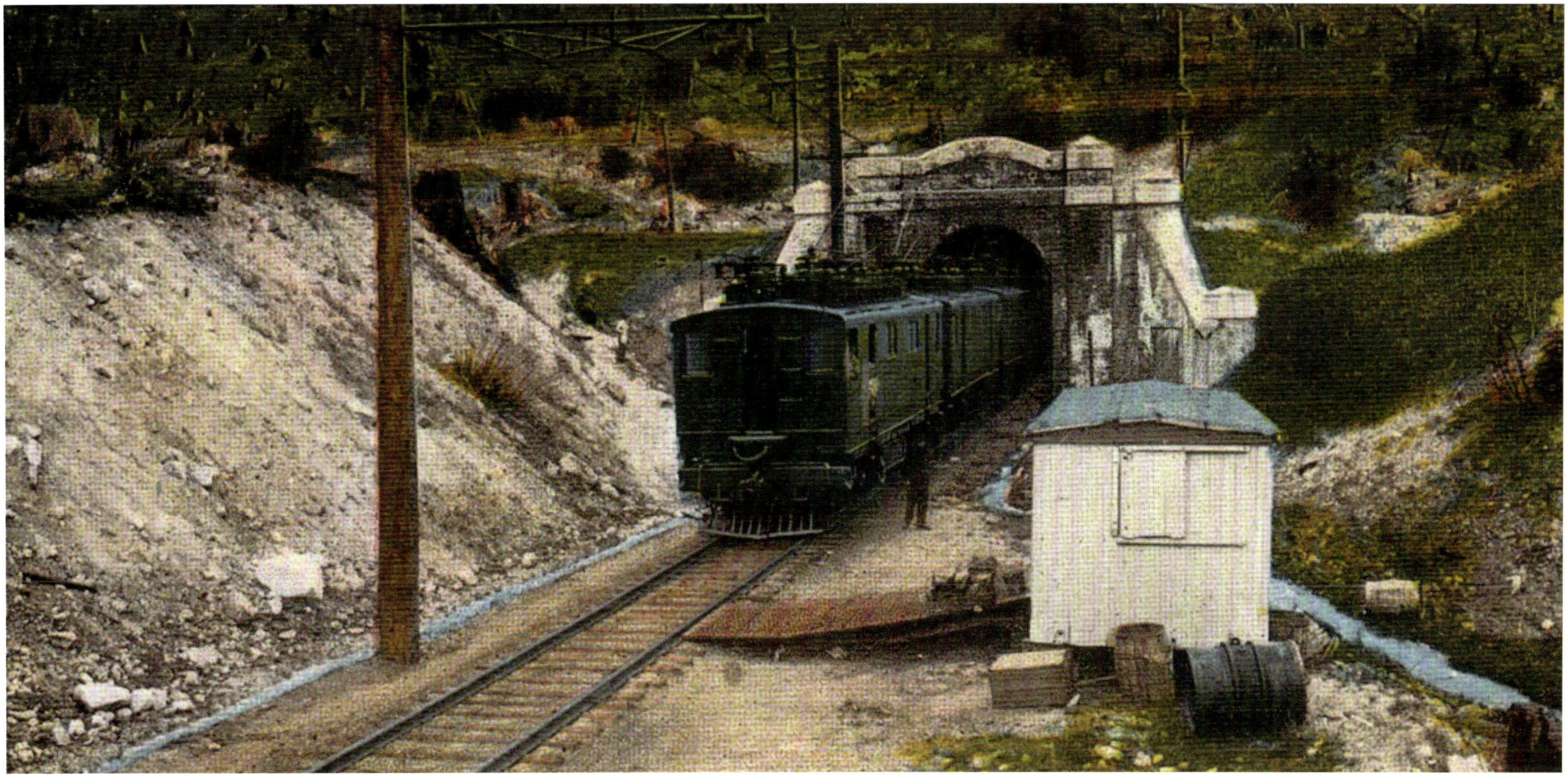

A statue of engineer John F. Stevens, who determined and executed a low-altitude route through Marias Pass in Montana, still stands near the site where the GN made its highest summit at 5,215 feet. The last spike was driven on January 6, 1893, at Scenic, Washington, marking the first time Seattle had been connected to the Midwest by rail.

The GN ran until 1970, when it merged with three other railways to form the Burlington Northern Railroad. At its peak, it operated more than 8,300 miles of track. Its regulars had their favorite trains, but the one closest to the hearts of GN veterans was the *Empire Builder*, which began running from Chicago to the West Coast in 1929.

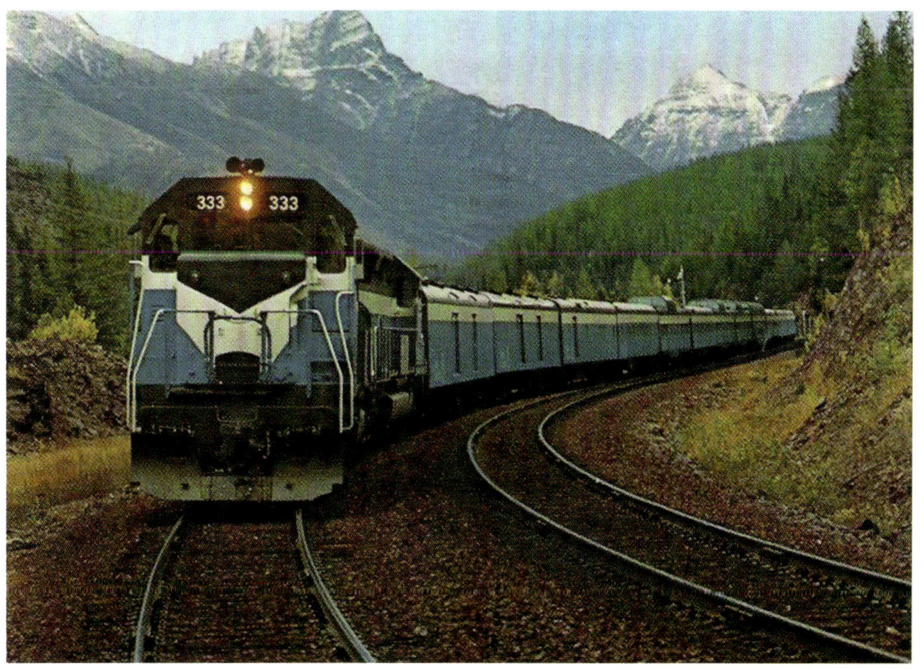

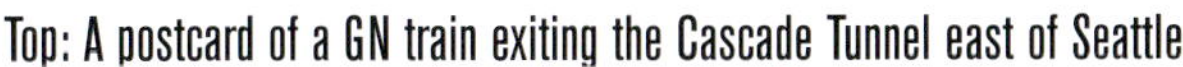

Top: A postcard of a GN train exiting the Cascade Tunnel east of Seattle.

Bottom: A GN train passing Glacier National Park.

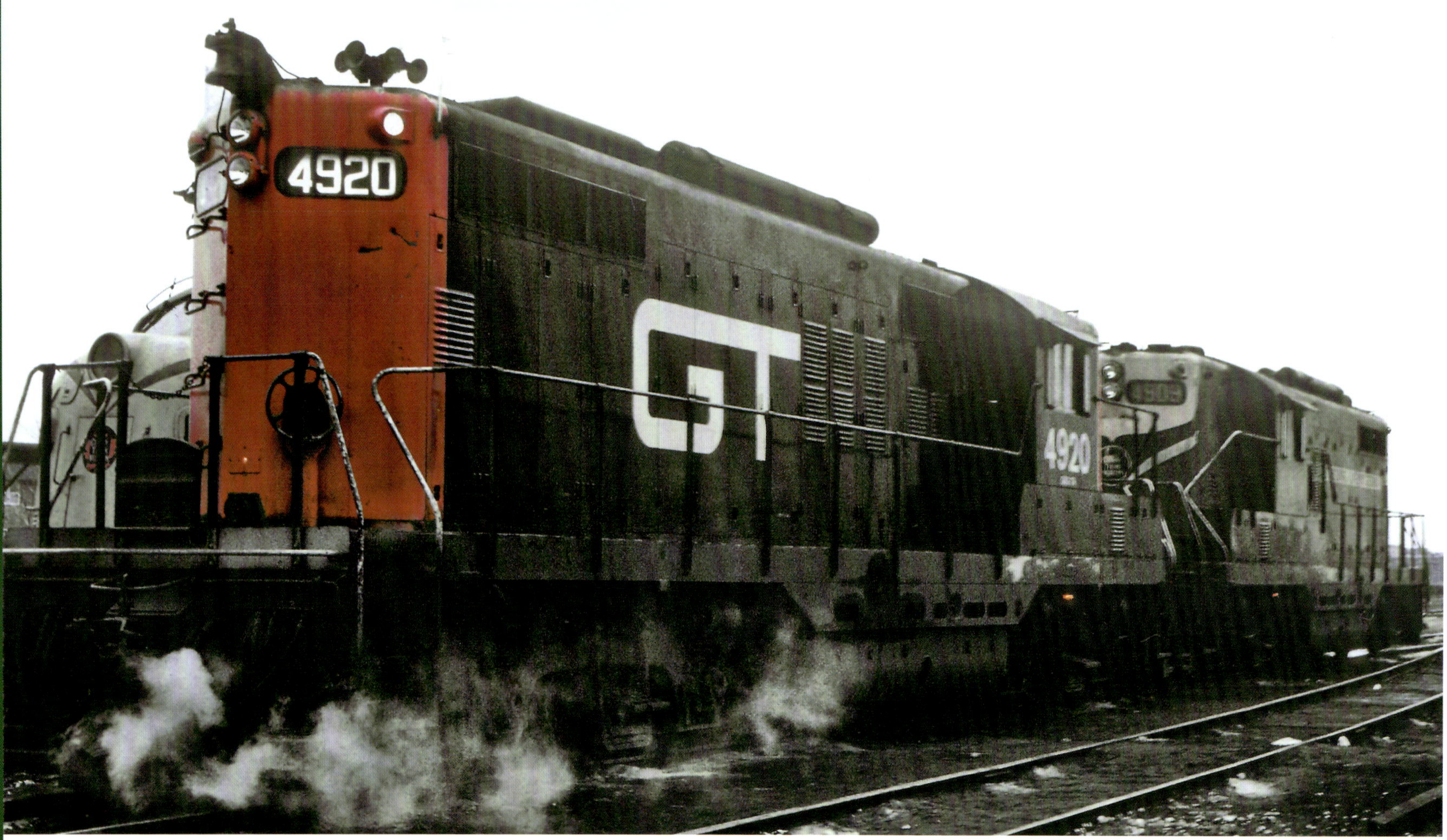

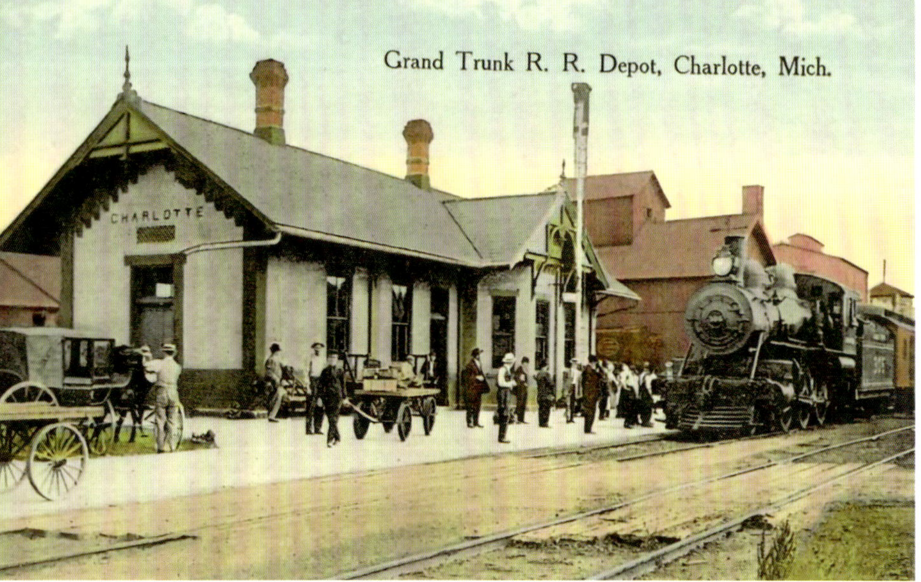

Top: A diesel locomotive at the Edison Yard in Chicago, Illinois.

Bottom: A postcard of the Grand Trunk Western Railroad at the Charlotte, Michigan, depot in 1912.

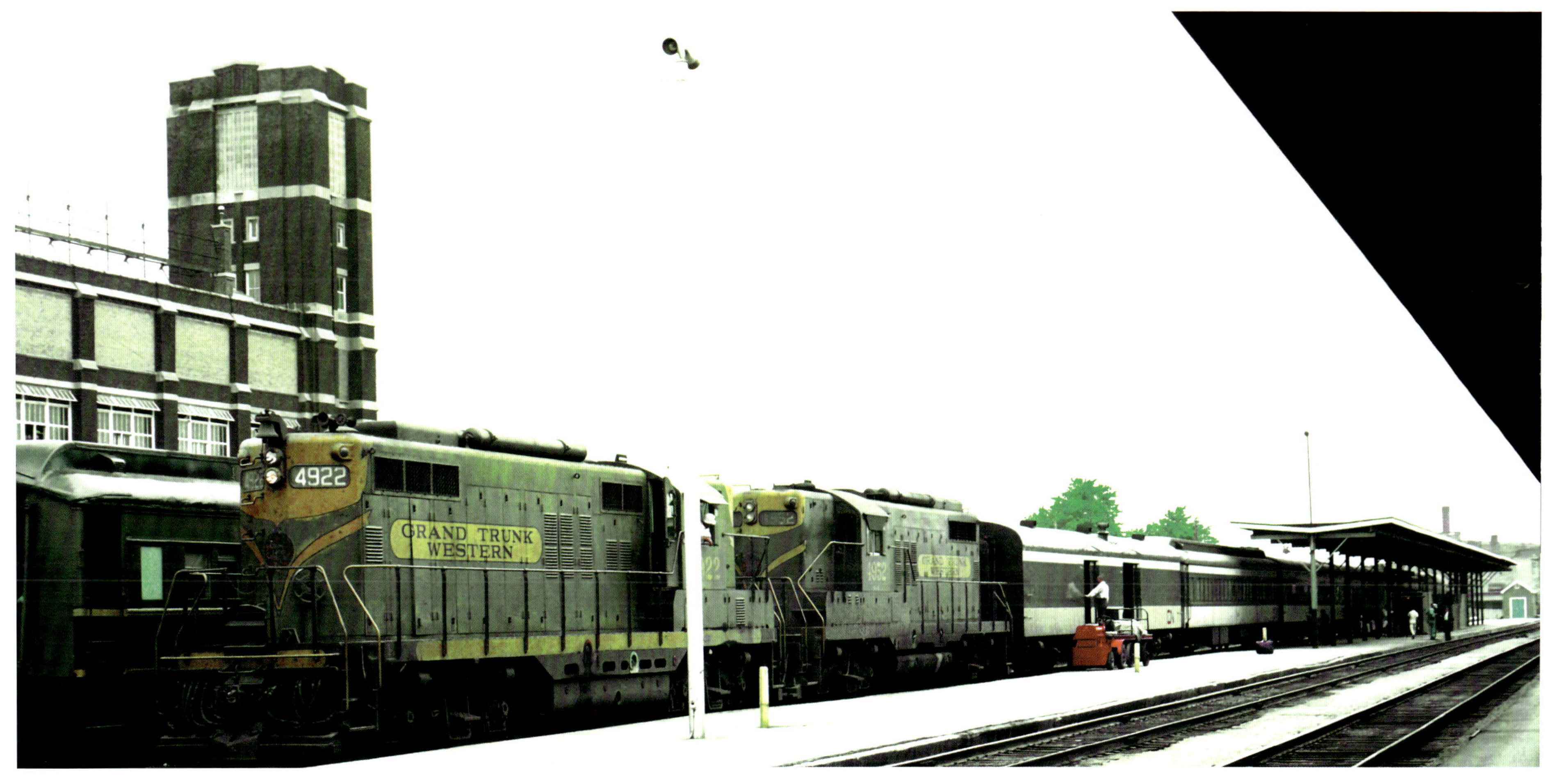

A diesel locomotive at the station in London, Ontario, in 1966.

To many railways, the automobile was the wheeled enemy. Cars and their rise in popularity contributed to financial difficulties for most railroads in the mid-twentieth century, as Americans were more inclined to get behind the wheel than hop aboard a train. For the Grand Trunk Western Railroad Company (GTW), the dynamic was quite a bit different.

That's because the GTW's success in Detroit, southern Michigan, and Chicago made it essential as a deliverer of high-value cars and parts as the American auto industry flourished. The railway still operates as an American subsidiary of Canadian National—the key cog in the company's Michigan division. As the auto industry thrived, the GTW enjoyed the fruits of that success.

Grand Trunk Western Railroad is sometimes confused with Canada's former Grand Trunk Railway (GTR), and there was indeed a connection. Though it was not called Grand Trunk Western at the time, GTW truly began in 1858, when the GTR was looking to link its Canadian lines to Chicago through lower Michigan. The first line in that quest—a 60-mile stretch from Port Huron, Michigan, to Detroit—opened on November 21, 1859.

Getting to Chicago proved to be no easy task for the GTR, as the Vanderbilts and their East Coast rail empire were chasing the same goal. The Canadian railroad weathered the storm, though, actually out-maneuvering the wealthiest railroad backers in the U.S. with some critical acquisitions and opening service to Chicago on February 8, 1880.

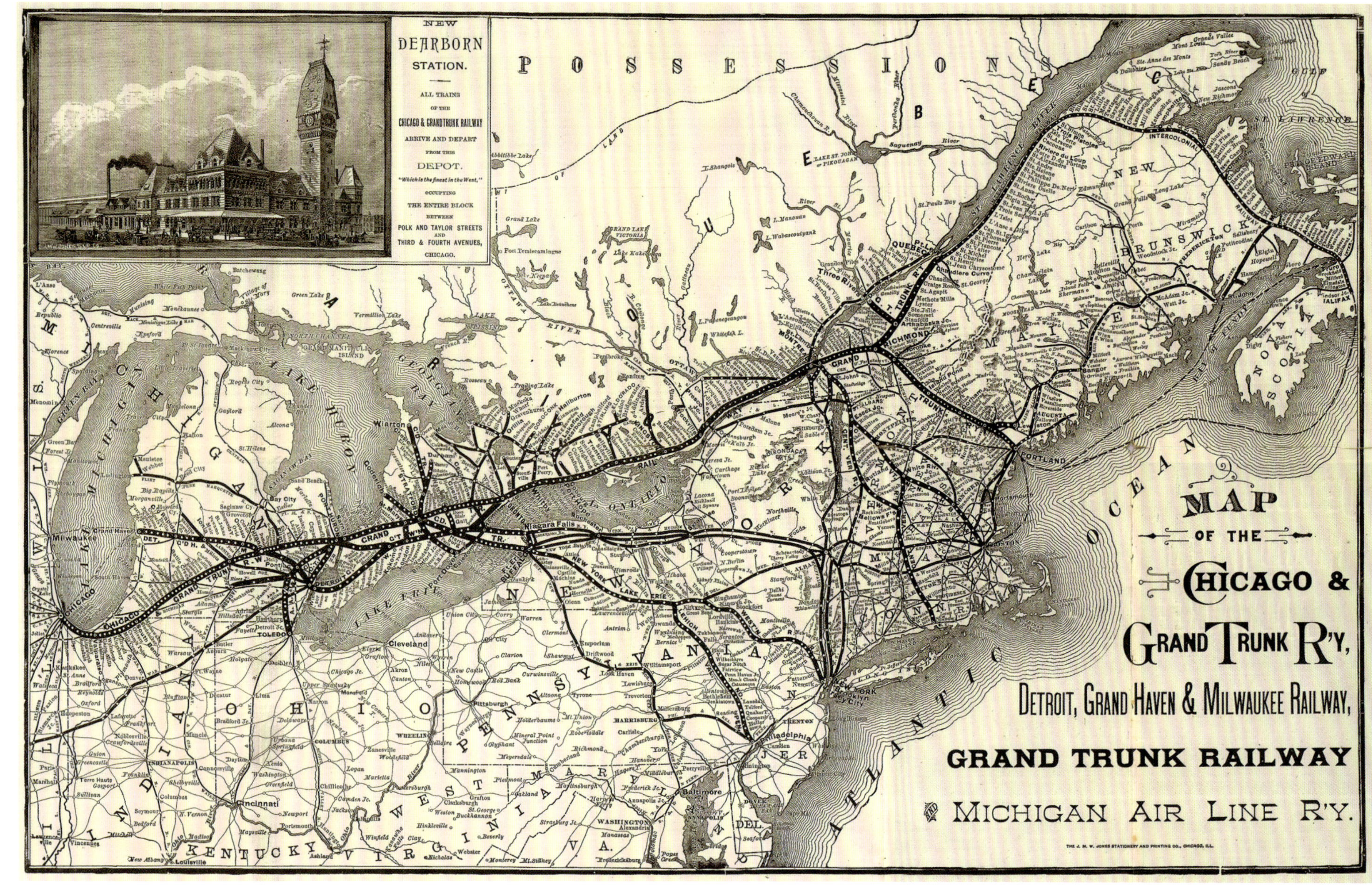

A map of the greater Grand Trunk Railway system from 1887.

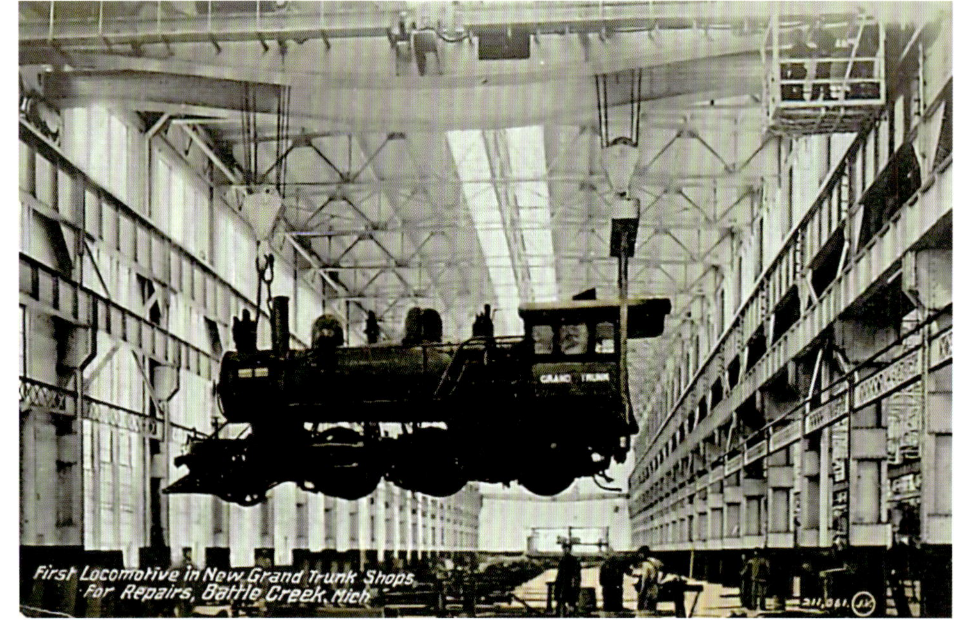

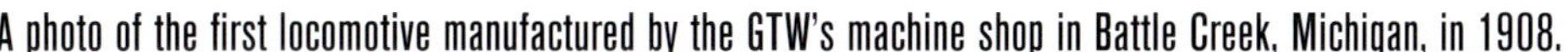
A photo of the first locomotive manufactured by the GTW's machine shop in Battle Creek, Michigan, in 1908.

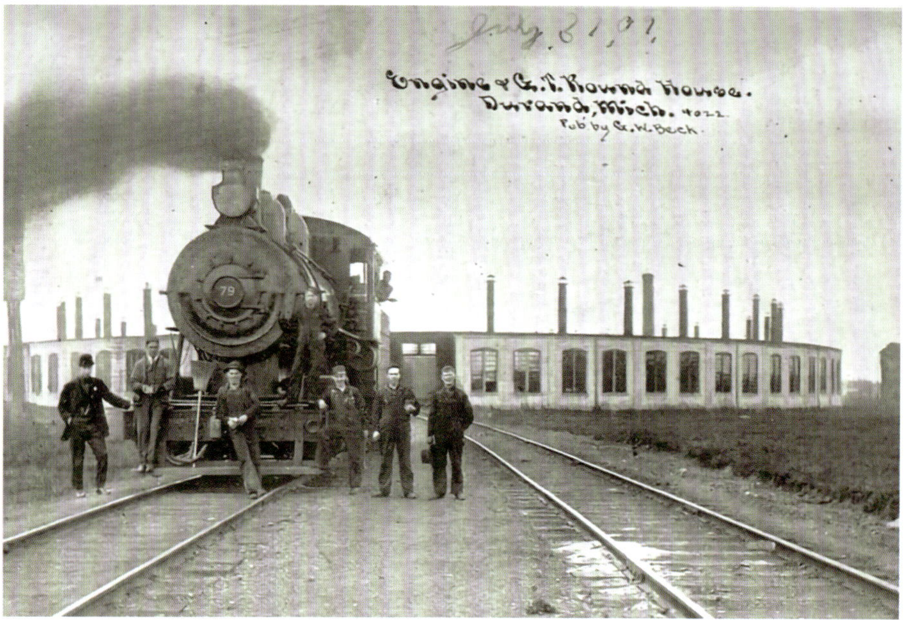

The GTW's roundhouse in Durand, Michigan, in 1909.

More mergers followed, and in 1900 the GTR opted to streamline its Midwestern operations by setting up the Grand Trunk Western as a subsidiary. It also set up the Grand Trunk *Pacific* Railway in its effort to compete against the Canadian Pacific and Canadian Northern in the quest to complete a transcontinental railway to British Columbia.

For its part, the GTW stuck to its Midwestern charge: connect Chicago and Detroit with cities and towns throughout the Midwest. That charge posed something of a catch-22 financially. The rise of the auto industry put high-value freight on GTW cars, helping the railway make ends meet. On the flip side, its hauls were short ones. That brought profits down compared to railways making cross-country treks.

Following the financial demise of the GTR, Grand Trunk Western was absorbed by Canadian National (CN) in the 1920s—an affiliation it retains to this day. CN, a publicly traded company, was instrumental in helping the GTW get through early twentieth-century troubles before World War II, and the growth of the auto industry also contributed to better days.

Through most of the twentieth century, the GTW generally followed its parent's path when it came to its identity and major decisions. It shared logos, color schemes, and equipment with CN, right down to the maple leaf background on its logo. In 1971, however, CN set up a new holding company—the Grand Trunk Corporation—to oversee the GTW. It allowed, in many ways, for the GTW to begin establishing its own identity.

The railroad began marketing itself as "The Good Track Road" in 1975, and it debuted red, white, and blue colors and a new, stylized "GT" logo on its cars. It increased its outreach to shippers, took greater care in upgrading its lines and equipment, posted impressive independent profitability, and earned several top safety awards in the 1980s.

However, the Grand Trunk Corporation was folded back into CN in 1992, which marked the beginning of the end for the GTW as a standout identity in the railroad industry. Its lines still make an impact in Michigan, Illinois, Indiana, and Ohio as a subsidiary corporation under the CN umbrella.

THE MOUNTAIN WEST

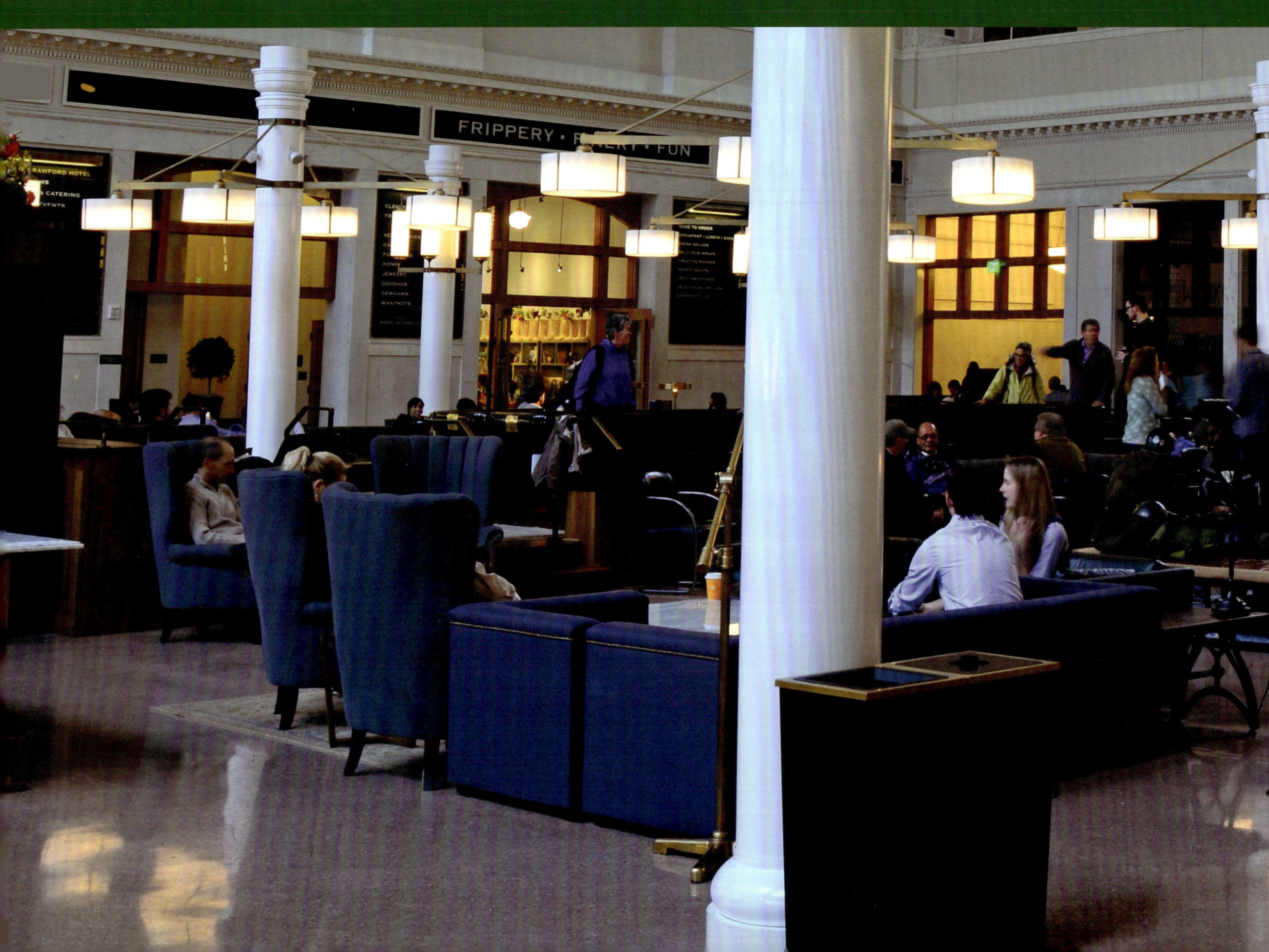

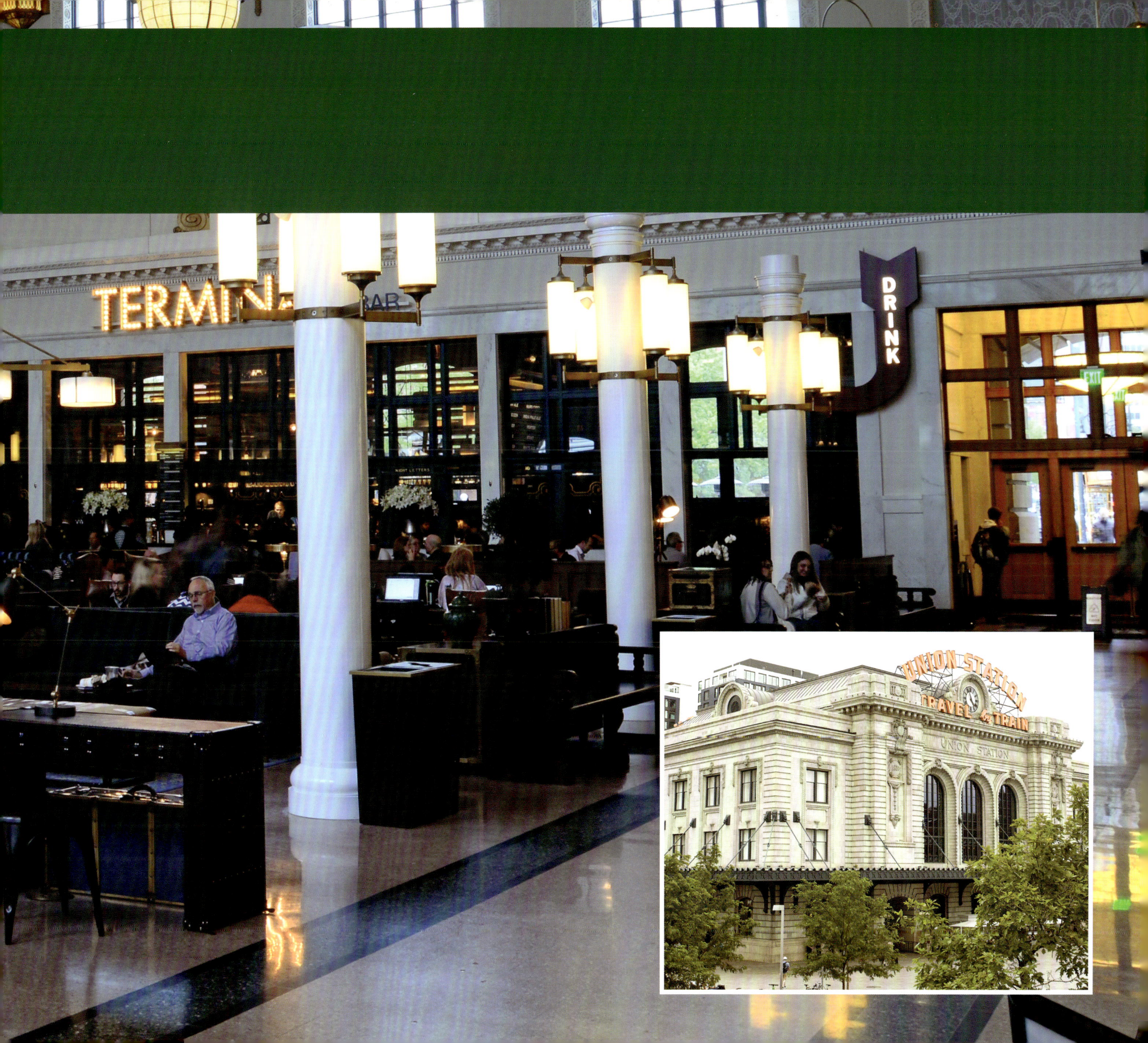
BAR
DRINK
EXIT
UNION STATION
TRAVEL
TRAIN
UNION STATION

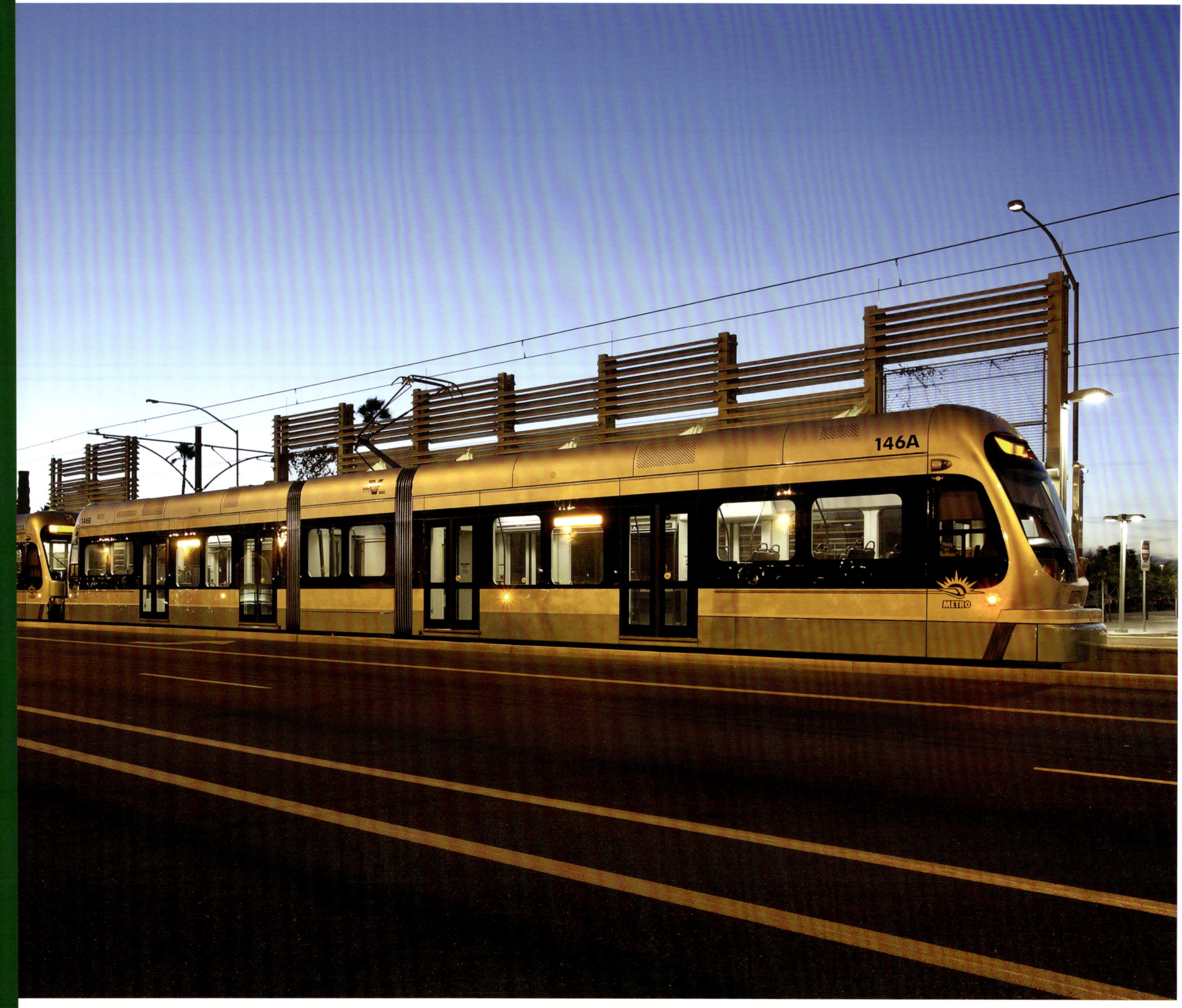

The Valley Metro Rail traveling east away from downtown Phoenix toward Tempe.

Voters in the Phoenix metropolitan area certainly made their desire for a light rail system evident. Beginning in the late 1980s, one local municipality after another began approving taxes to support a revamped local transportation system. The Arizona State Legislature created the Regional Public Transportation Authority (RPTA) in 1985, and it was determined by that group in 1993 that Valley Metro would be the name of a comprehensive plan to use that massive funding to get folks from point to point across the valley.

With a distinct purple and green color scheme, Valley Metro Rail, Inc., was formed in 2002 and tasked with designing, building, and operating a high-capacity transit system. Construction began in 2005, and on December 27, 2008, light rail began serving the cities of Phoenix, Mesa, and Tempe. There have been several expansions since Valley Metro Rail opened, and more extensions in the works will connect to South Phoenix and the State Capitol.

Even at its current capacity, Valley Metro has enjoyed great success by many standards. Spanning almost 30 miles with 41 stations, the light rail serves over 30,000 riders daily and has revolutionized the way people in the area get around. Valley Metro has garnered awards for achievements ranging from safety to environmental excellence to special needs services—accolades its leaders are justifiably proud of.

Powered electrically by overhead lines, Valley Metro makes its 30-mile trek in a little over 90 minutes, including stops. Its maximum speed is 58 miles per hour. While the light rail does not run all the way to Sky Harbor International Airport, the PHX Sky Train operates a free service to and from the 44th Street/Washington light rail station, keeping locals and visitors conveniently on schedule.

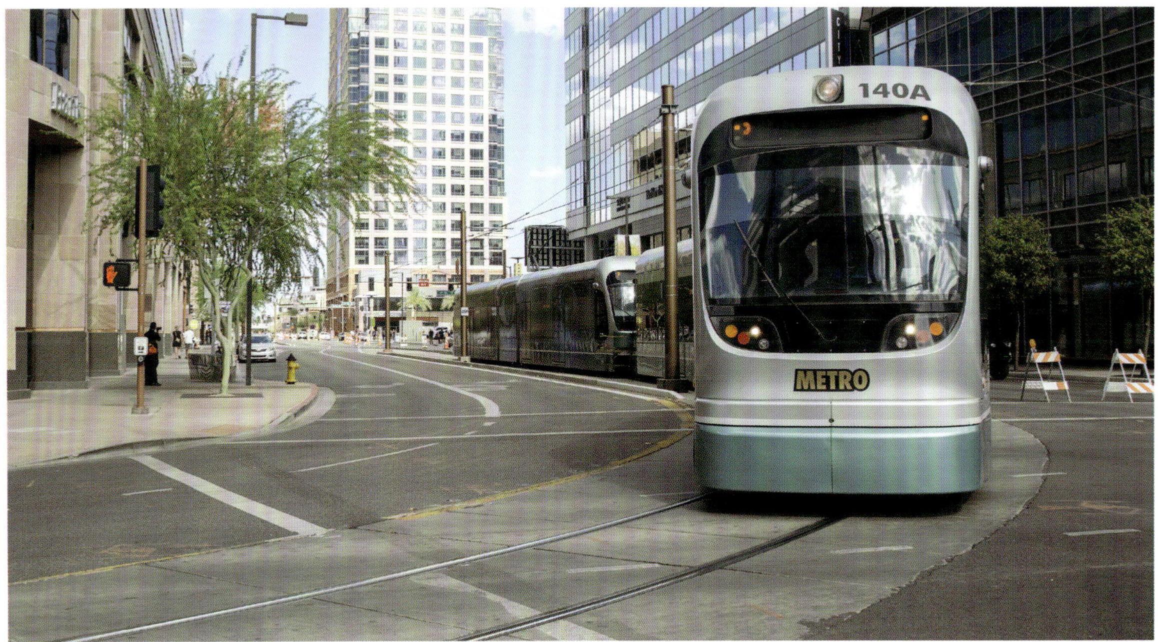

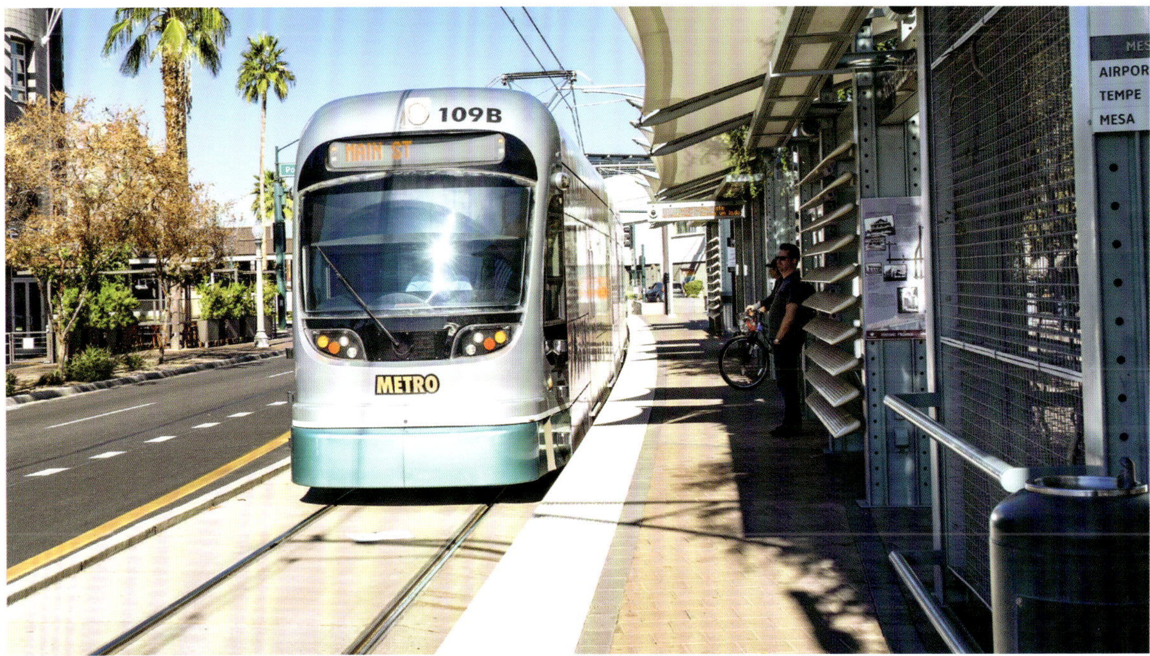

Top: The Valley Metro Rail in downtown Phoenix.

Bottom: Multiple extensions to the Valley Metro Rail are currently being planned, with the line extending into the city of Glendale to Arizona State University's West Campus by 2044.

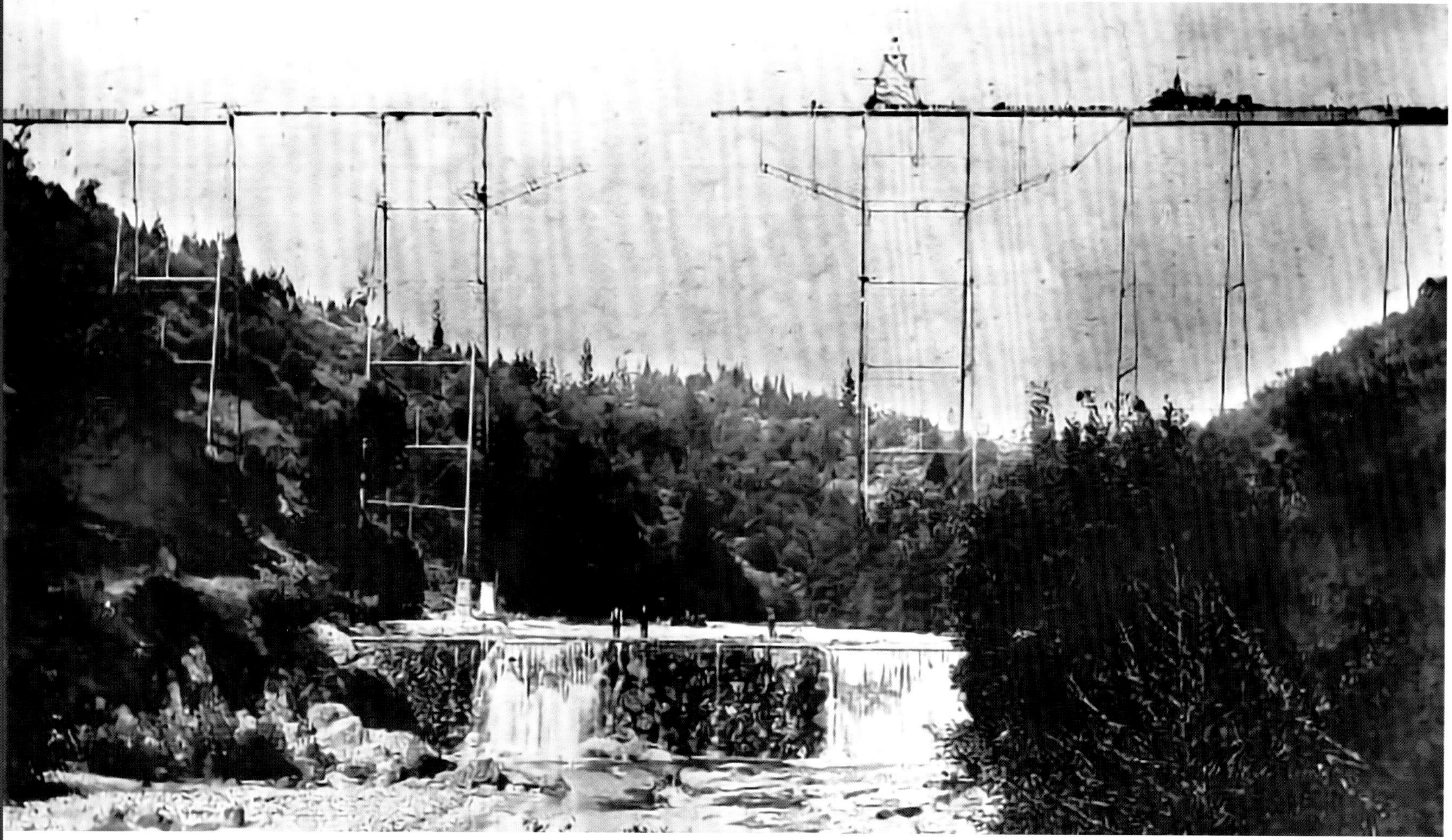

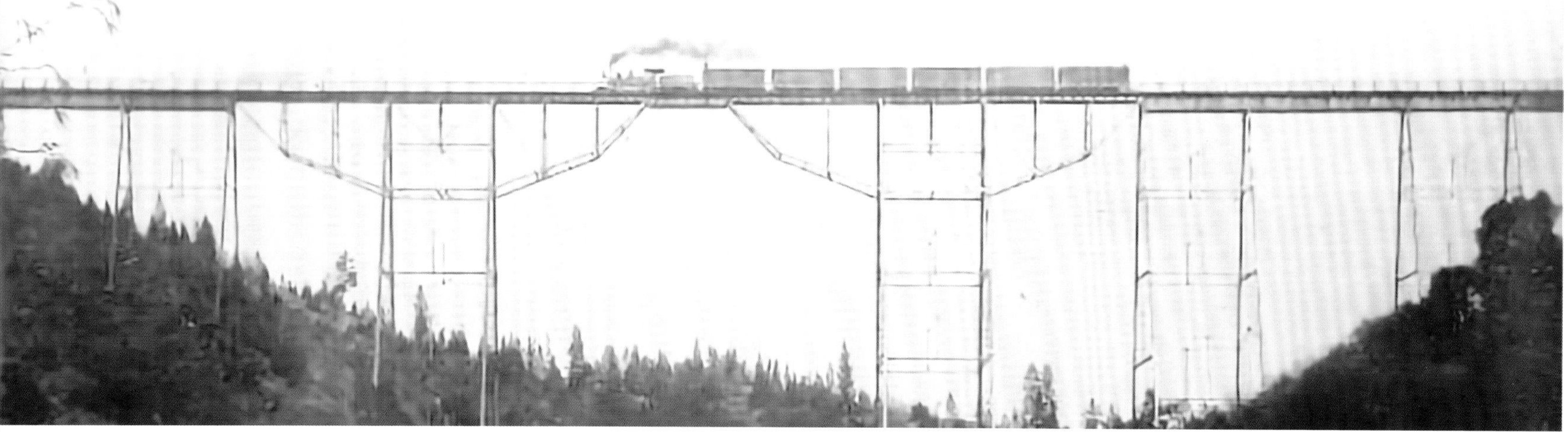

The Bear River Bridge under construction during 1908 (top), and a Nevada County Narrow Gauge locomotive crossing it in December of 1908 (bottom).

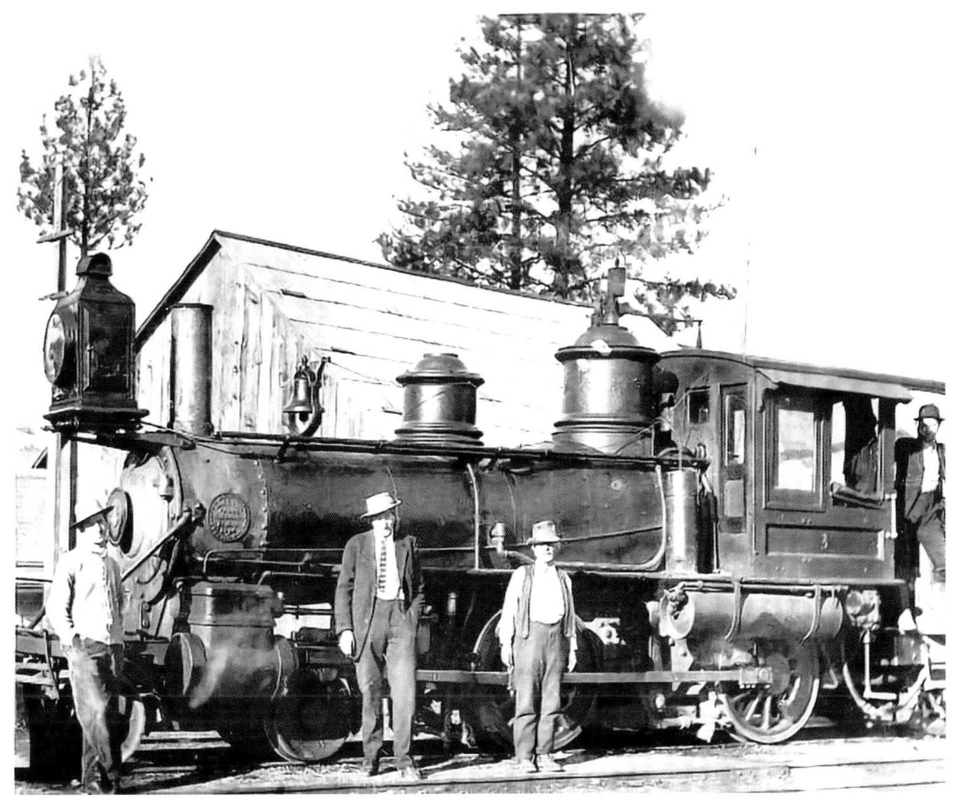

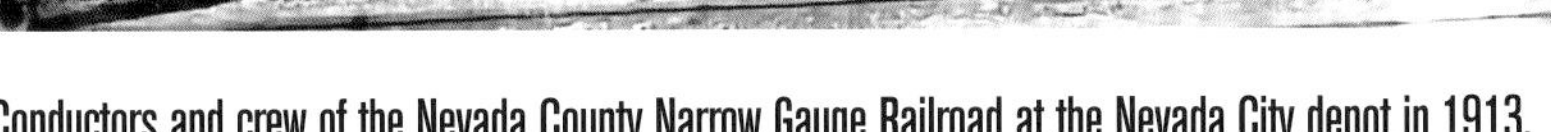

Conductors and crew of the Nevada County Narrow Gauge Railroad at the Nevada City depot in 1913.

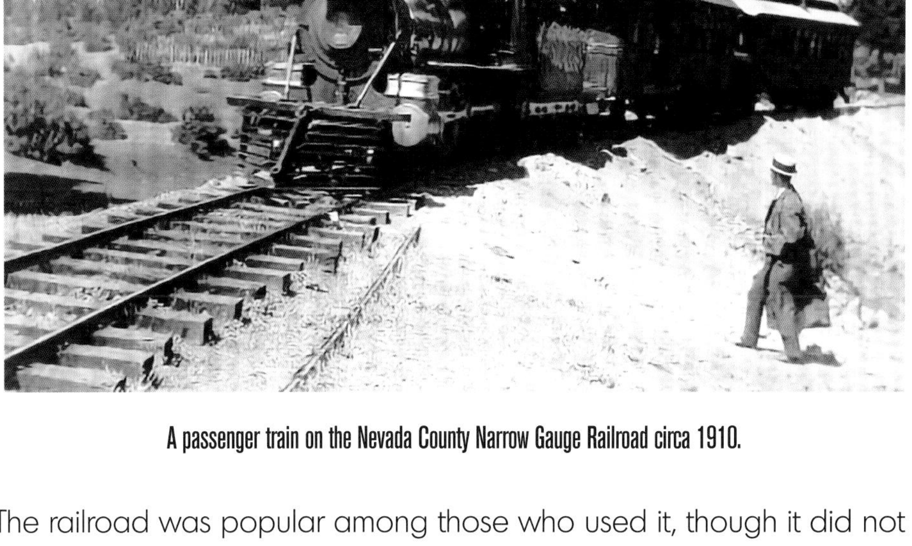

A passenger train on the Nevada County Narrow Gauge Railroad circa 1910.

The influx of go-getters to Northern California thanks to the Gold Rush of the mid-nineteenth century brought about a unique dilemma—how to get all these new folks from one place to the next. The answer, for those who lived between Colfax and Nevada City, was the Nevada County Narrow Gauge Railroad. Twenty-two miles at its peak, it was nicknamed "Never Come, Never Go" because of its NCNG initials. It connected with the Central Pacific Railroad and was renowned for having the highest railroad bridge in California. The Bear River Bridge stood 200 feet above the Bear River and stretched some 900 feet long.

It was 1874 when the California legislature and Governor Newton Booth approved the building of a railway between Colfax and Nevada City. The only bid that came in was for a half-million dollars, so M. F. Beatty was hired. He employed 600 men for the job, which included the construction of two bridges, two tunnels, and five trestles.

The railroad was popular among those who used it, though it did not get much fanfare at the time. Perhaps its lasting legacy is that of Sarah Kidder. She managed the Nevada County Narrow Gauge Railroad between 1901 and 1913, becoming the first woman to manage a major American railroad. Her family's estate was the largest home in the area. Kidder's railroad shipped more than $270 million in gold without ever losing an ounce.

By 1913, though, financial troubles hit hard. That's when Kidder sold her interests and retired in San Francisco. By 1926, the railroad was worth just a single dollar. Earl Taylor bought it for that amount and sold it in 1942 for a quarter-million. That marked the year the final NCNG train chugged over the scenic tracks of the Sierra Nevada Mountains.

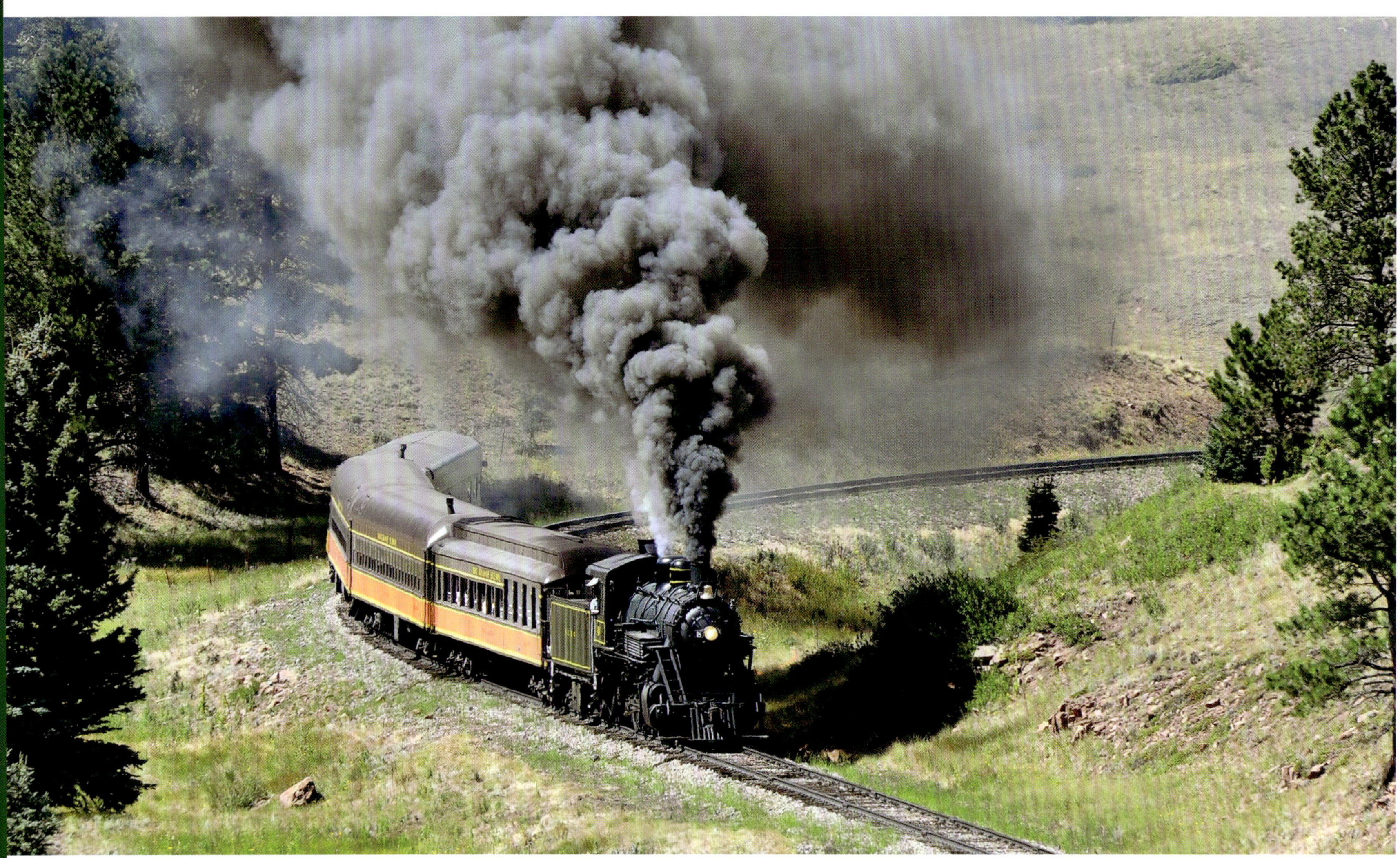

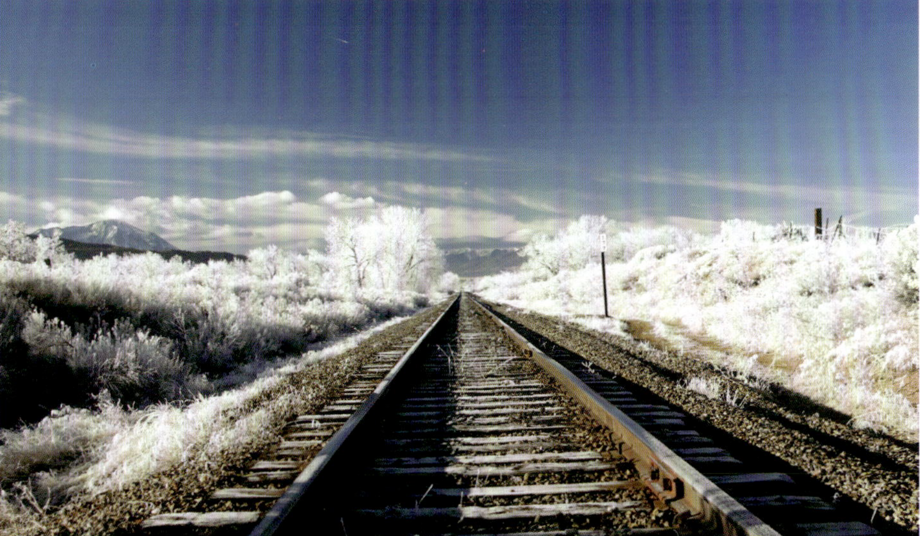

Top: The Rio Grande Scenic Railroad operates both steam and diesel engines.

Bottom: Icy tracks crossing the Veta Pass (9,220 feet) in the Sangre de Cristo Mountains. The Denver and Rio Grande Western Railroad's narrow-gauge track used a different route to cross over the pass. When the track was converted to standard gauge, the route was moved about seven miles southeast of the old route.

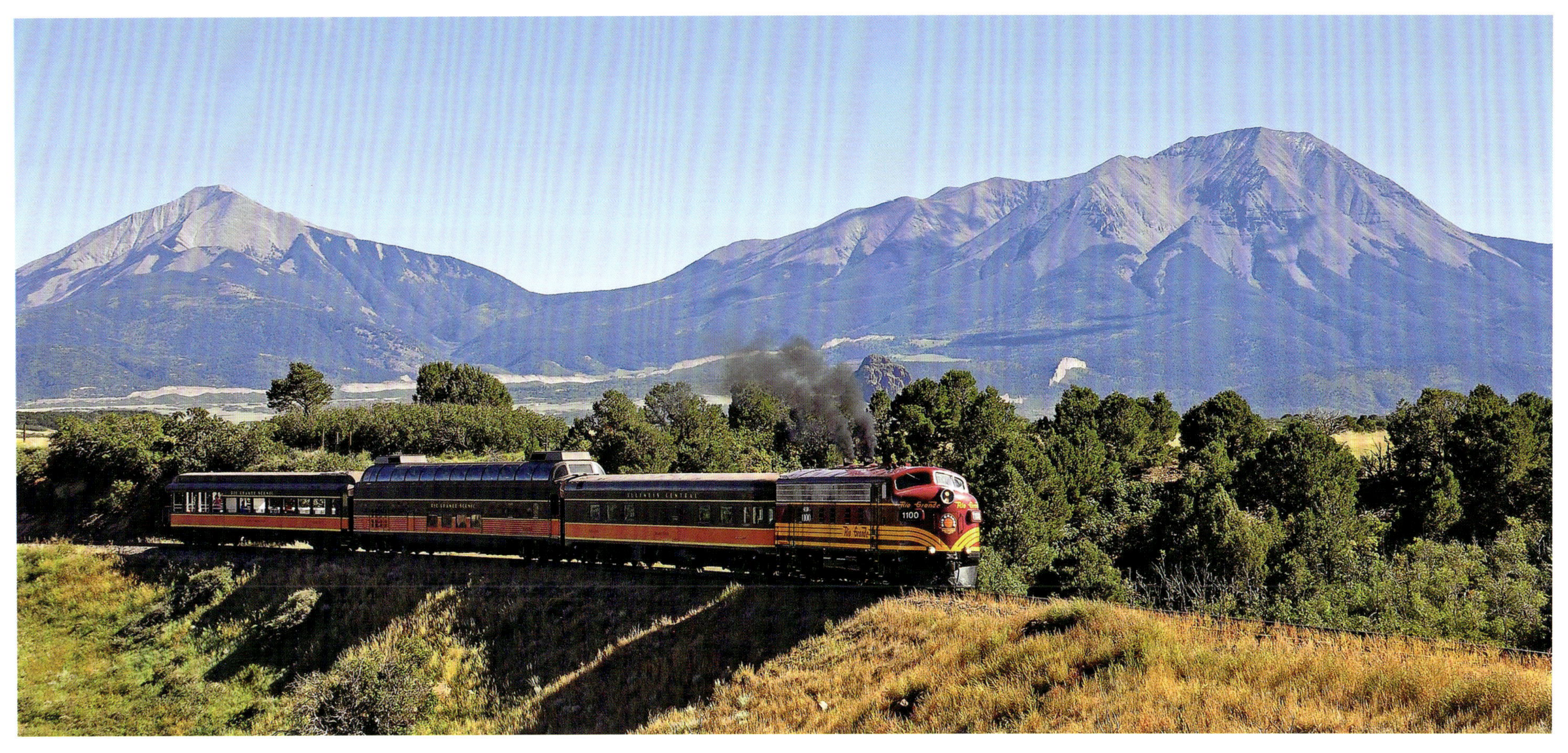

The Rio Grande Scenic Railroad just outside of La Veta with the Spanish Peaks behind it.

Located 200 miles south of Denver, Colorado, in the larger area of the San Luis Valley, the Rio Grande Scenic Railroad opened in 2006 as a tourist train. Traveling east between Alamosa, Colorado, and Walsenburg, Colorado, located in the smaller La Cuchara Valley below the Spanish Peaks, the Rio Grande Scenic Railroad travels for 60 miles along what was originally the San Juan Extension of the Denver and Rio Grande Western Railroad (D&RGW).

The D&RGW was incorporated in 1870 with the intent of working its way south along the Rio Grande toward El Paso, Texas. The narrow-gauge railway reached Colorado Springs by 1871, moved south toward Pueblo, and made it to the Veta Pass and into Alamosa by 1878. The D&RGW is credited with opening the San Luis Valley up to the rest of the world, allowing for the area's rich resources to be easily transported out of the area and enter into the world's economy. The city of Alamosa was literally built in one day as buildings were transported into town on the railway the day the station opened. Alamosa was the narrow-gauge hub of North America, with more narrow-gauge tracks operating out of the city than any other city. Though, the original narrow-gauge tracks that operated between La Veta and Alamosa were replaced with standard gauge in 1899 to compete with larger freight companies.

The Rio Grande Scenic Railway is owned by the Colorado Pacific Rio Grande Railroad (CXRG), which bought 154 miles of D&RGW tracks in 2003. The CXRG operates three lines radiating out of Alamosa. The Rio Grande Scenic Railway runs along tracks from Alamosa to a junction with the Union Pacific in Walsenburg, Colorado. Another train operated by the CXRG runs south from Alamosa to Antonito, where it meets with the Cumbres and Toltec Scenic Railroad.

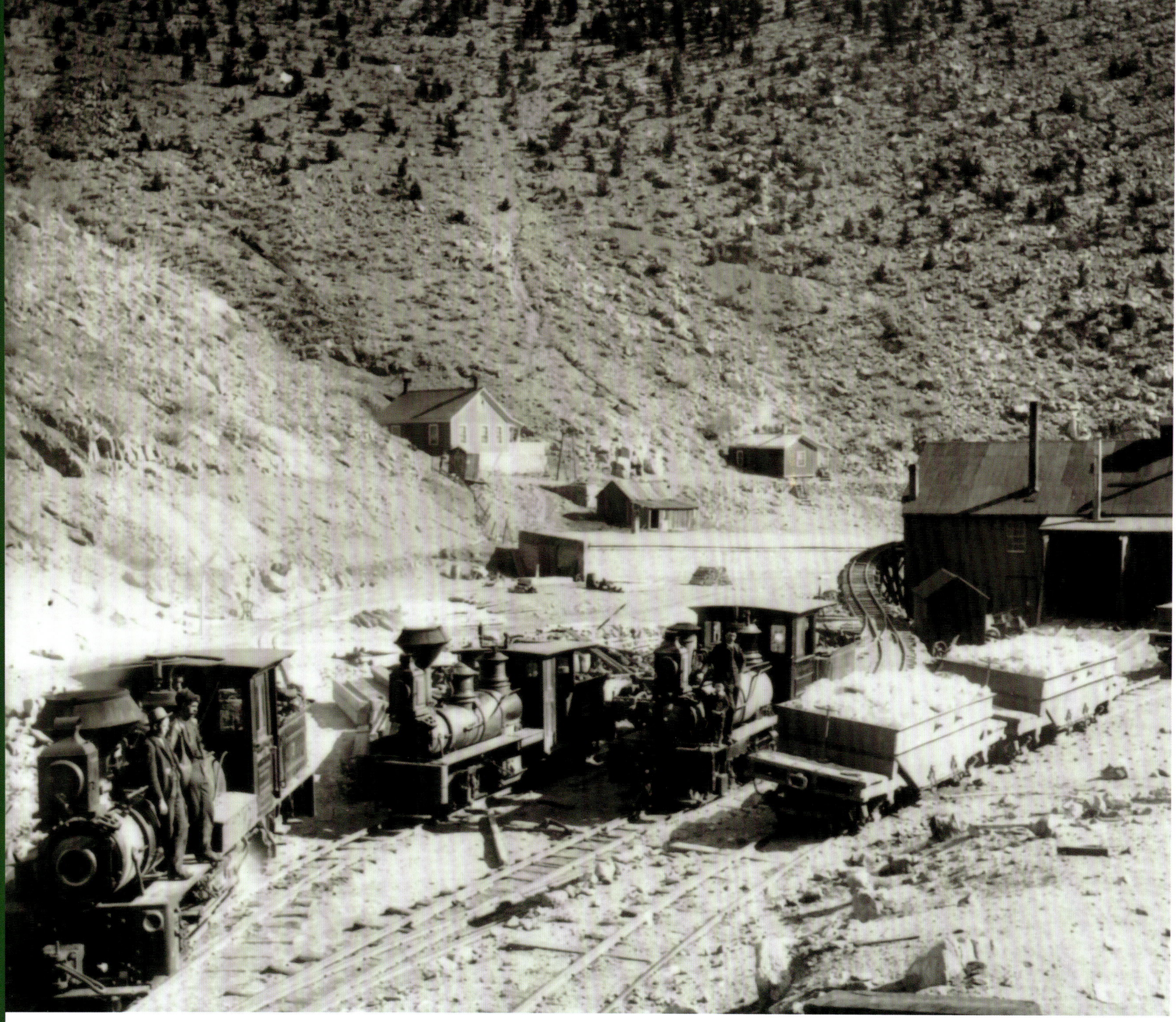

Pictured here is the Gilpin Tramway. The Gilpin Tramway was a two-foot gauge tramway used to transport gold ore from the mines in the mountains down into the mining towns in the valleys.

A Gold Rush-era railroad that conquered some steep grades in the Rocky Mountains, the Colorado Central Railroad ran over three-foot narrow-gauge tracks in Colorado and southeastern Wyoming in the late nineteenth century. Some of its "back stories" involved heated battles between local interests and outside investors at a time when the West was just taking shape, and a rivalry between the cities of Denver and Golden for Colorado supremacy.

At its heart, however, this railway was built to haul gold out of the mountains. It was founded as the Colorado and Clear Creek Railroad Company by Golden entrepreneur William Loveland in 1865. Golden was the capital of the Colorado Territory at the time. Loveland, and his fellow investors wanted to supply rail service to the mining towns along Clear Creek, while also connecting Boulder and Denver.

Outside investors from Union Pacific (UP) also wanted a piece of the action, and the railway went through two early reorganizations that led to its renaming as the Colorado Central Railroad in 1868. In the face of off-track politicking, the Colorado Central experienced several delays but eventually made its way through the challenging Front Range. A line between Golden and Jersey Junction, near Denver, opened in 1870. Trains could reach Forks and Black Hawk by the end of 1872. Idaho City became accessible in 1877, and Central City the following year.

Loveland and Union Pacific went back and forth in the courts in their battle for ownership of the railroad, which had been resourceful in its navigating of tough terrain but not so successful in the financial books. UP jumped on the Colorado Central's troubles and leased the company in 1879. By 1893, though, the UP collapsed. The Colorado Central became the Colorado and Southern Railway (1899), but service began to be discontinued and tracks abandoned in the early 1900s.

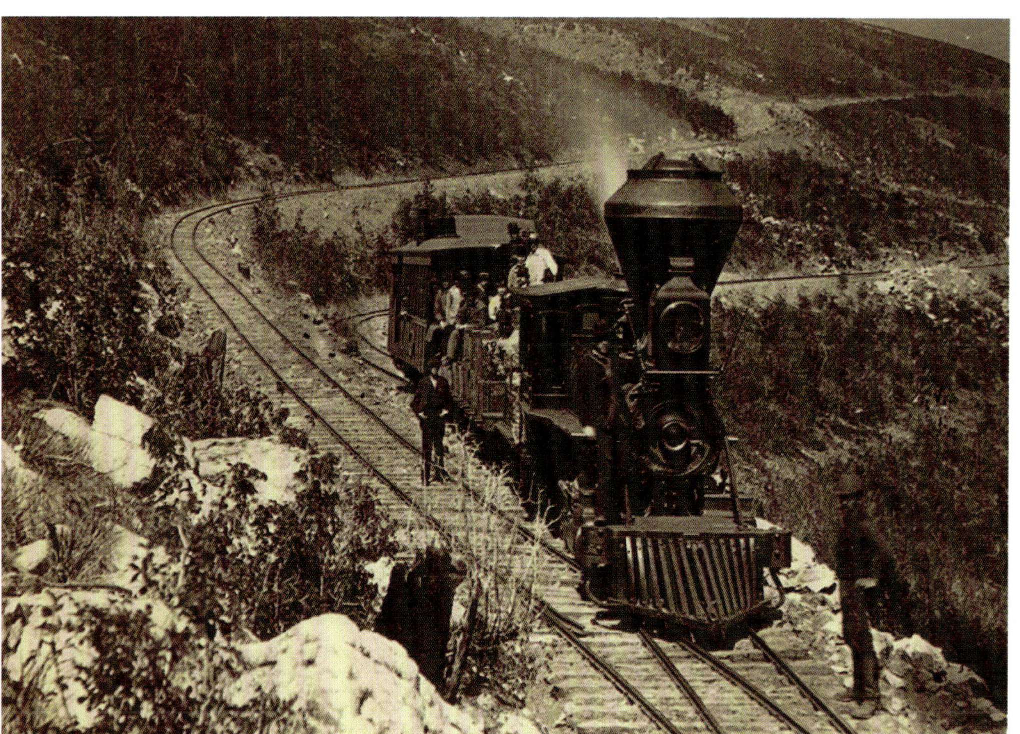

Top: A Colorado Central locomotive in the Colorado Rockies' Snowy Range just north of Cameron Pass.

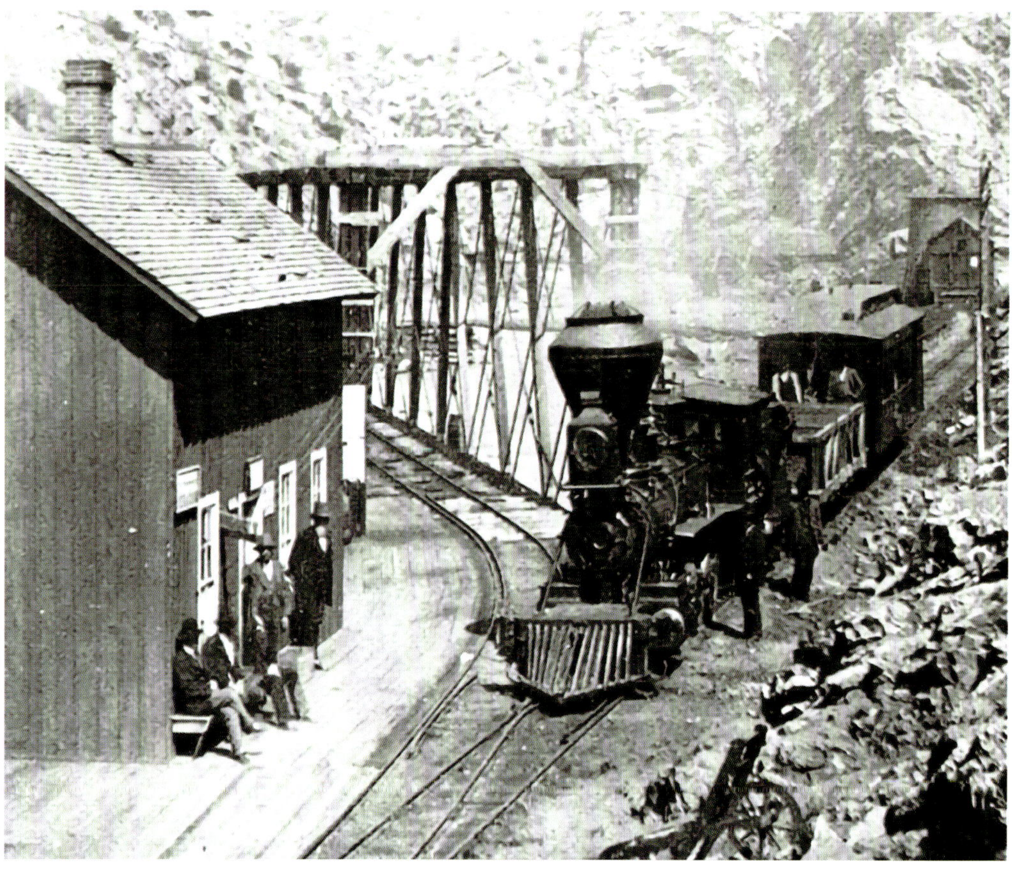

Bottom: The Colorado Central at Clear Creek heading east from Black Hawk, Colorado, toward Golden, Colorado.

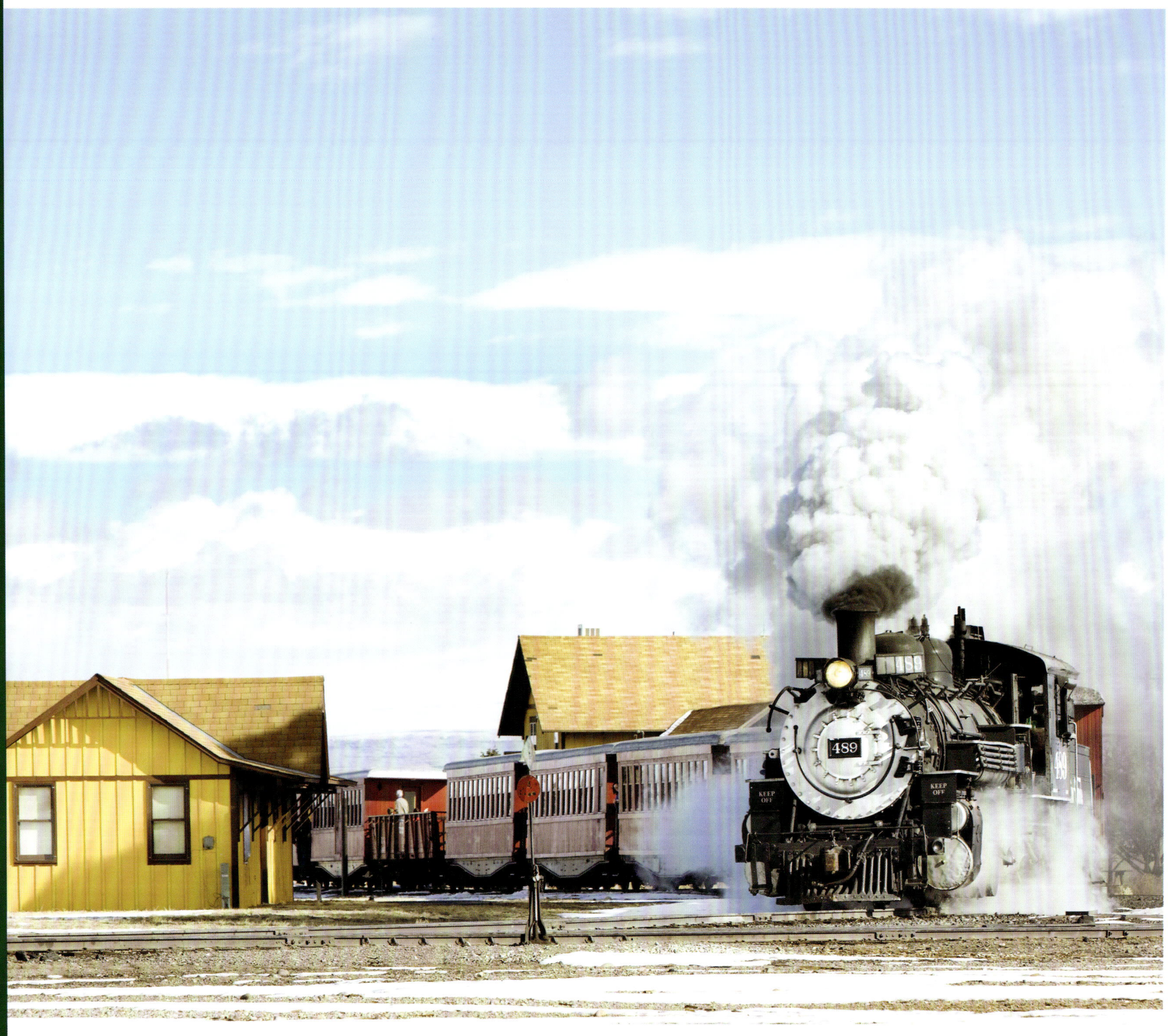

All the steam locomotives on the Cumbres and Toltec Scenic Railroad formerly belonged to the Denver and Rio Grande Western Railroad.

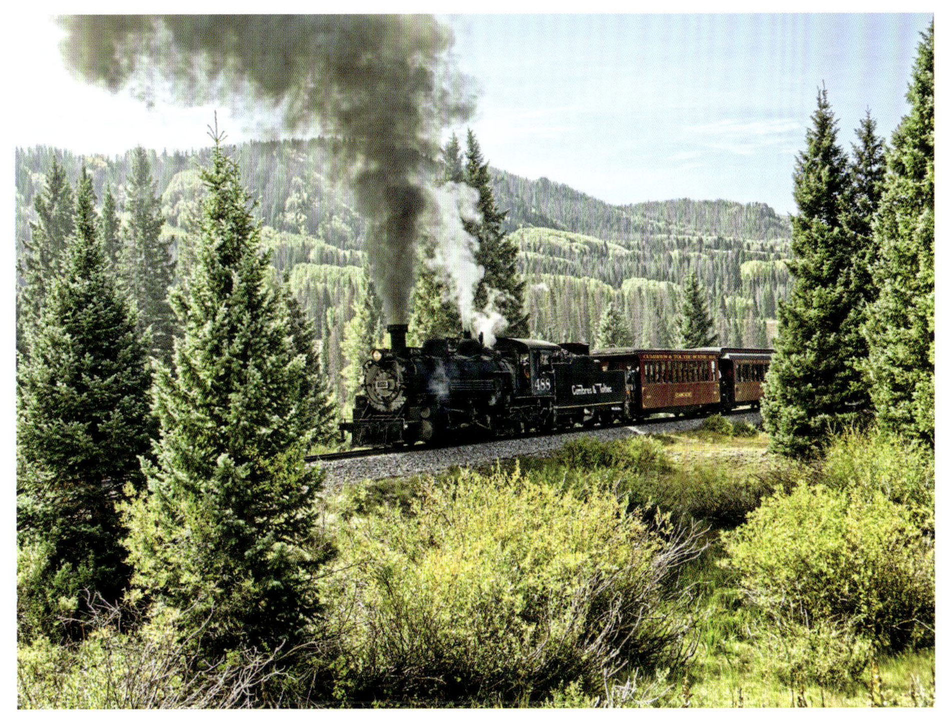

The C&TSRR operates two trains a day from the Chama and Antonito terminuses.

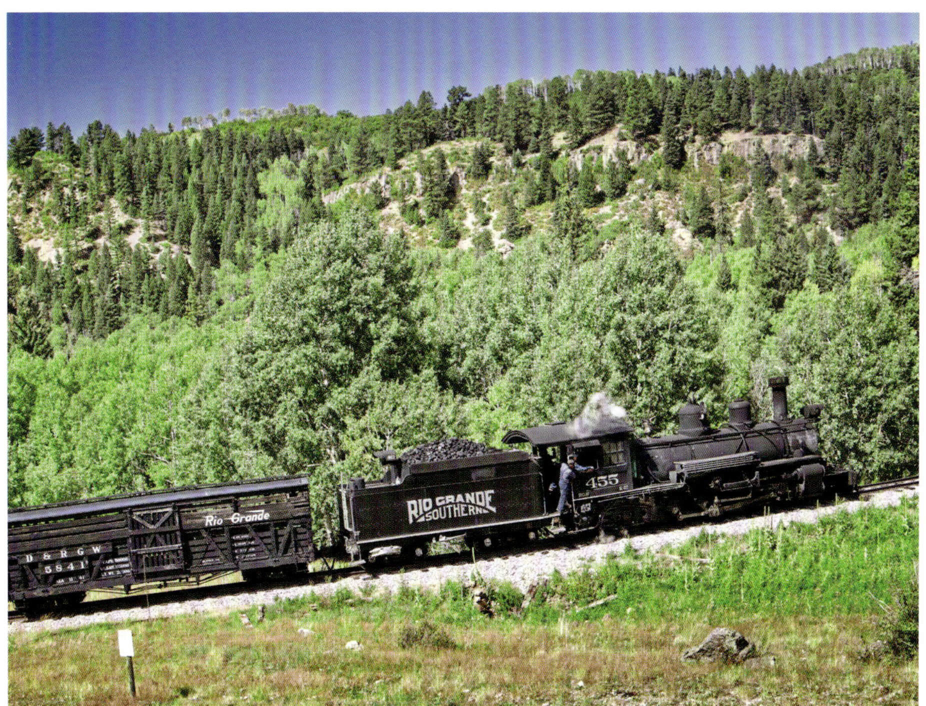

Heading east from Chama, New Mexico, and beginning its ascent to Cumbres Pass.

A narrow-gauge heritage railway traveling between Antonito, Colorado, and Chamas, New Mexico, the Cumbres and Toltec Scenic Railroad (C&TSRR) gets its name from two of the major geographic landmarks it passes en route. On its 64-mile journey, it travels over the 10,000 foot Cumbres Pass of the San Juan Mountains and the Toltec Gorge of the Rio de los Pinos in the Rio Grande National Forest. The Denver and Rio Grande Western Railroad (D&RGW) built the track in 1881 as part of the Alamosa–Durango Line that ran between Alamosa and Durango, Colorado. The D&RGW was a pioneer in American mountain locomotion, traveling through the Rockies to connect Denver, Colorado with Salt Lake City, Utah. It was instrumental in transporting the rich mineral resources of Colorado to the rest of the nation.

The C&TSRR is one of the few portions of track still used today that was part of the D&RGW's San Juan Extension. Only the Durango and Silverton Narrow Gauge Railroad, the standard gauge Colorado Pacific Rio Grande Railroad running out of Alamosa, Colorado, and the narrow-gauge C&TSRR remain. The D&RGW began to abandon the C&TSRR in 1968 due to declining freight operations, but rail enthusiasts prompted the states of New Mexico and Colorado to buy a 64-mile section of the track for preservation. Both states are now part owners of the track and have developed the Cumbres and Toltec Scenic Railroad Commission to operate the railroad as a passenger excursion service. The Friends of the Cumbres and Toltec Scenic Railroad was established in 1988 to help create educational programs about the railroad's history.

The C&TSRR operates a handful of historic steam locomotives with a mixture of flat-roofed and clerestory carriages. There are four tiers of service on the route, including Coach, Deluxe, Parlor, and Premium services. The railroad was listed on the National Register of Historic Places in 1973, and the boundaries of that designation were expanded in 2007. In 2012, the railroad was officially listed as a National Historic Landmark. In recent years, service has been cancelled for short amounts of time due to roadbed damage and threats from forest fires. In 2010, the Lobato Trestle, just east of Chama, was severely damaged by fire, causing a shortened route for trains traveling from each direction. The trestle was repaired in 2011 and is back in use.

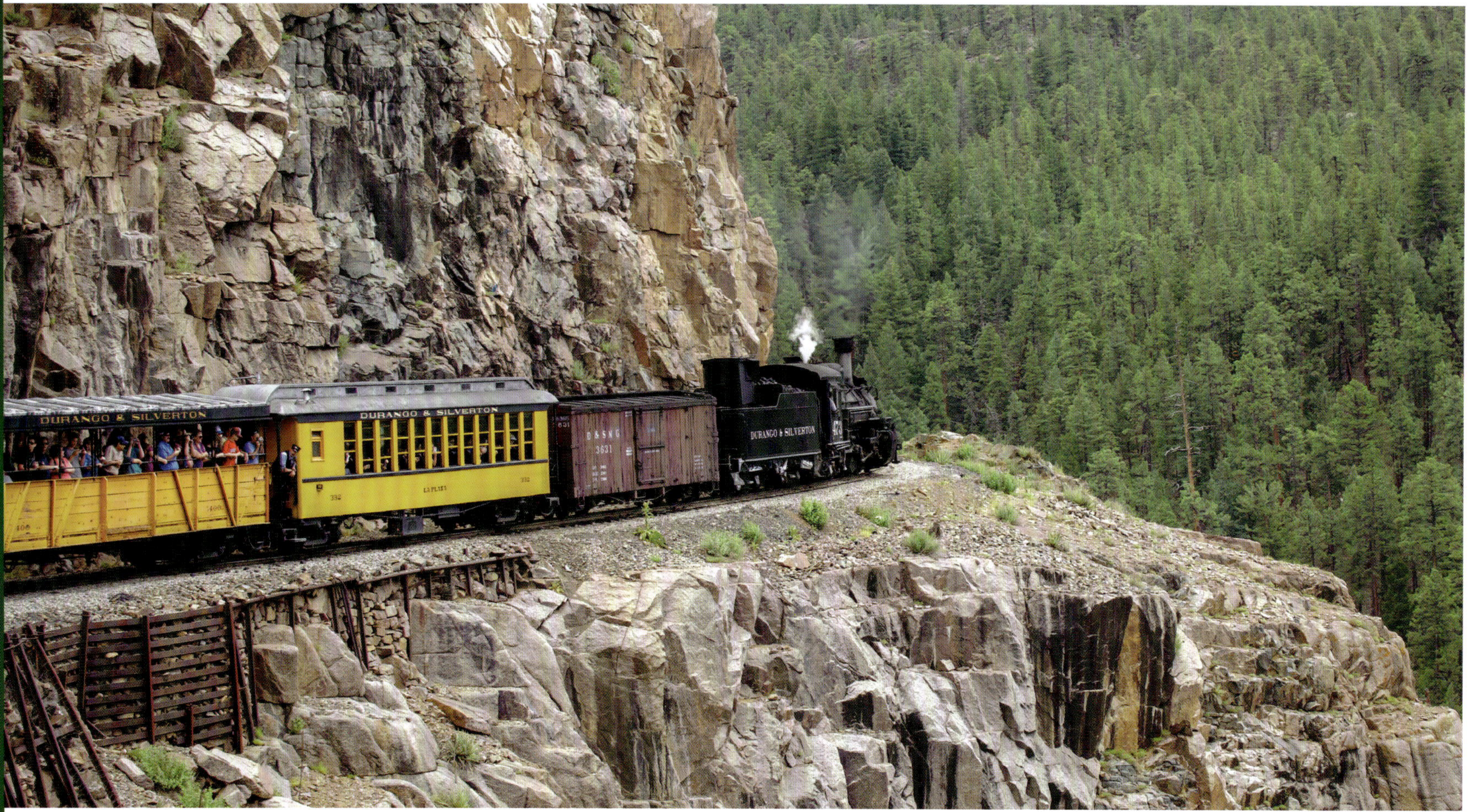

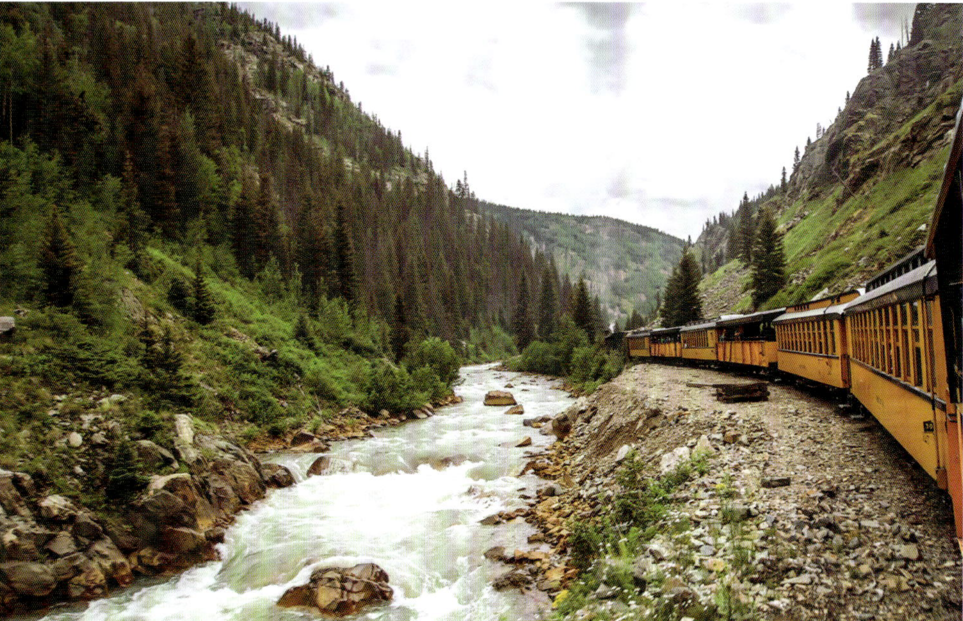

Top: There are historic museums commemorating the D&SNG in both Silverton and Durango.

Bottom: The D&SNG crosses the Animas River five times in the 45-mile stretch between Silverton and Durango.

A relic from the mining days of the Wild West, the Durango and Silverton Narrow Gauge Railroad (D&SNG) has been in operation for over 130 years, although it carries more tourists today than the silver and gold ore it was built to transport. The portion of track that is today known as the D&SNG was originally a part of the San Juan Extension of the Denver and Rio Grande Western Railroad (D&RGW). From Denver, the San Juan Extension traveled south to Santa Fe, New Mexico, and then west toward Durango. The Silverton Branch of the San Juan Extension then went north from Durango toward the abundant mining areas in the San Juan Mountains around Silverton, which is the same track the D&SNG operates on today.

The tracks were laid into Durango on August 5, 1881, and then expanded into Silverton by July 10, 1882. It was not long before the Silverton Branch was hit hard by economic factors. The Panic of 1893, the oversupply of silver, the decreased mining ventures in the area, and declining revenue from passengers all brought the Silverton Branch's service nearly to a halt. And with all economic factors aside, environmental factors, like mudslides and flooding, were also huge hurdles that nearly stopped the railway in its tracks.

Since the 1980s, the D&SNG has been bought and sold by various private owners, and it continues to be one of only a few rail services that still operate steam locomotives in the nation. It is also only one of two remaining narrow-gauge tracks still operating in what was once Colorado's very extensive narrow-gauge network. It is a federally designated National Historic Landmark and a Historic Civil Engineering Landmark designated by the American Society of Civil Engineers.

The narrow gauge of track that the D&SNG is named after is only three feet wide, one foot and eight and a half inches narrower than today's standard railroad gauge.

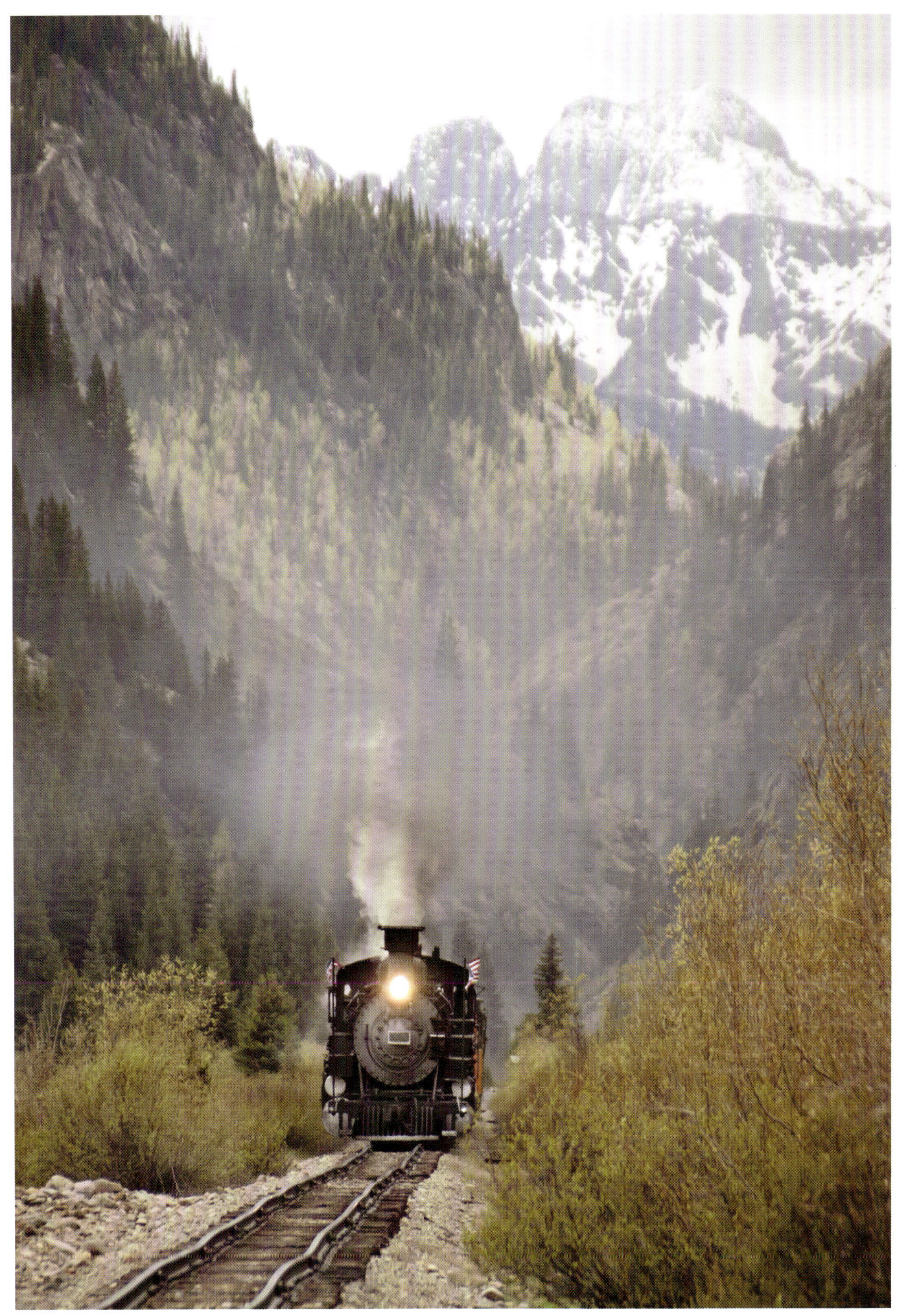

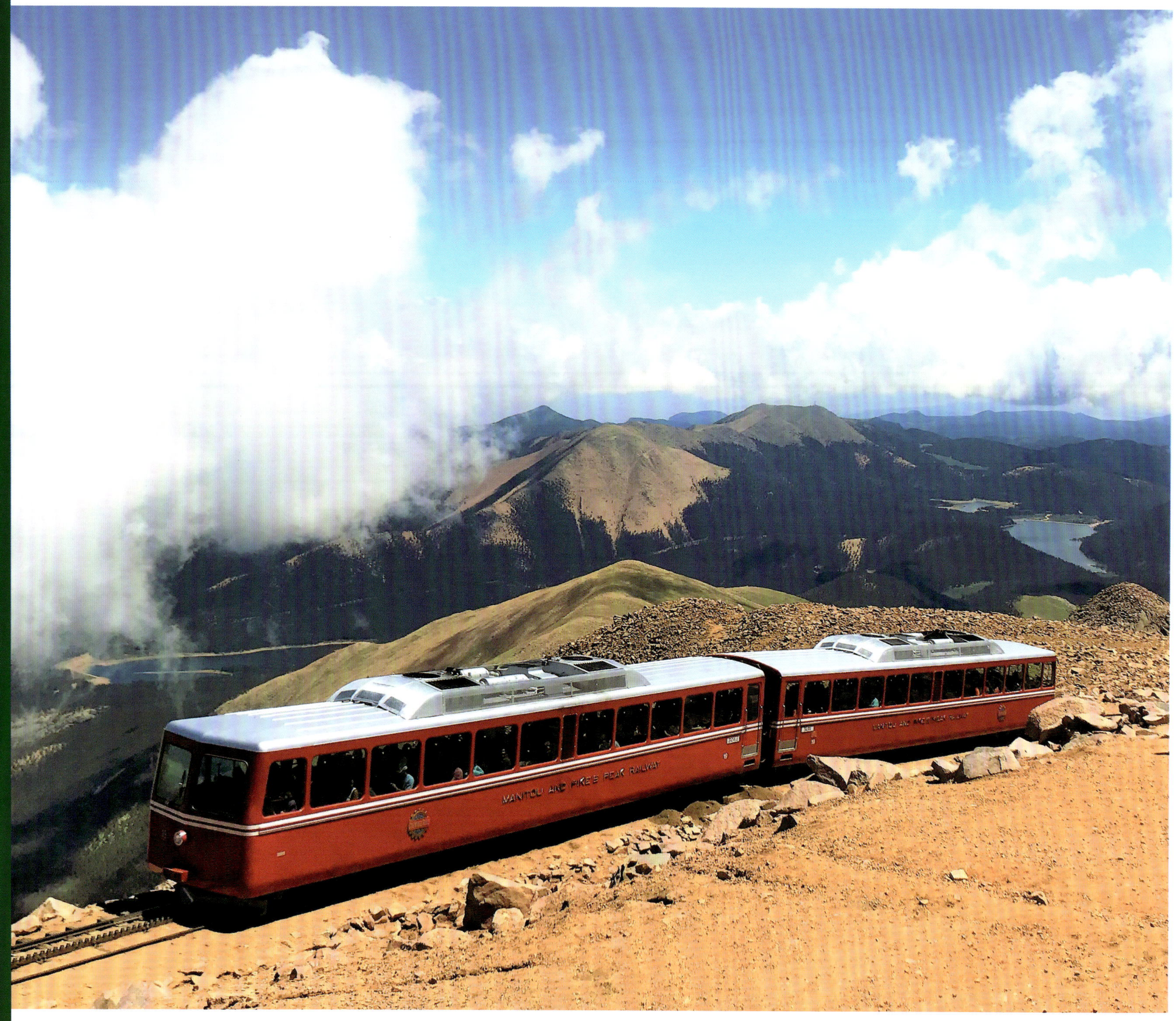

Diesel railcars built by the Swiss Locomotive and Machine Works company in 1975 ascending to the summit of Pikes Peak.

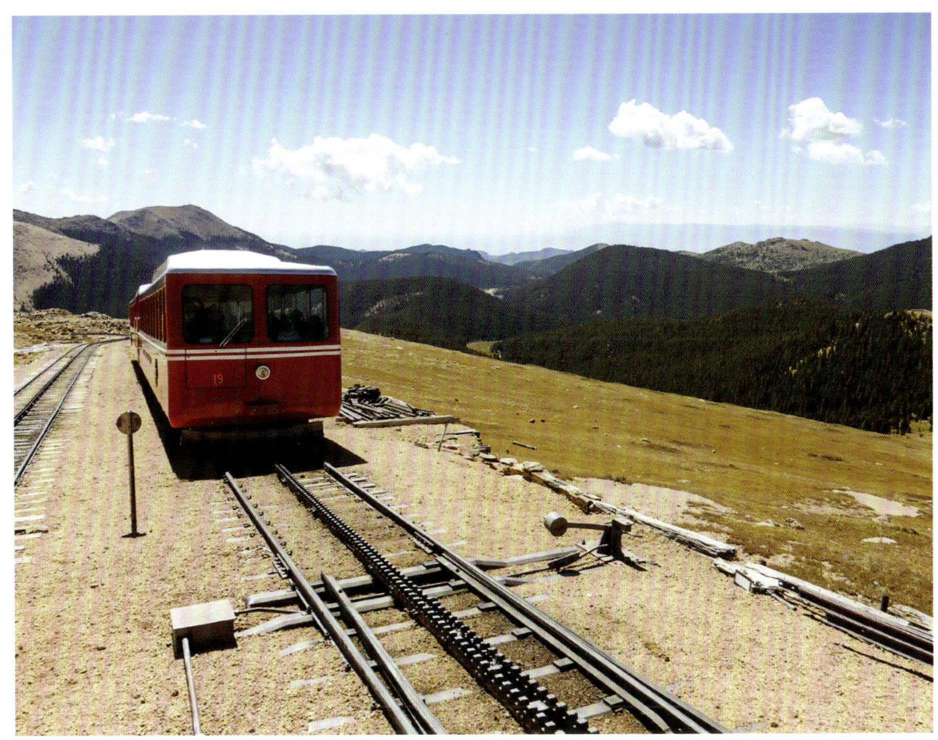

Pikes Peak in a registered National Historic Landmark and is named after Zebulon Pike, an American explorer who was unable to reach the summit of the mountain in 1806 after two days of foodless hiking through waist-deep snow.

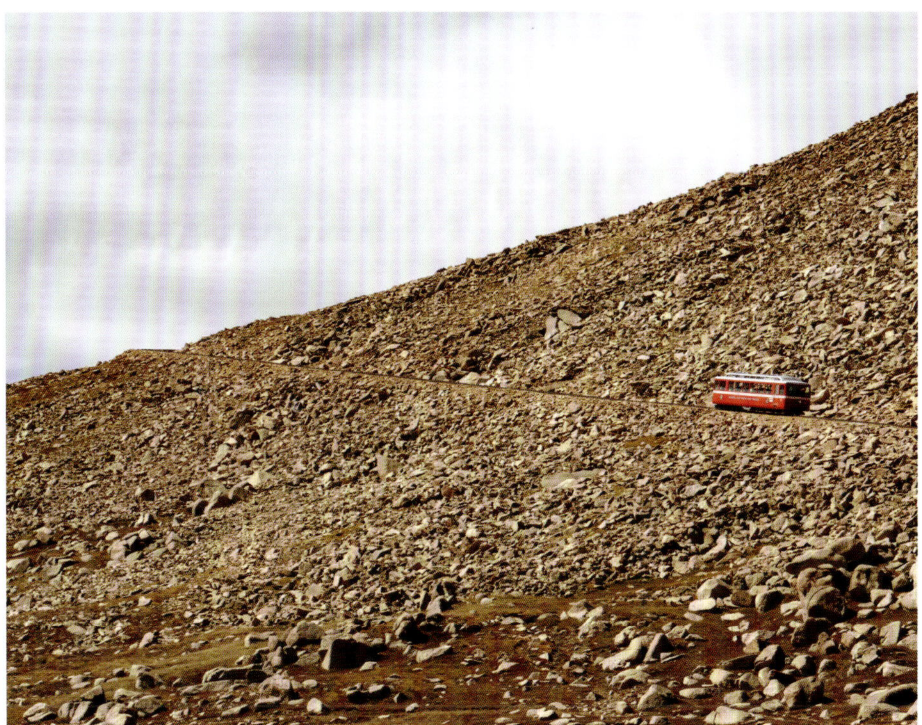

Because of its high elevation and polar climate, snow can be expected at any time of year at the top of Pikes Peak.

Unlike the many train routes in Colorado that were built for mining and other causes that promoted westward expansion, the Pikes Peak Cog Railway was built solely for the entertainment of tourists. It is the highest railway in North America, starting from the Manitou Springs Train Station at 6,412 feet above sea level and climbing to the summit of Pikes Peak at 14,115 feet above sea level. The route is almost nine miles long with an average gradient of 16 percent, which was surmounted with an Abt rack system that was in place from 1889 until 2019.

Zalmon G. Simmons, of Simmons Bedding Company fame, founded the railway in 1889. At that time, limited service was provided to the Halfway House Hotel, located in Ruxton Park, Colorado, about halfway between Manitou Springs and Pikes Peak. By 1891, the railway reached the summit, and service was opened to the public. The railway used steam locomotives for the first few decades of service, which can be incredibly arduous and difficult work for a railway that operates on such a gradient. Gasoline- and diesel-powered railcars were purchased for the railway by 1938, but steam-powered service was still provided until the 1960s. By the 1970s, the railway increased its capacity to support the growing number of tourists that were attracted to the destination, which led to the railway acquiring four 214-passenger railcars in 1976. In addition to those railcars, the Pikes Peak Cog Railway also operates several smaller Swiss-built railcars, GE locomotives, several diesel-electric locomotives, a snowplow, and an original Wasson wooden coach.

The Pikes Peak Cog Railway provides service six to eight times a day during the peak months of the summer. In 2006, the railway started operating year-round—as long as snowplows are able to clear the tracks. Service was put on hold in the winter of 2017 when scheduled maintenance uncovered larger maintenance issues that required extensive updates. The line reopened on May 20, 2021, and it debuted several new Stadler Rail trainsets from Switzerland that are powered by diesel-electric locomotives. Other track upgrades included radio-controlled switches, a new Strub rack system, and metal ties replacing the old wooden ties. A new visitor center at the summit welcomes travelers to the highest summit of the southern Front Range of the Rocky Mountains.

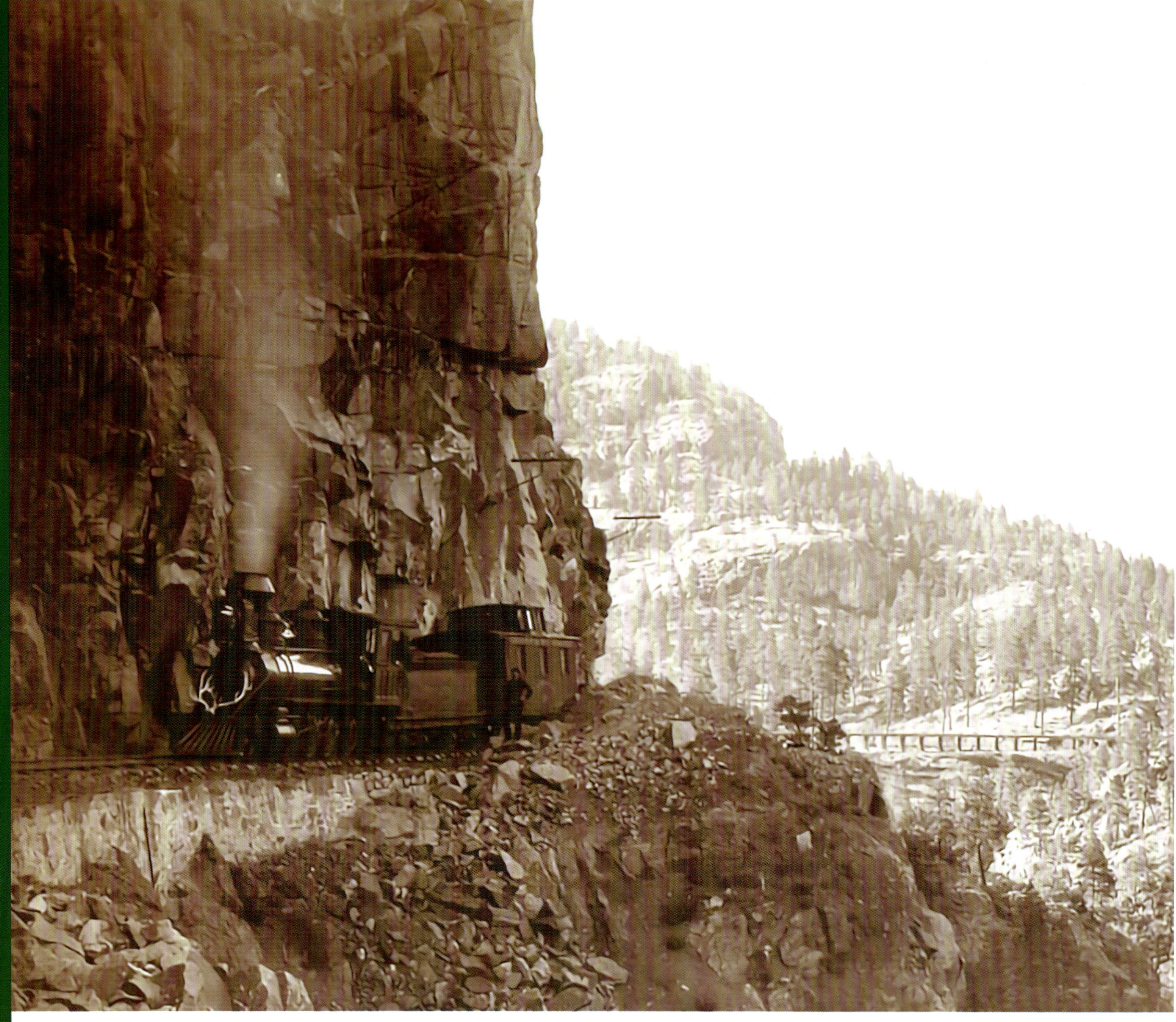

The Rio Grande Southern Railroad running along the Animas River, a tributary of the San Juan River in Colorado.

It seemed the odds were always stacked against the Rio Grande Southern Railroad (RGS), a small and relatively short-lived operation on the Western frontier. Her intentions were great. Her timing was anything but. The Colorado mining towns of Silverton and Ouray were accessible by branch rail lines in the 1800s, but the rugged San Juan Mountains made it too difficult for some of the smaller railways of the time to connect the two directly. Entrepreneur Otto Mears founded the RGS in 1889 to take on that task—and then some. Not only would his fledgling company connect Silverton and Ouray, it would go south of Silverton to Durango and north of Ouray to Ridgway, negotiating the tough terrain.

Construction began in 1890 and the 162-mile route was finished right before the Panic of 1893 gripped the nation, which forced closure on the mines the railway planned to serve. The RGS struggled to stay afloat through those early hard times, and again during the Great Depression of the 1930s. Compounding its problems was the fact its rolling stock was largely secondhand, so mechanical issues were an unfortunate way of life.

Back taxes owed by the company, the loss of a mail contract, and the growing popularity of cars also contributed to the demise of the RGS in 1953. In spite of its struggles, or perhaps because of them, railway aficionados still celebrate the run of the RGS. Several tourism trains in Colorado use refurbished rolling stock from the railroad, other cars are preserved in local museums, and small-scale replicas are popular among model railroaders who seem to hold a special place in their hearts for this little railroad that tried to buck the odds.

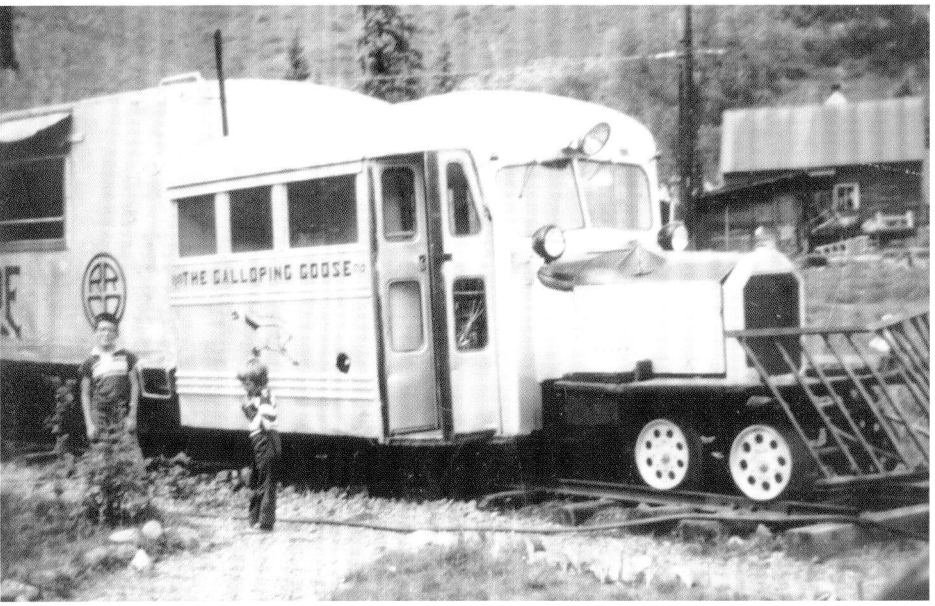

Top: A photo of the RGS's locomotive the *Galloping Goose*, which was used to save the railway money while providing mail service on the route.

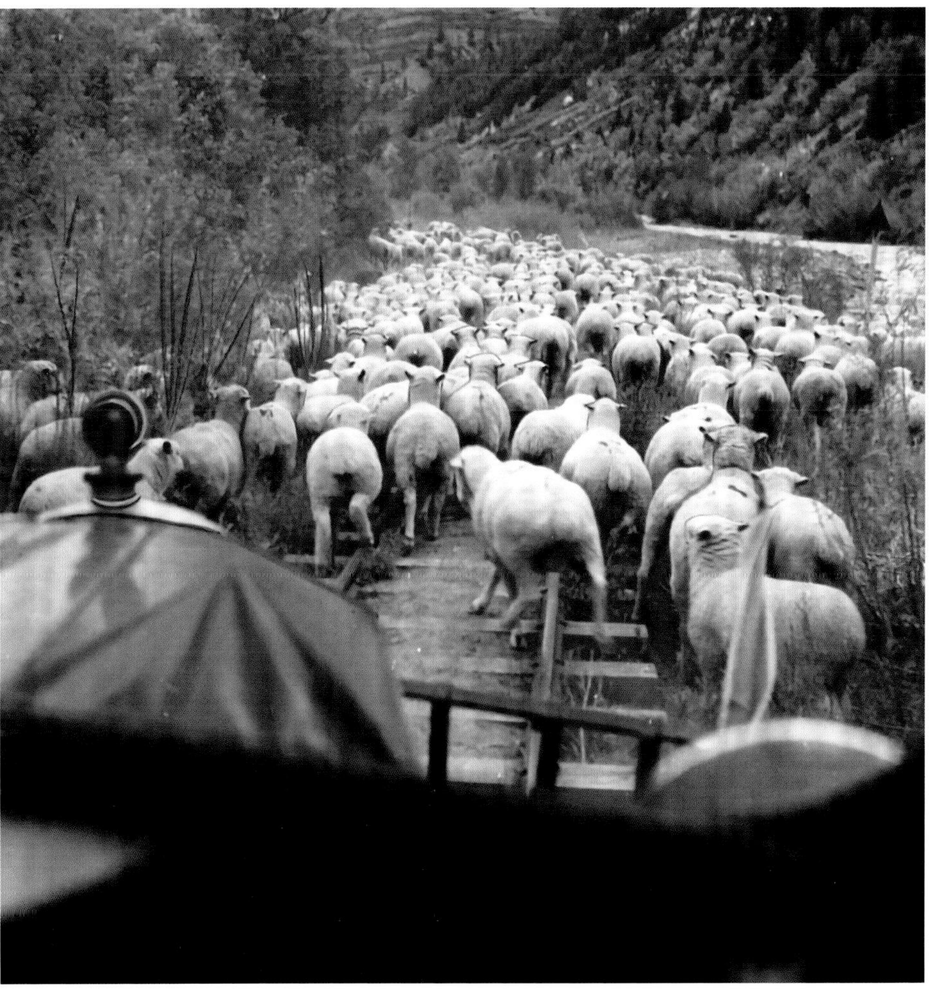

Bottom: A herd of sheep in front of the *Galloping Goose*. The train would have to stop completely in order to clear the herds from the tracks.

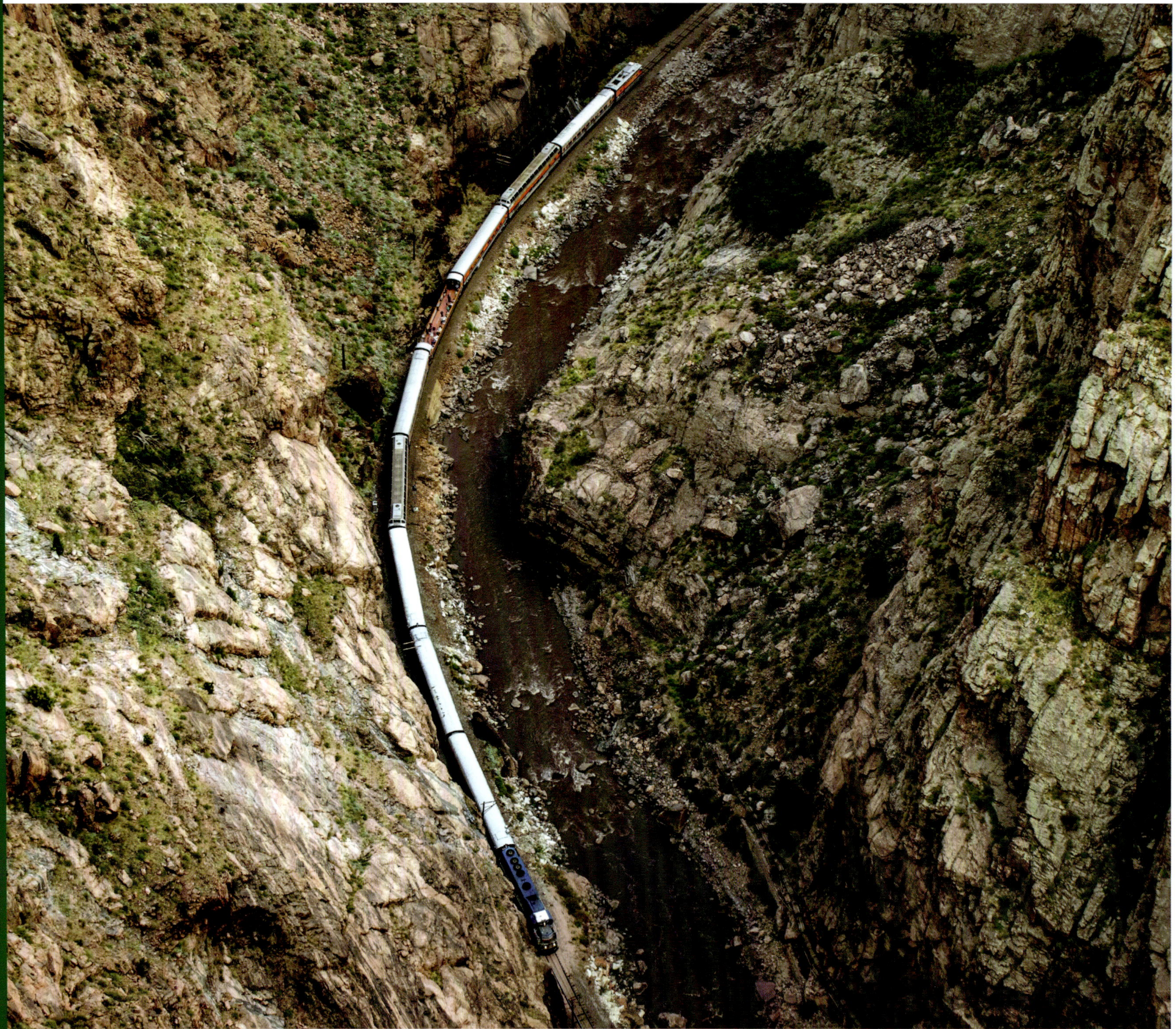

The Arkansas River is the sixth-longest river in the U.S., and second longest river in the Mississippi-Missouri river system. It originates in the Rockies just outside of Leadville, Colorado.

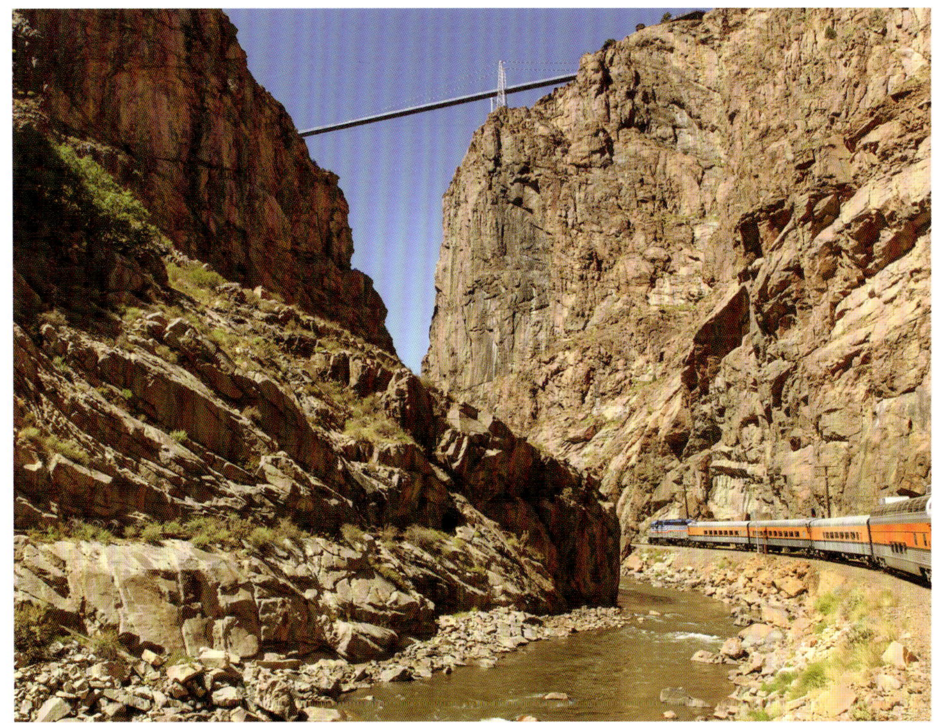

The Royal Gorge is six miles long, walled with cliff faces of dense igneous rock, and can reach depths of 1,000 feet. It was almost an impossible barrier to overcome when constructing the railway.

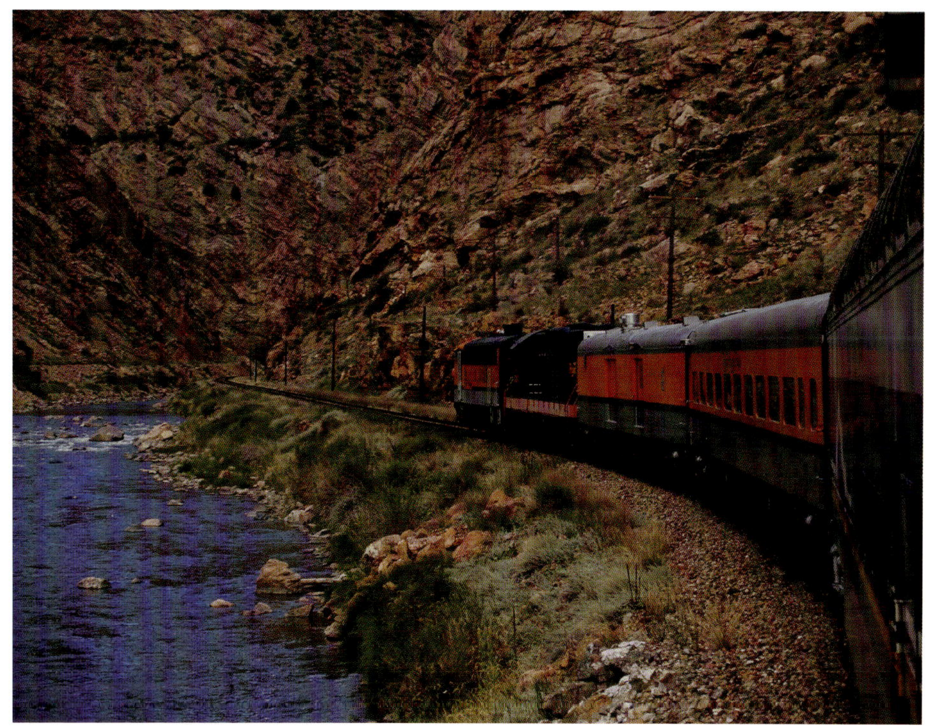

The train offers three classes of service—Coach, Deluxe, and Vista Dome—with first-class dining and a wide selection of drinks.

Also a vestige of the famed Denver and Rio Grande Western Railroad (D&RGW), the Royal Gorge Route Railroad of Colorado passes through what some may say is the most exquisite route the former Rocky Mountain railroad company built. In the late 1870s, miners flocked to the Arkansas River valley in Colorado in hopes of finding carbonate-rich lead and silver ore, leading the railway companies to follow.

With the influx of mining operations in the Arkansas River valley, both the Sante Fe Railroad and D&RGW looked to build a line through the narrow pass at Royal Gorge in order to reach Leadville. After taking their dispute to the United States Circuit Court, the Santa Fe Railroad was given the right to construct the railway in the gorge, while the D&RGW was given secondary rights to construct their own tracks as long as they did not interfere with Santa Fe's tracks.

The D&RGW began laying tracks west of the gorge as Santa Fe began working its tracks through the gorge. Both crews were victims of sabotage committed by the competition. And Santa Fe even hired Doc Holliday and other hired guns to take over train stations along the D&RGW route.

Rights to the land swayed back and forth between the two companies, and the legal battle went on for years, but the dispute was eventually resolved out of court. Santa Fe conceded the rights of use to D&RGW and received $1.8 million in payment for the track it built in the gorge. By 1880, the tracks made it up the friendly one percent water-grade ascent through the gorge to Leadville. Passenger service began that year and continued until 1967. Freight continued to be hauled on the line until 1989.

Sold off and merged a handful of times since the days of D&RGW ownership, a 12-mile section of the gorge railway was bought from Union Pacific by the newly incorporated Royal Gorge Express in 1997. Trains that pass through the gorge depart year-round from Cañon City, Colorado, toward its terminus of Parkdale, Colorado.

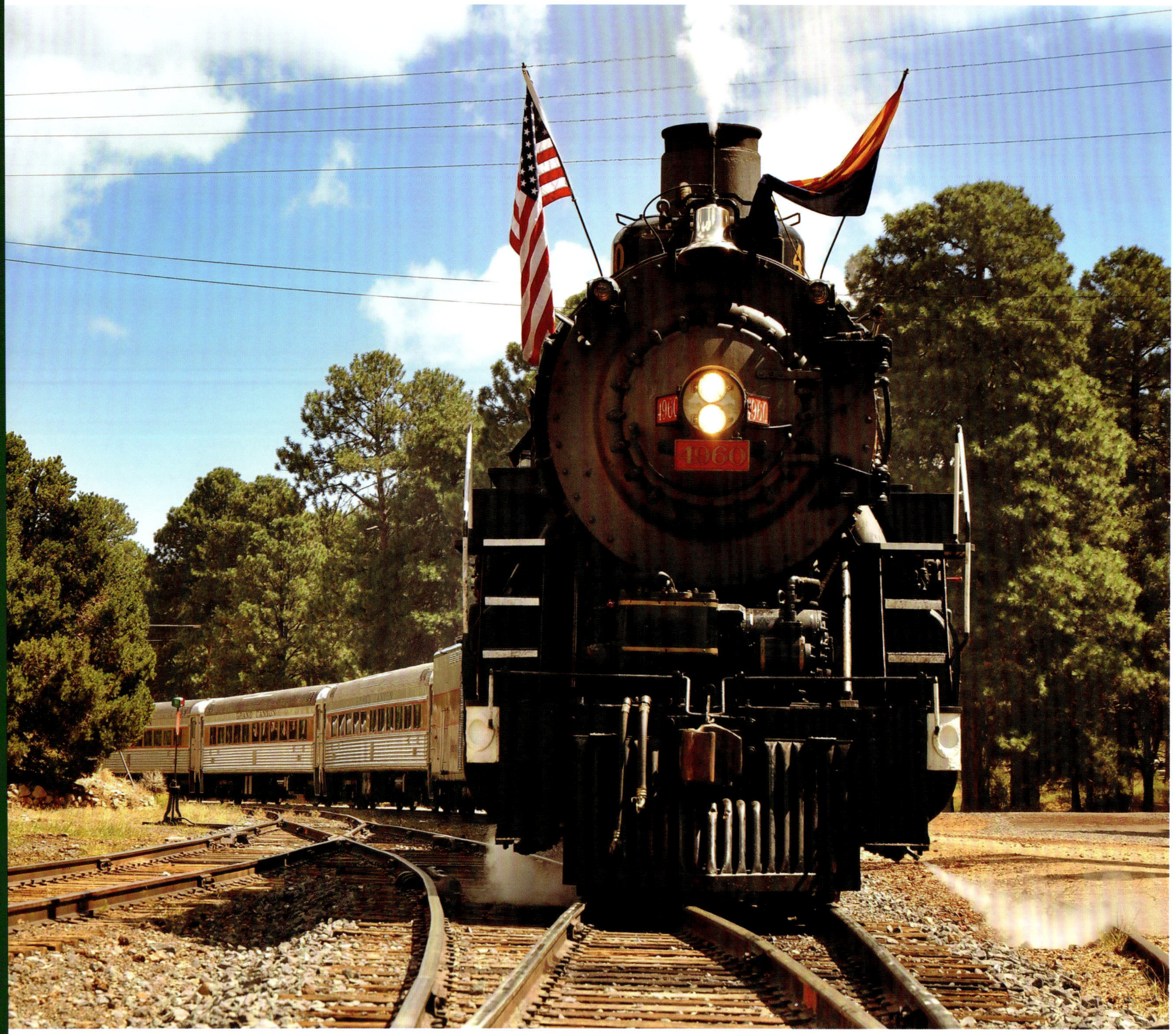

Although train service was once discontinued because of the popularity of cars and falling ridership, today the Grand Canyon Railway keeps nearly 50,000 cars away from this natural wonder every year.

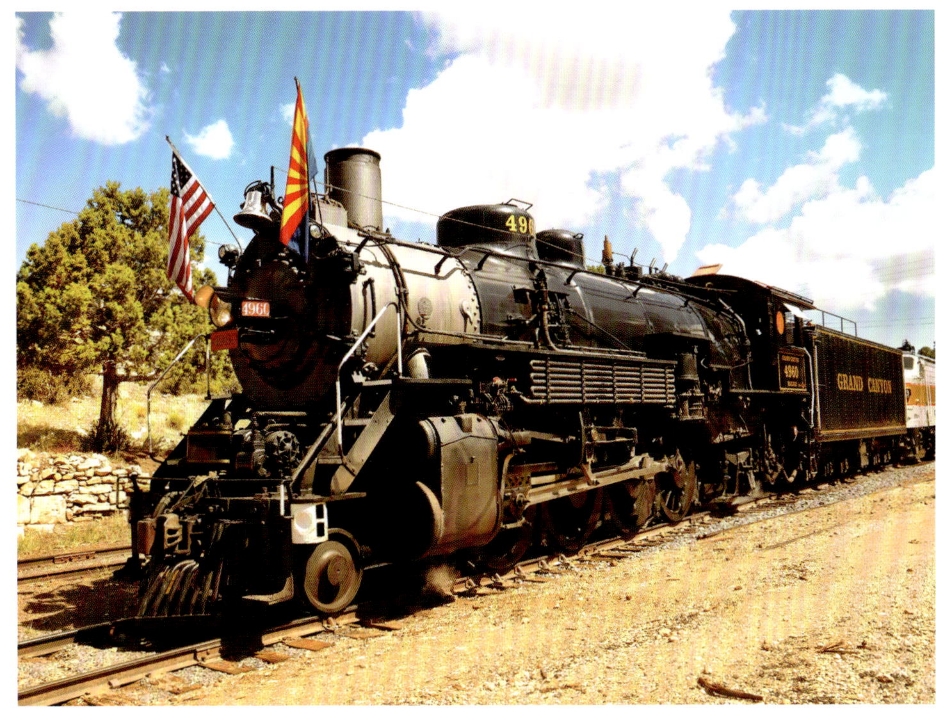

More than 4,500,000 people visit the Grand Canyon every year, so be sure to plan ahead when trying to visit one of the nation's busiest national parks.

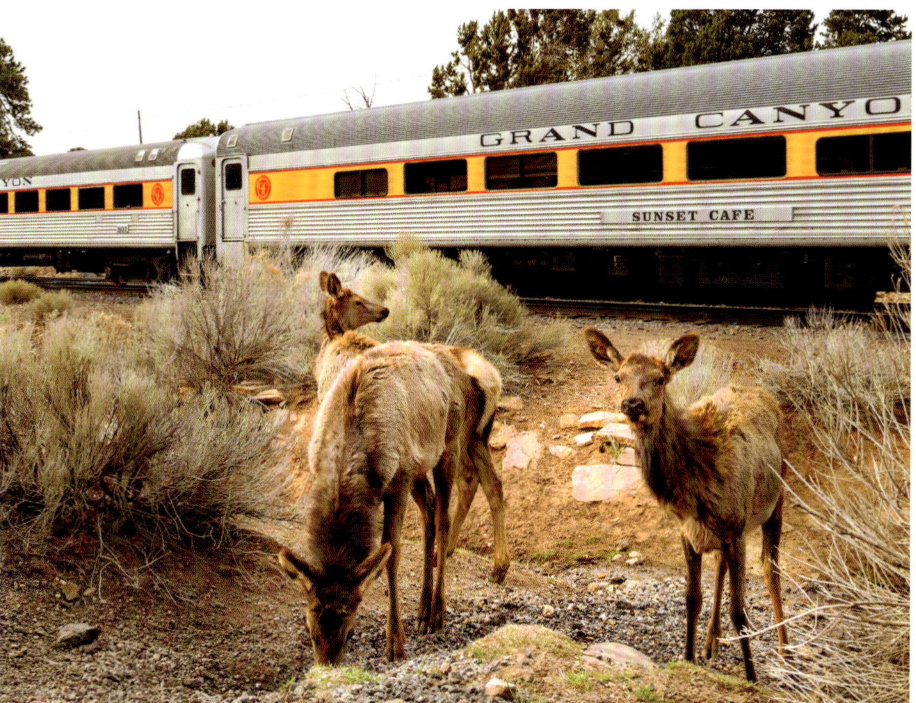

Young elk near the Grand Canyon. The Kaibab National Forest is home to various types of mammals, including elk, mule deer, white-tailed deer, pronghorn, coyote, turkey, cougars, bobcats, and black bears.

Passenger service to the Grand Canyon hasn't always been consistent since the Grand Canyon Railway opened in 1901. Built by the Santa Fe Railroad as a branch between Williams, Arizona, and the Grand Canyon Village, the Grand Canyon Railway offered tourists a reliable way to visit this natural wonder hidden in the high desert of northern Arizona for less than four dollars. But tourists weren't the main driver for building the railway at first. Investors wanted a way to transport silver, lead, copper, and gold ore to the Santa Fe Railway, which connected the country by rail from Chicago to Los Angeles, from the Anita Mines just 40 miles north of Williams. But as mining at the site proved unsustainable, investors were still able to make a return on their money through tourism.

With the growing popularity and ubiquity of the automobile in the mid-twentieth century, train services throughout the nation were struck with shrinking passenger numbers, and the Grand Canyon Railway was no exception. The train's original stint of passenger service lasted until 1968, while the Santa Fe Railroad continued to use the route as a freight service until 1974. But the route was completely abandoned by 1977, with a few unsuccessful attempts to resurrect the service to no avail. But by 1988, the continued operation of the railway was beginning to look hopeful when the Biegert family acquired the railway. By 1989, the Biegerts had the railway carrying passengers again to the South Rim of the Grand Canyon and continued to do so until 2006 when Xanterra Travel Collection bought the railway. Under Xanterra ownership, the traditional steam engines were no longer used in light of the growing energy crisis of the twenty-first century. Today, diesel engines make a majority of the runs, but engines that run on vegetable oil can be seen on the tracks from time to time.

On the 64-mile trip from Williams to the Grand Canyon, passengers float through the high desert of Arizona thousands of feet above sea level. The route passes through the Coconino National Forest filled with ponderosa pines and up to the Kaibab National Forest that lines the South Rim with juniper and sagebrush. The trip also offers scenic views of the San Francisco Peaks, the highest point in the state of Arizona, which lie just outside of Flagstaff. Today, the entire area is completely serene compared to the seismic, volcanic, and geographic forces that so violently shaped much of the majestic American Southwest.

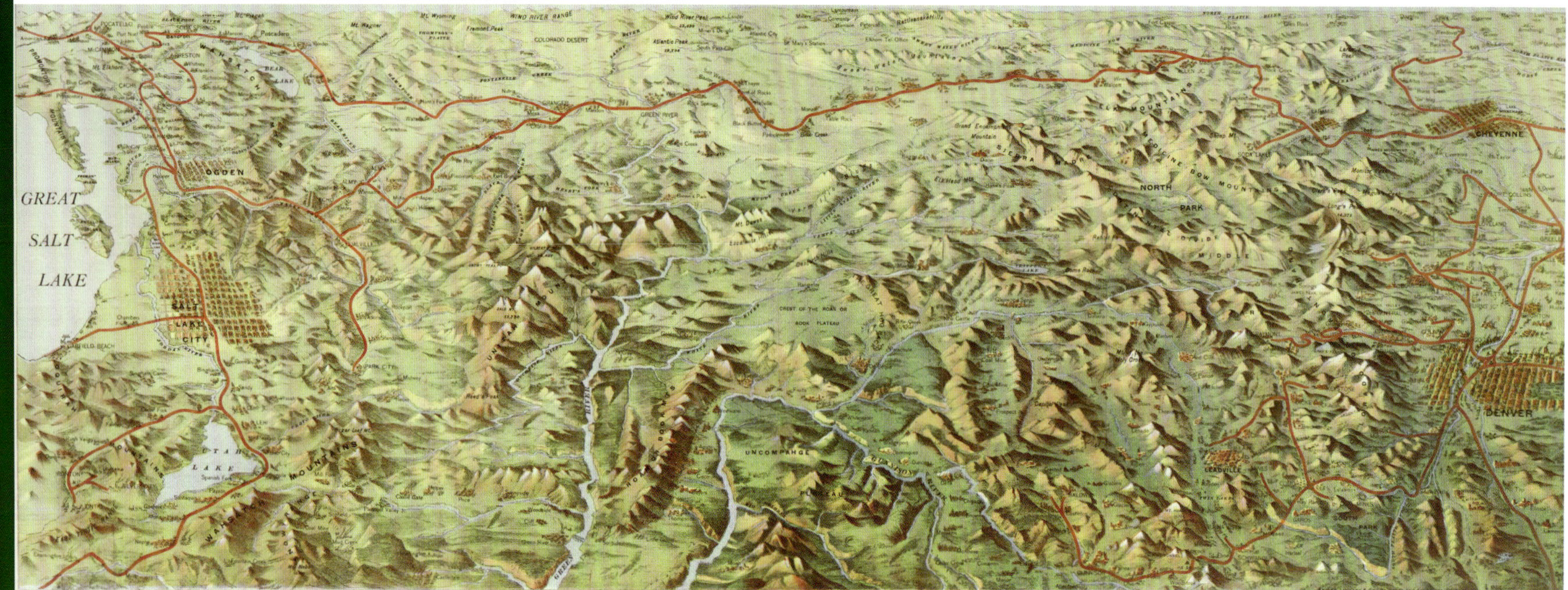

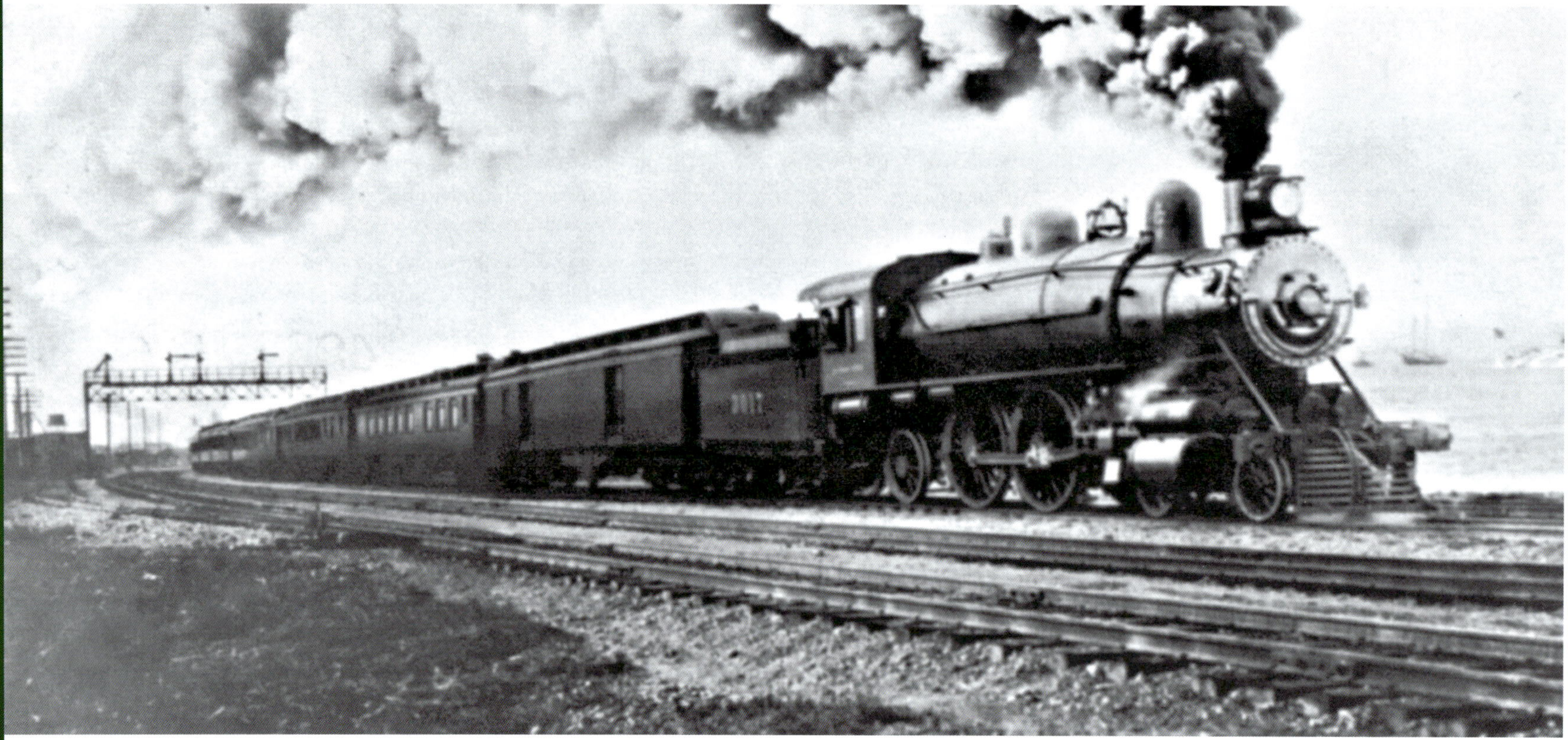

Top: A map of the Overland Route between Cheyenne, Wyoming, and Ogden, Utah, from the 1890s.

Bottom: The *Overland Limited* departing from Oakland, California, circa 1906.

The Overland Route, known originally as the Pacific Railroad, was the primary transcontinental rail line operated by the Union Pacific Railroad (UP) in partnership with the Central Pacific Railroad (CPRR). The "Overland" name comes from a stagecoach line that operated along part of the same route in Utah and Nevada during the 1860s.

Construction of the railway began after the passage of the Pacific Railroad Act of 1862, which authorized the UP to build westward from Omaha, Nebraska, while the CPRR built eastward from Sacramento, California. The two railroads met at Promontory Summit, Utah, on May 10, 1869, marking the completion of the first transcontinental railroad in the United States. This railway dramatically reduced cross-country travel time from months to just days, revolutionizing transportation and commerce in the American West.

By the late nineteenth century, UP had gained control of additional lines to solidify its dominance over transcontinental rail travel. The Overland Route passed through Nebraska, Wyoming, Utah, Nevada, and California, serving as the backbone of long-distance passenger and freight traffic. Key cities along the route included Cheyenne, Laramie, Ogden, Reno, and Sacramento, many of which grew significantly due to the route's presence. In 1885, the Southern Pacific Railroad took control of the CPRR and jointly operated the Overland Route with UP.

The Overland Route became famous for its passenger services, particularly with the introduction of the *Overland Limited* in 1896, a luxury train that ran between Chicago and San Francisco in about 63 hours each way. This train featured upscale amenities, dining cars, and sleeper cars, making it one of the premier passenger trains of the early twentieth century. The *City of San Francisco*, a train introduced in the 1930s as part of UP's streamlined "City" fleet, offered even faster and more modern service, moving passengers along the same route as the *Overland Limited* in about 39 hours each way.

The route remained key for passenger and freight traffic throughout the twentieth century, even as passenger rail service declined with the rise of automobiles and airlines. By 1971, Amtrak took over most passenger rail services in the U.S., operating the *California Zephyr* over much of the historic Overland Route. However, that train passes through Denver and Salt Lake City instead of Cheyenne and Ogden, and the tracks crossing the famous Promontory Summit have long since been removed.

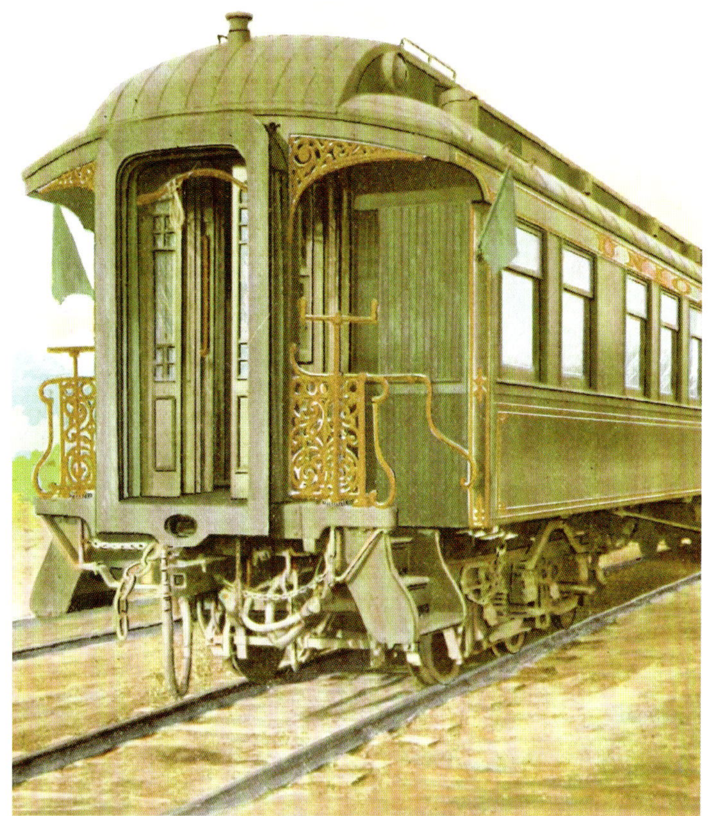

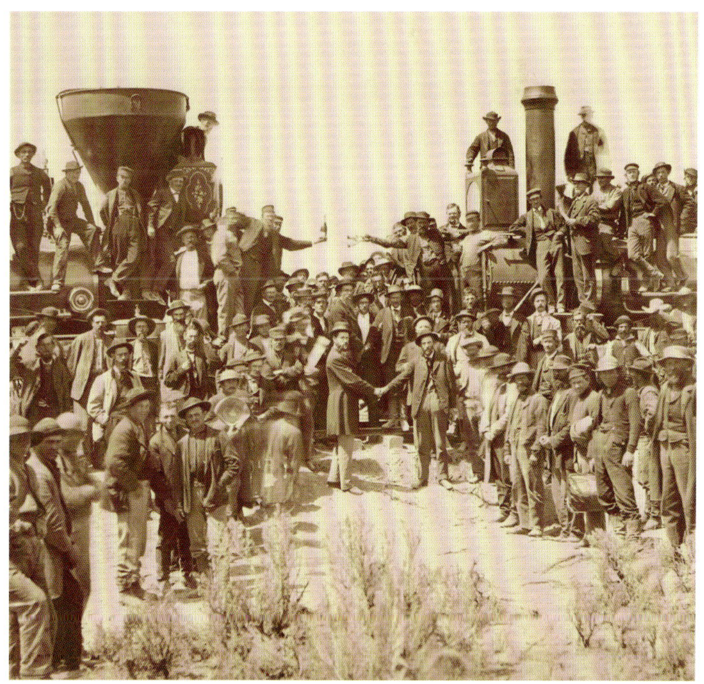

Top: An exterior view of a train car on the *Overland Limited* from 1896.

Bottom: On May 10, 1869, engineers, business leaders, and workers from the Union Pacific and Central Pacific Railroads held a ceremony driving the golden spike where the railroads connected at Promontory Summit, Utah.

THE WEST COAST

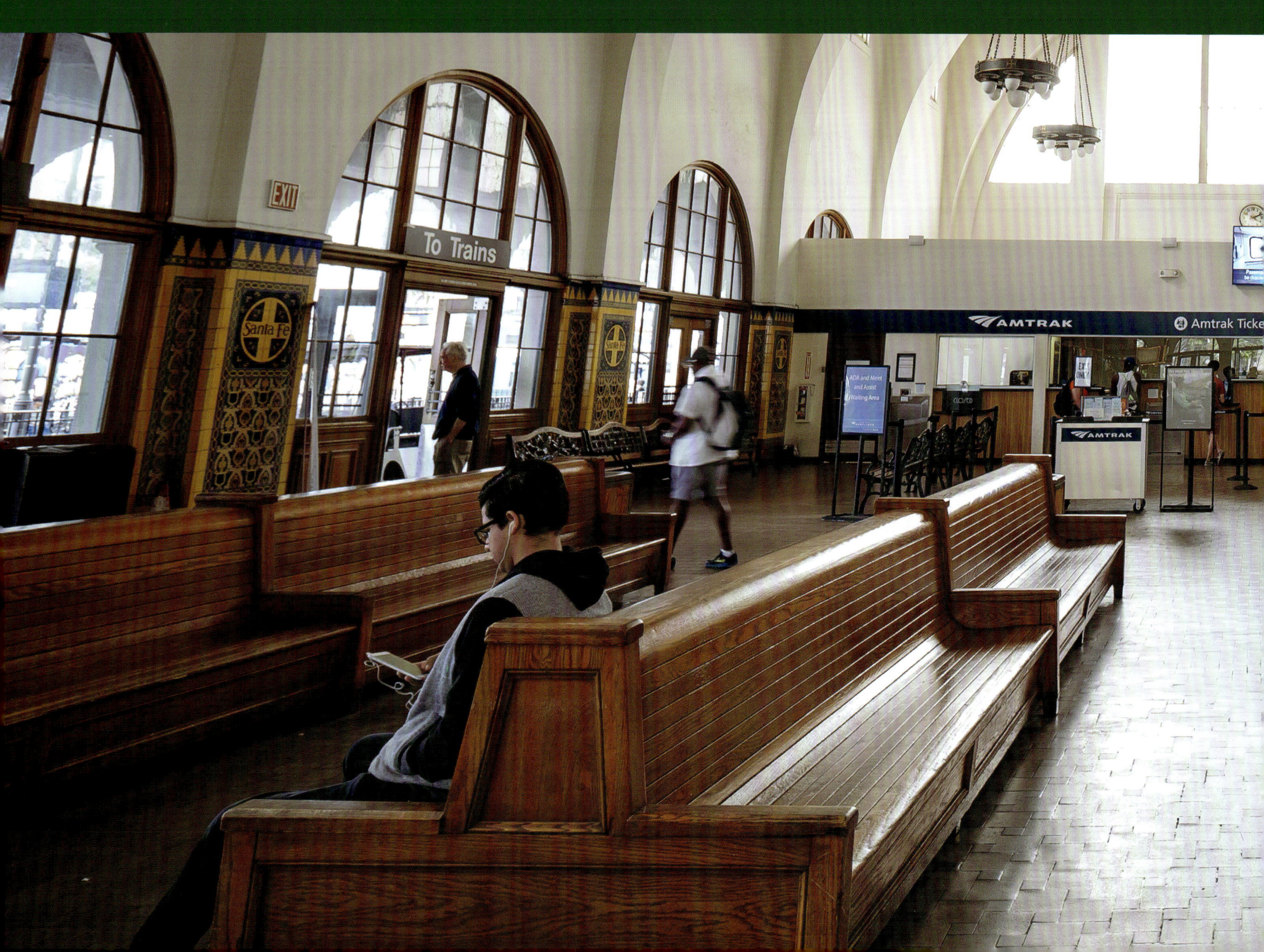

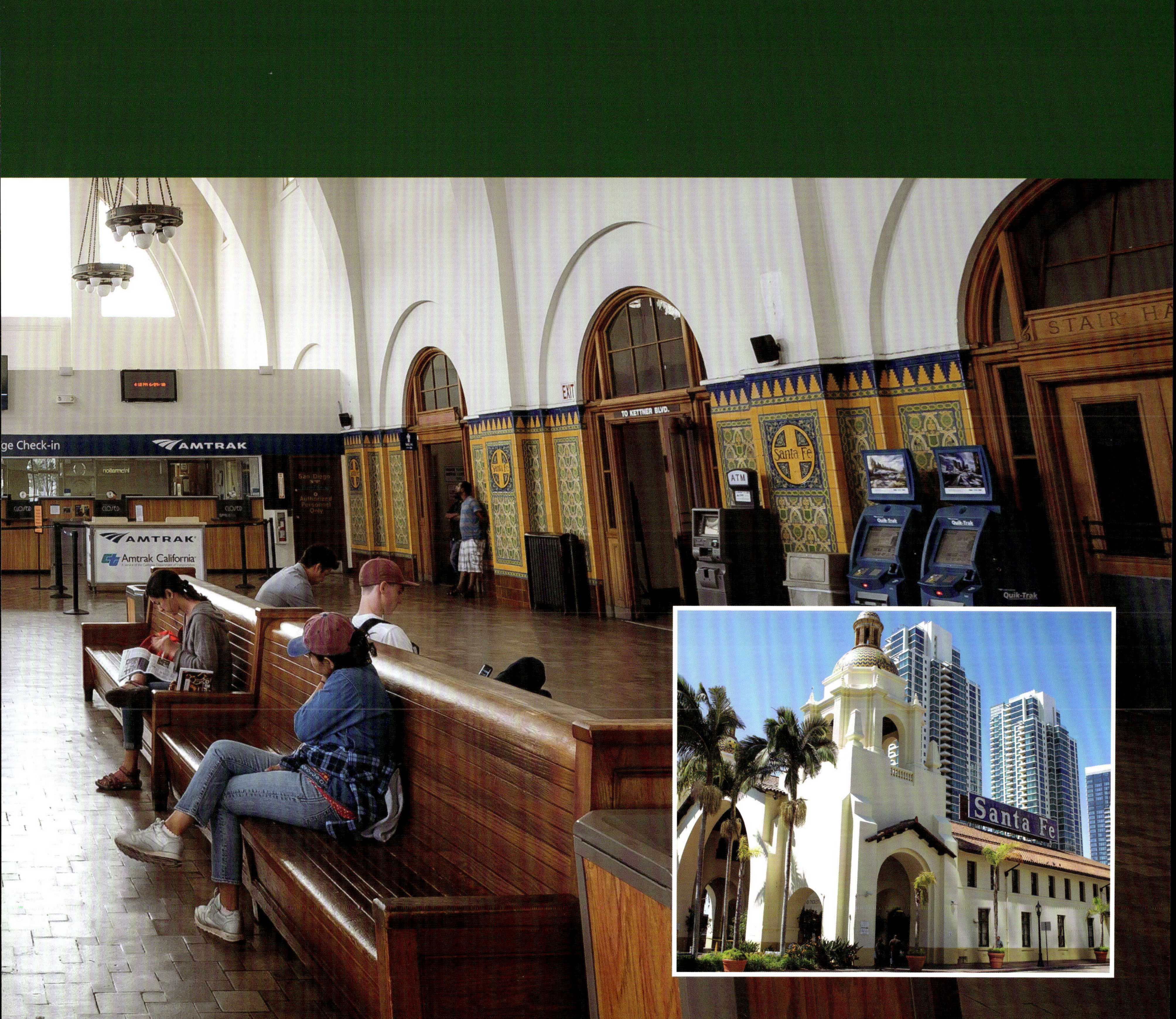

AMTRAK
AMTRAK
Amtrak California
EXIT
TO KETTNER BLVD.
ATM
Santa Fe
Quik-Trak
Santa Fe

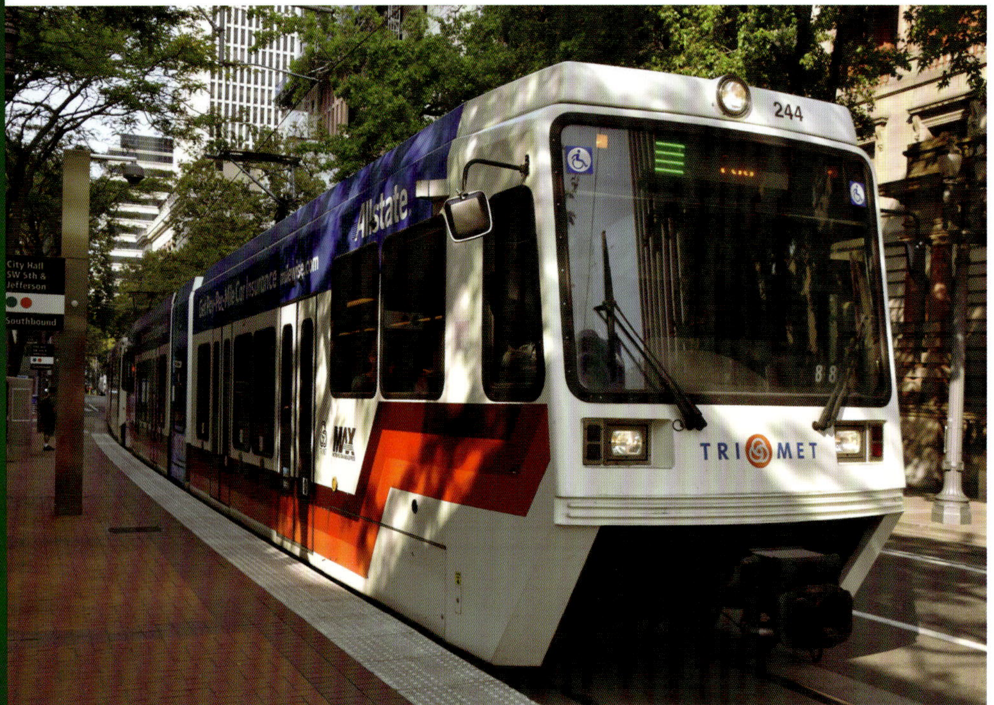

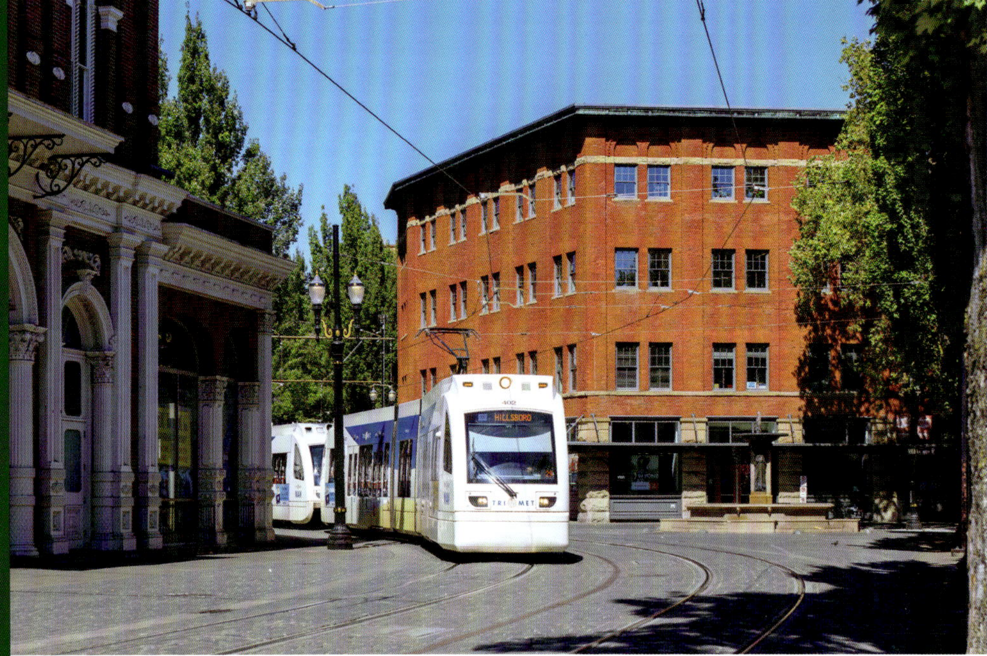

Top: The MAX Light Rail in downtown Portland.

Bottom: The MAX Light Rail traveling through the Skidmore neighborhood of Portland.

Portland commuters have the 1972 mayoral election to thank for one of the best light-rail systems in the country. The MAX (Metropolitan Area Express) Light Rail system was born of a freeway revolt. Anti-freeway candidate Neil Goldschmidt beat freeway proponent Frank Ivancie in an election that turned into a referendum on transportation in Oregon's biggest city. With his victory, Goldschmidt oversaw the cancellation of a proposed freeway project and was able to route federal funds to the light-rail line. The MAX Light Rail was off and running.

Construction began in 1982 and the MAX Blue Line, Portland's first light rail, hit the tracks in 1986. Since then, four more lines have been added. The Orange Line was the latest addition in 2015. It increased the total mileage of the MAX Light Rail to 60 and the number of stations to 97. Three stations closed in 2020, leaving the current number at 94 with the tracks covering the same mileage. The Blue Line remains the standard, serving 51 of those stations. Most stations reside at street level, although Washington Park—the system's only underground station—holds the distinction of being the deepest transit station in the U.S. at 260 feet underground.

So, what makes MAX Light Rail so special? Operated by TriMet, it seems to have thought of everything. It gets high ADA marks for serving riders with disabilities. Its safety marks are just as impressive. It actively solicits and implements feedback from riders. It's bicycle-friendly. It has a popular mobile app. And perhaps most important to commuters, it's reliable. Trains run every 30 minutes for most of the day and every three minutes during rush hours, reaching an average daily ridership of more than 75,000 people.

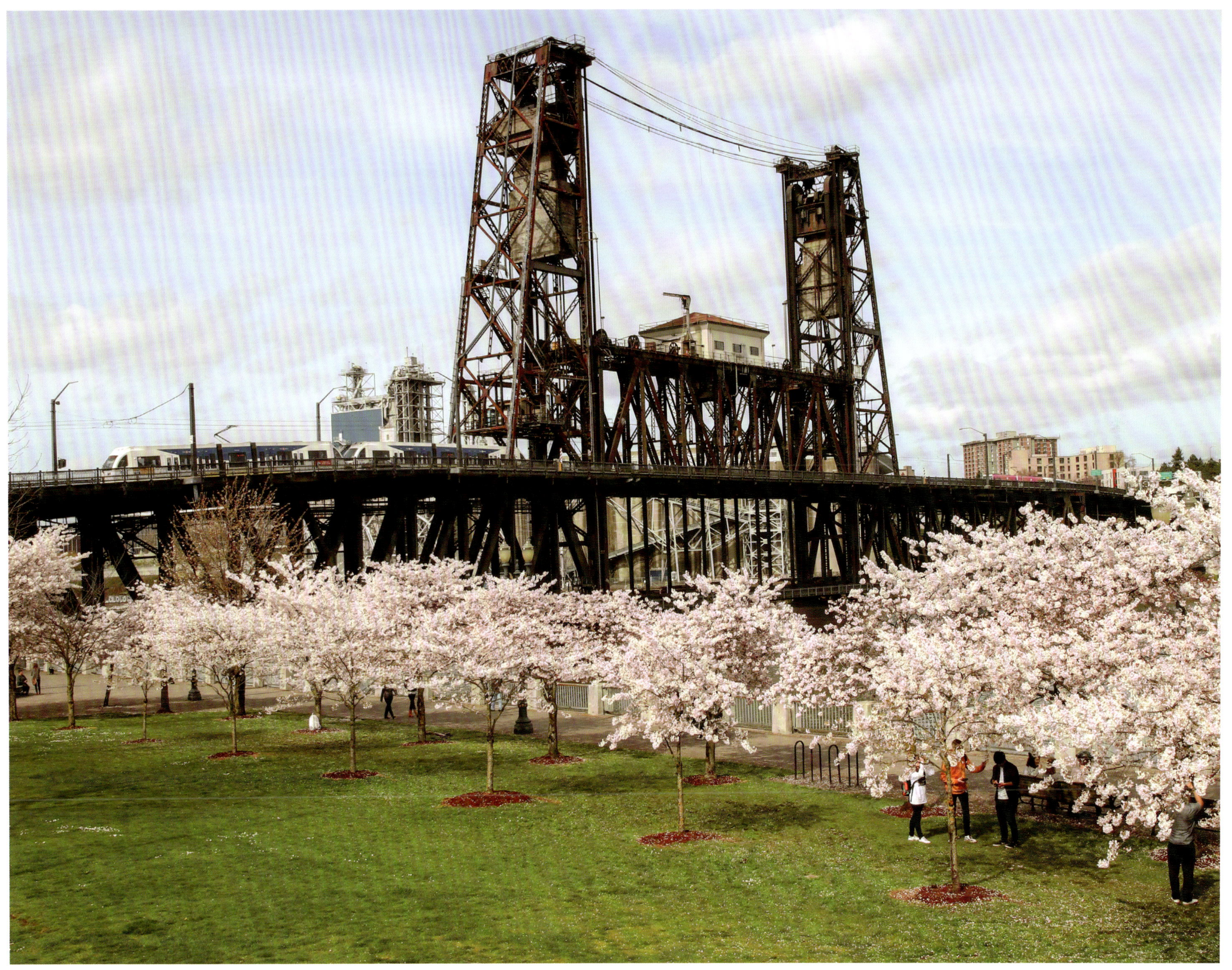

The MAX Light Rail crossing the Willamette River on Steel Bridge.

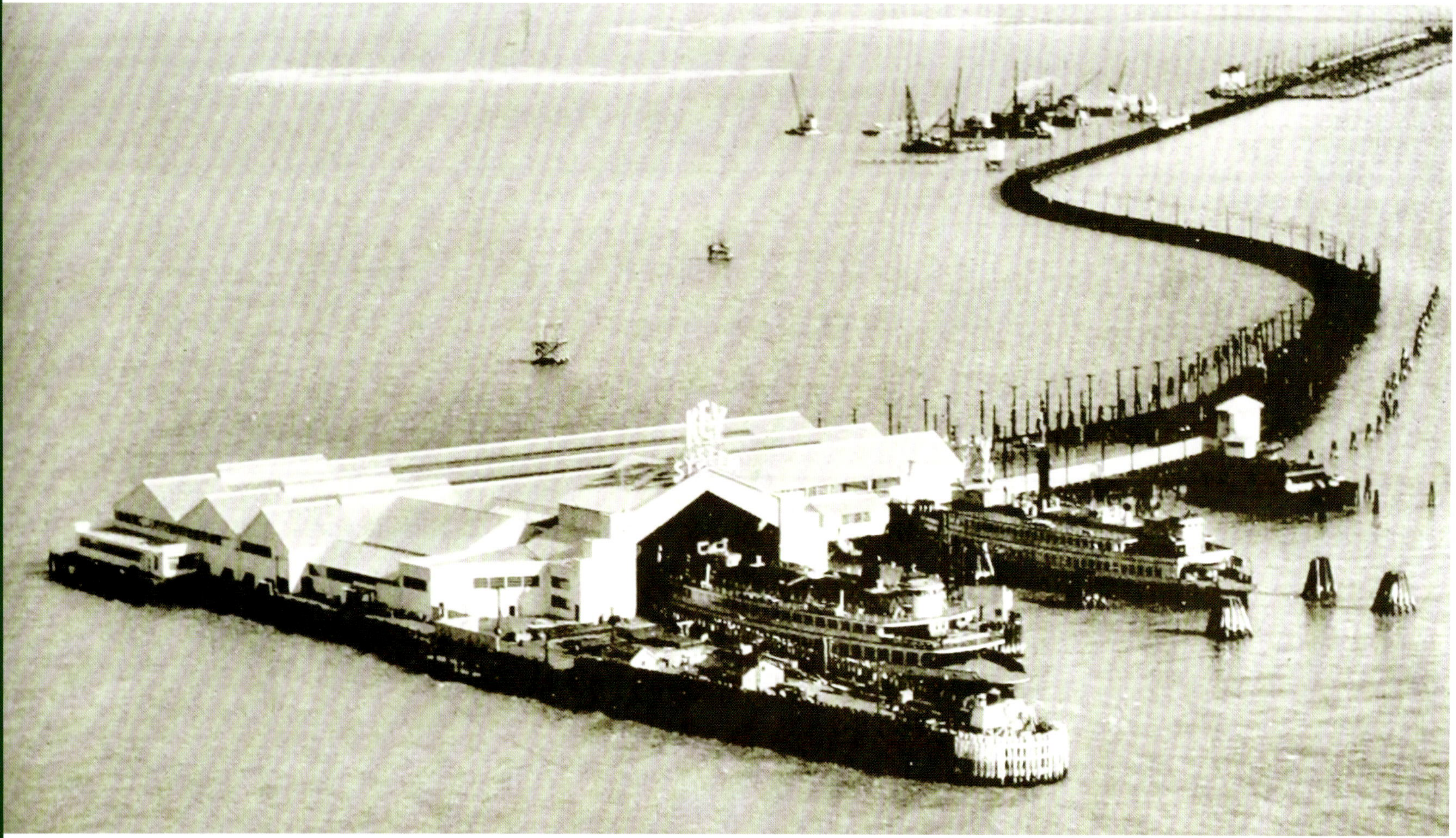

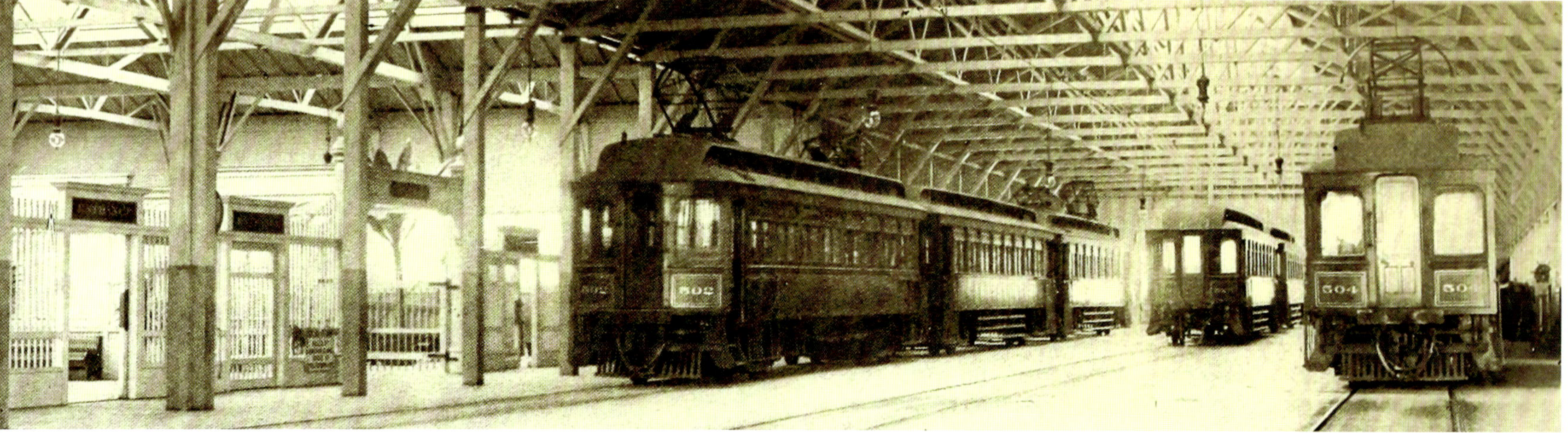

Top: The Key System's mole that served as a transit center for rail riders to transfer to a ferry for service to San Francisco.

Bottom: A Key System depot in 1909.

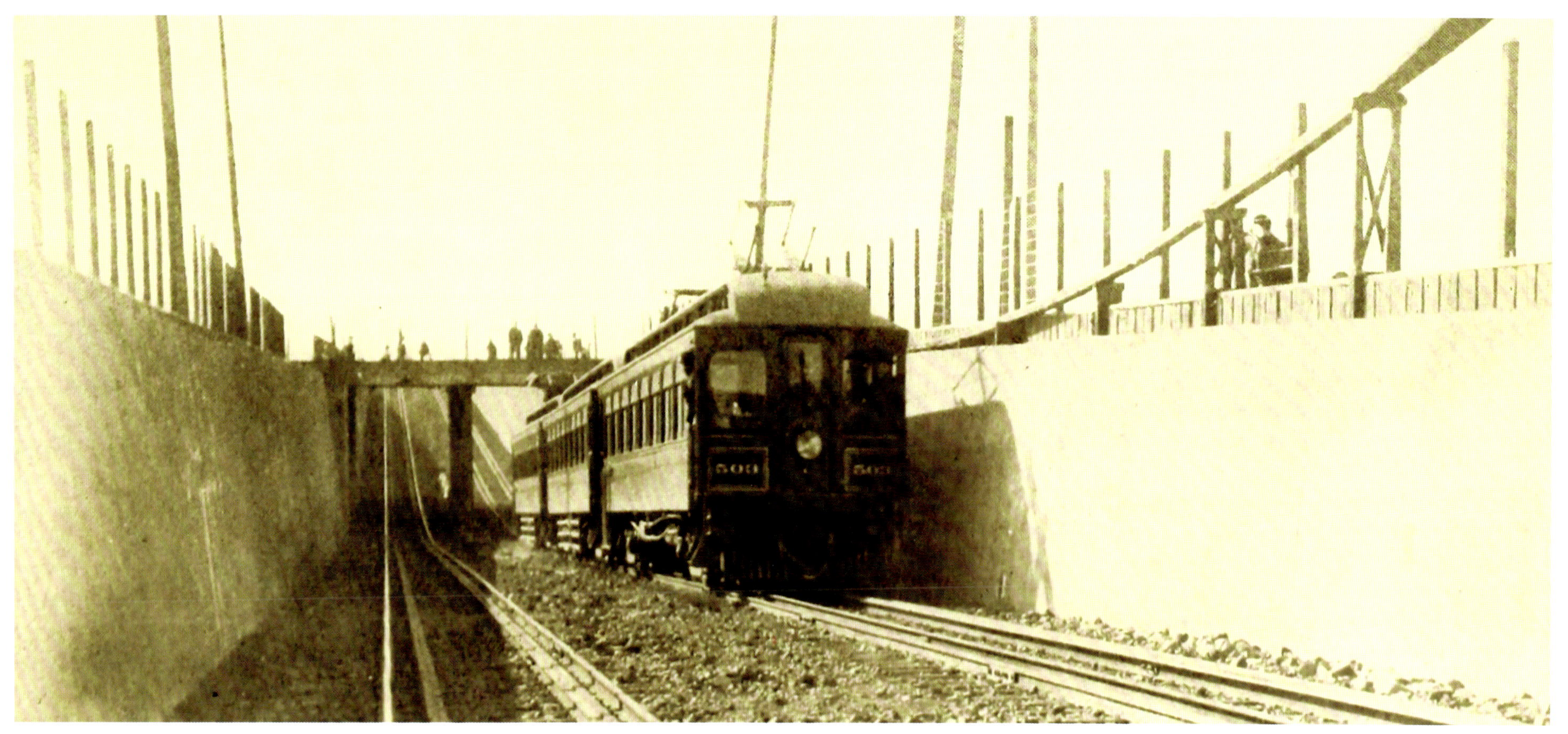

A photo from 1909 of a train in Emeryville heading toward the mole.

In the early 1900s, the San Francisco, Oakland, and San Jose Consolidated Railway was a mouthful. So, when someone noticed that a map of the public transportation network primarily serving the East Bay in California's Bay Area resembled an old-fashioned key, it became commonly known as the Key Route or Key System.

The network of independently owned entities started when Francis Marion Smith started accumulating railroads, streetcars, and real estate in the East Bay in 1893. In 1902, he founded the aforementioned railway, and began electric train service the following year. Together with ferry boats, streetcars, and the rail lines, the system treated local residents to amazing transit service for the time.

In addition to easily making it to and from Berkeley, Piedmont/Claremont, and Oakland—the three "loops" on the key handle—passengers could quickly access the San Francisco Bay's ferry system, whose slits represented the notches on the key. Commuters had it made. Many of the paths and stairways in Oakland today were designed to give residents easy access to the Key System. And for those with upscale tastes (and budgets), there were dining cars with linen tablecloths and fresh-cut flowers.

Ferry service ended due to a fire in 1933, but six years later the Key System started moving passengers between San Francisco and Oakland on the lower deck of the Bay Bridge. Financial issues turned out to be the ultimate demise of the system. Bills racked up during World War II and the growing popularity of the automobile in the postwar years contributed to hard times. The Key System's last transbay train ran on April 20, 1958. Two years later, the Alameda–Contra Costa Transit District took over the remaining Key System bus lines. Still today, transit lines in the East Bay follow routes first run by a system that was, in many ways, ahead of its time.

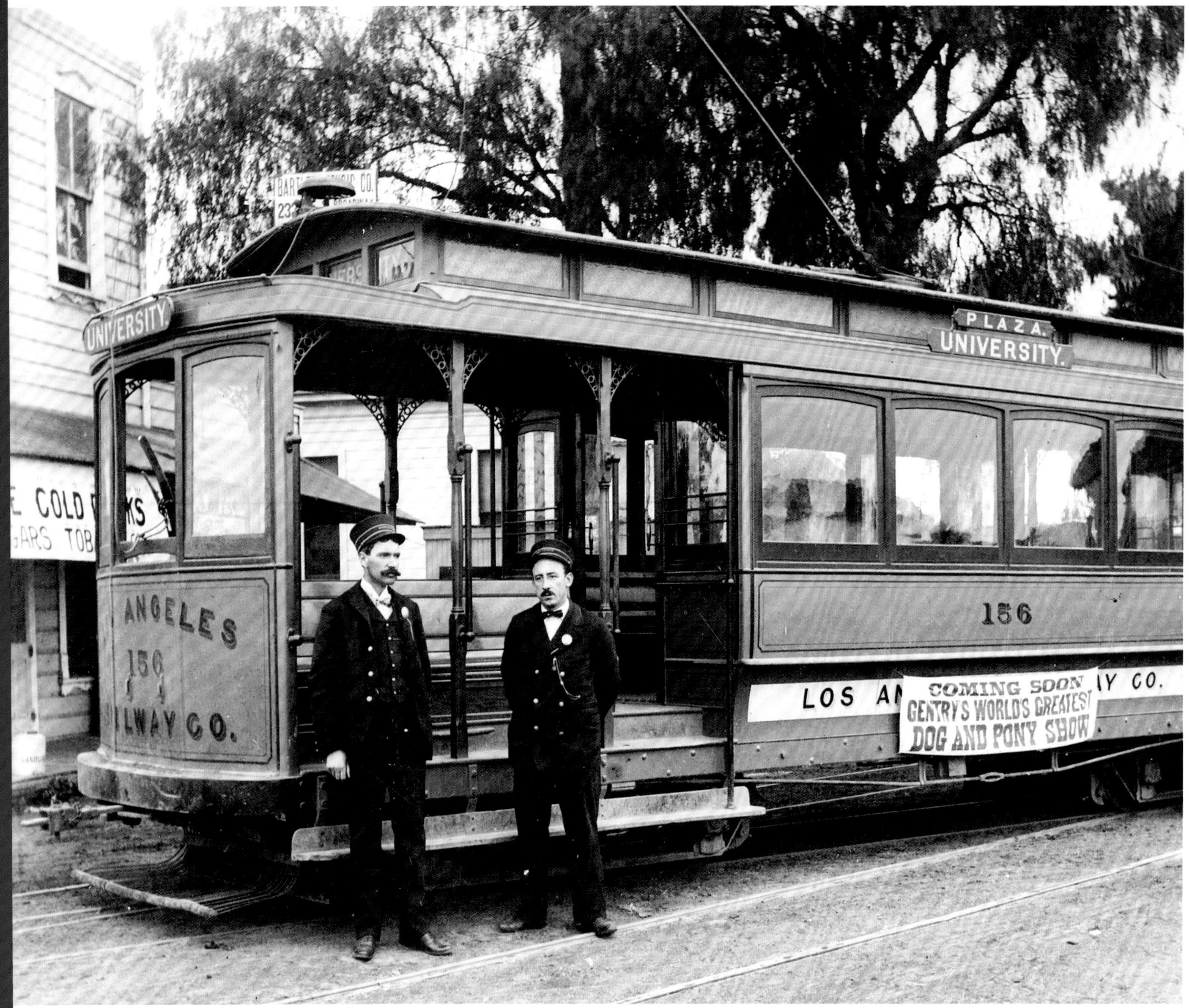

A photo of the University Plaza Line circa 1900.

The Los Angeles Railway had L.A. commuters seeing yellow over the first six decades of the twentieth century. The Yellow Cars, nicknamed for their hard-to-miss paint scheme, comprised Los Angeles's first local streetcar system. The system was largely the doing of Henry Huntington, a real estate, railroad, and utility tycoon. He bought it in 1898 and opened the streetcar system in 1901, sensing the needs of Los Angeles residents to get around their growing city. His timing could not have been better, as L.A.'s population surged from 100,000 to more than 300,000 in the first decade of the 1900s.

The Pacific Electric Railway's Red Cars served a wider area of Los Angeles, but the Yellow Cars carried more passengers. Huntington's system operated 20 streetcar lines and more than 1,200 trolleys at the height of its popularity, running through the heart of the city and into many growing neighborhoods. Huntington's death in 1927 did not slow the success of the Los Angeles Railway. His estate took over the system until selling it to National City Lines in 1944. It was then renamed the Los Angeles Transit Lines, largely to downplay the use of rail lines.

Approximately one million L.A. residents lived within a half mile of the bus or streetcar lines during the 1940s. By the 1950s, riders could go as far as Beverly Hills to the west and Hawthorne to the south. The Hawthorne Line, at 13 miles, was the longest. However, the rise of the automobile in the post-World War II era had a significant impact on ridership, as L.A.'s crowded freeways would soon prove. Some of the rail lines were converted to bus routes in the 1940s and 1950s, and in 1958 the Los Angeles Metropolitan Transit Authority purchased the remaining lines to complete that quest.

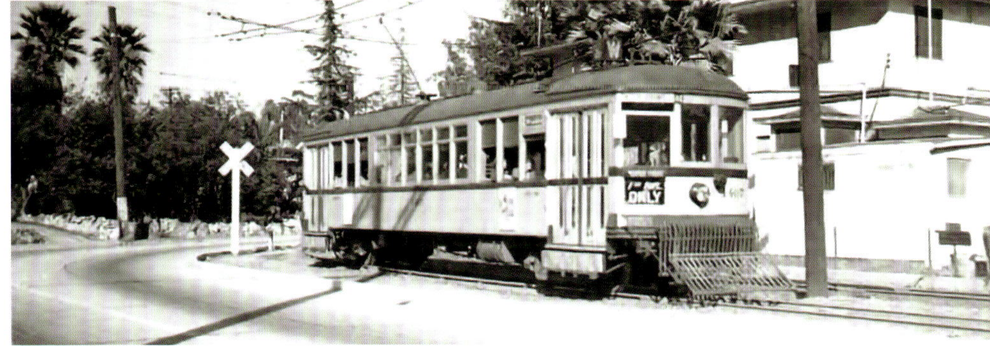

Left: A railcar passing Marmion Way in 1923.

Right: The cover of the Los Angeles Railway's map from 1942.

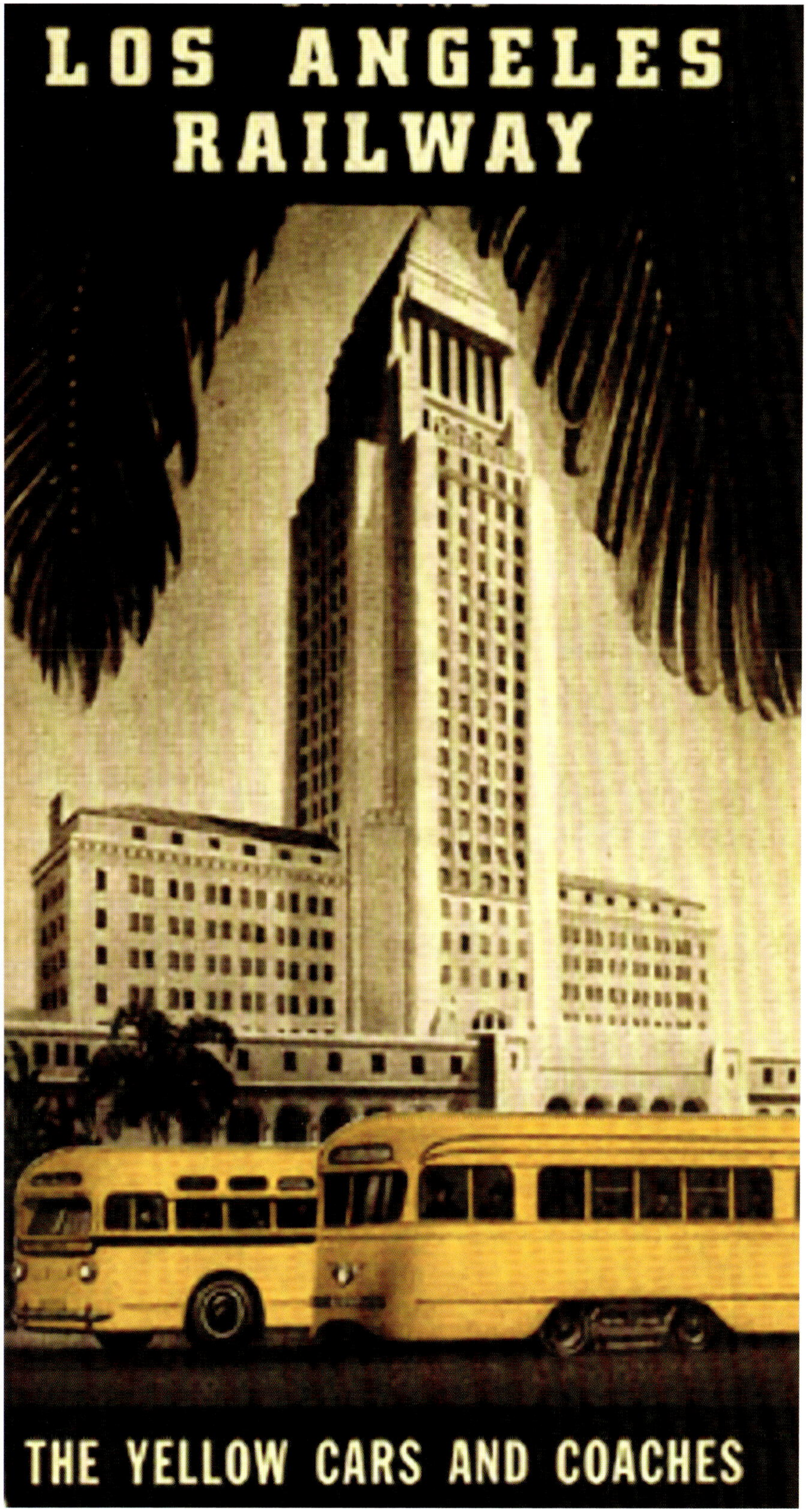

LOS ANGELES RAILWAY

THE YELLOW CARS AND COACHES

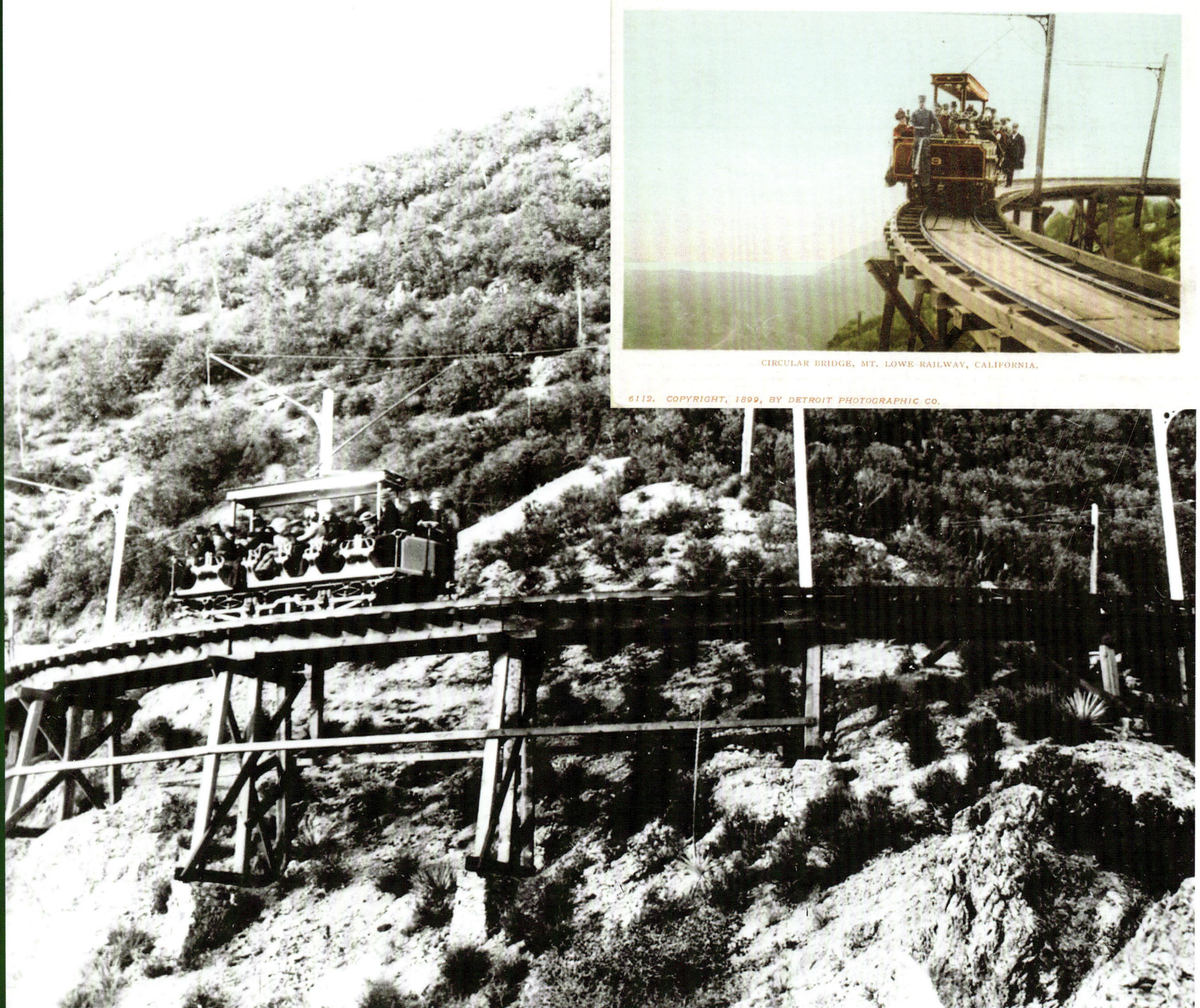

Top: A view of the sheer drops negotiated by the Mount Lowe Railway's Circular Bridge.

Bottom: A railcar traveling over the Circular Bridge on the Alpine Division branch.

It became known as "Earth's Grandest Mountain Ride," and it wasn't an exaggeration. The Mount Lowe Railway, the visionary project of aeronaut, scientist, and inventor Thaddeus Lowe, was the only scenic, electric-traction mountain railroad in United States history. It operated from 1893–1938 in the San Gabriel Mountains (then called the Sierra Madres). And though Lowe operated the railway for less than 10 years, his role in getting people into the mountains by rail resulted in the renaming of Oak Mountain to Mount Lowe.

Southern Californians had been pondering the possibility of running a train to the top of the mountain as early as the 1880s. As it was, getting there required an all-day trek on horseback or foot. There were a couple of fledgling efforts to start a railway that could climb the San Gabriels, but nothing that lasted long enough to get the job done until Lowe—a New Hampshire native and former chief aeronaut of the Union Army during the Civil War—took up residence in Southern California and decided to give it a shot.

Partnering with Cornell-educated civil engineer David Macpherson, who had earned acclaim while with the Santa Fe Railway, Lowe incorporated the Pasadena and Mount Wilson Railway Company in 1891. It would later become known as the Mount Lowe Railway Company. Their initial plan was to connect the Pasadena area to the peak of Mount Wilson, but property and rights-of-way issues forced a rethinking of the route. Lowe and Macpherson settled on a new route, and began building the rails shortly after the company came together. On July 4, 1893, the railway went into service. Some seven miles of track connected Mountain station in Altadena with the top of Echo Mountain, where a Victorian hotel, chalet, astronomical observatory, casino, and dance hall awaited passengers along with magnificent natural scenery.

The climb to the top was known as the Great Incline of the three-route railway. The Mountain Division branch transferred passengers from Altadena to the foot of the mountain where they would connect to the Great Incline. The Alpine Division branch opened in 1896 and provided passengers with three miles of spectacular canyon views along the ride to the upper terminus at Crystal Springs, where the railway's Ye Alpine Tavern offered tennis, mule rides, and wading pools.

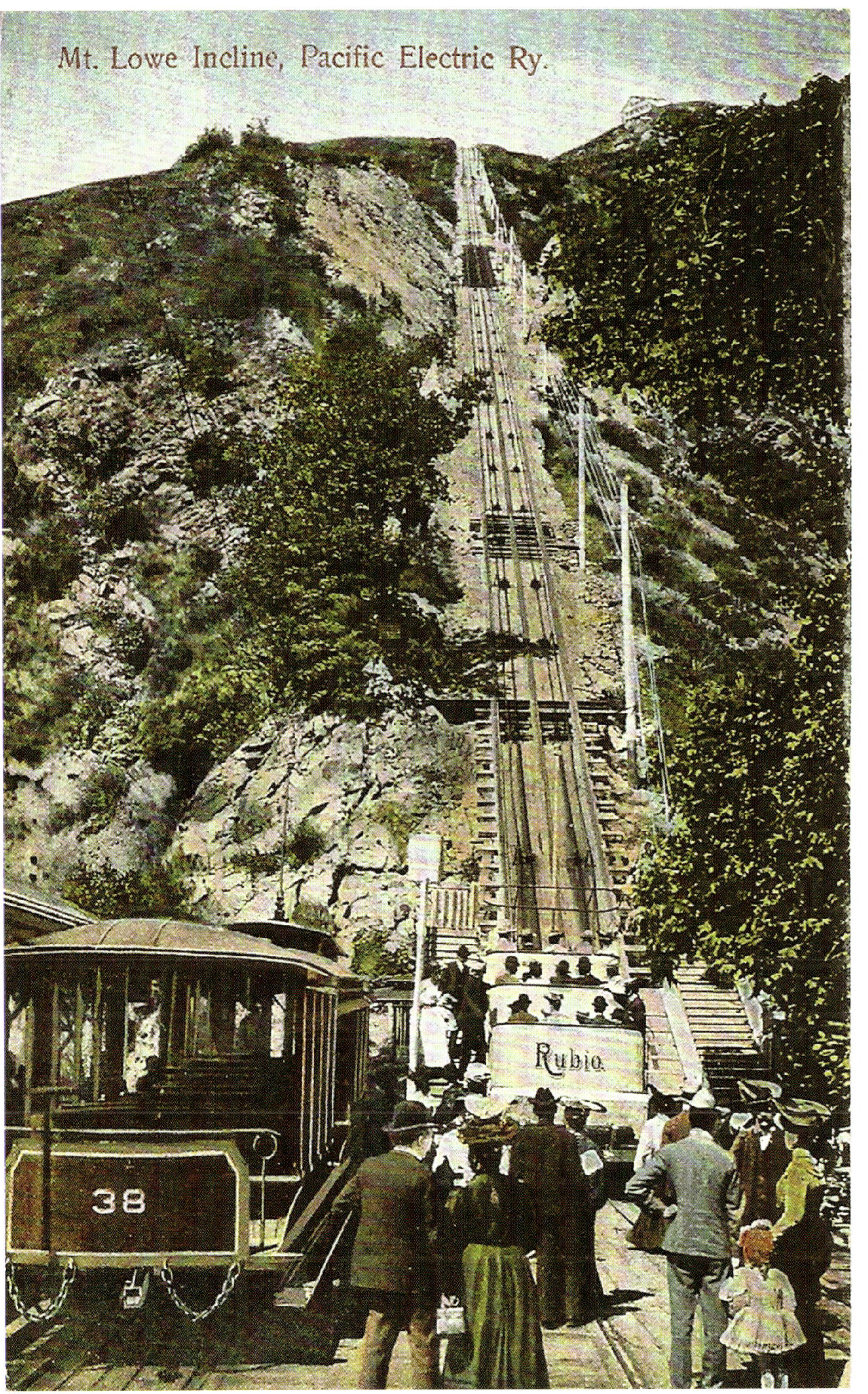

A postcard of Mount Lowe's Great Incline from 1908.

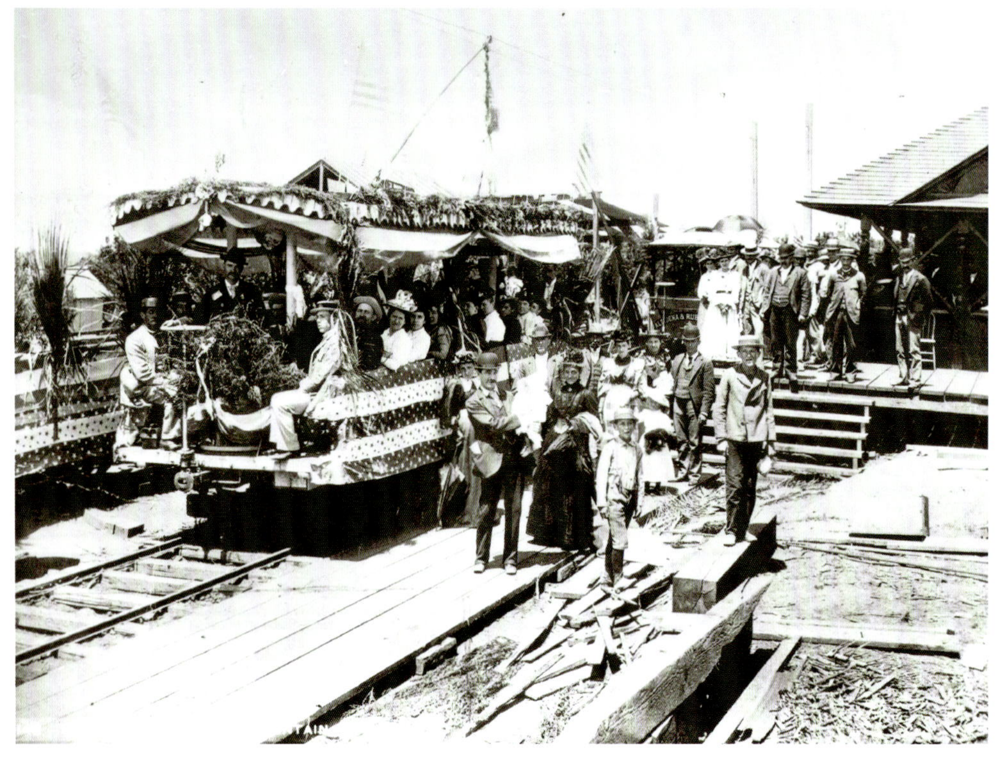

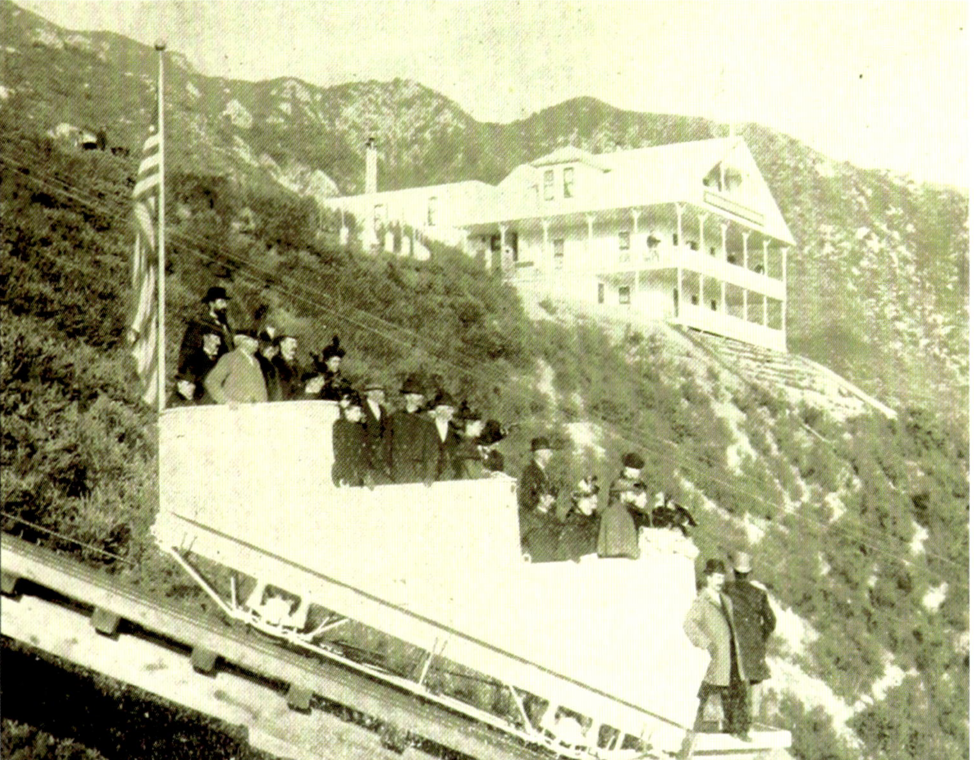

The Great Incline was truly a spectacle to behold, and certainly to ride. Using technology designed by San Francisco cable car inventor Andrew Smith Hallidie, the railway climbed grades between 48 and 62 percent over almost 2,000 feet of ascent. It was built using three rails, with a fourth rail as a "passing track" halfway up. A trestle that spanned a 150-foot deep chasm was expertly designed and implemented and became known as the Macpherson Trestle.

Mount Lowe Railway was widely celebrated, but a series of setbacks ultimately knocked it off the rails. Lowe could not keep up with expenses, had to liquidate his personal assets, and was left in financial ruin by the turn of the century. Henry Huntington took over the operation, making it part of his Pacific Electric Railway in 1902. Fires in 1900 and 1905 destroyed almost every facility on Echo Mountain. A flash flood in 1909, gale-force winds in 1928, and an electrical fire in 1936 all took their tolls. Finally, when heavy rains in 1938 devastated the mountainside, the Mount Lowe Railway was abandoned.

Residents and visitors can still enjoy its legacy, however. The railway was placed on the National Register of Historic Places in 1993, and the Mount Lowe Preservation Society was formed in 2000. Ruins of the railway, along with spectacular views, can be seen on a challenging but popular 9.8-mile hike called the Mount Lowe Railway Loop.

Top: A photo of the first railcar to leave the Altadena station on the Mountain Division branch on July 4, 1893.

Bottom: The specially built funicular railcar descending the Great Incline.

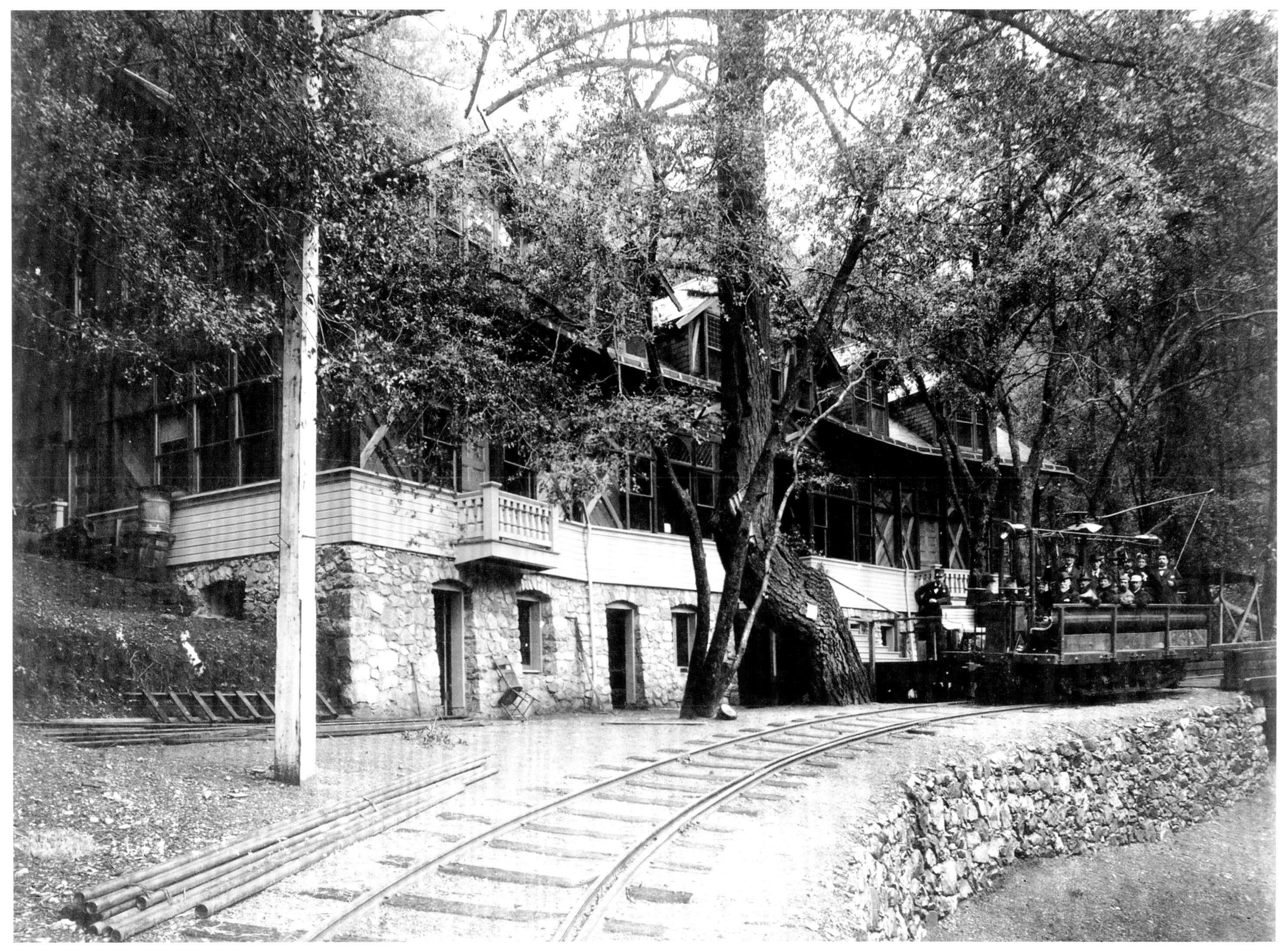

The Ye Alpine Tavern at Crystal Springs, the terminus of the Alpine Division branch. The Ye Alpine Tavern was renamed the Mount Lowe Tavern in 1925 and burned down in 1936.

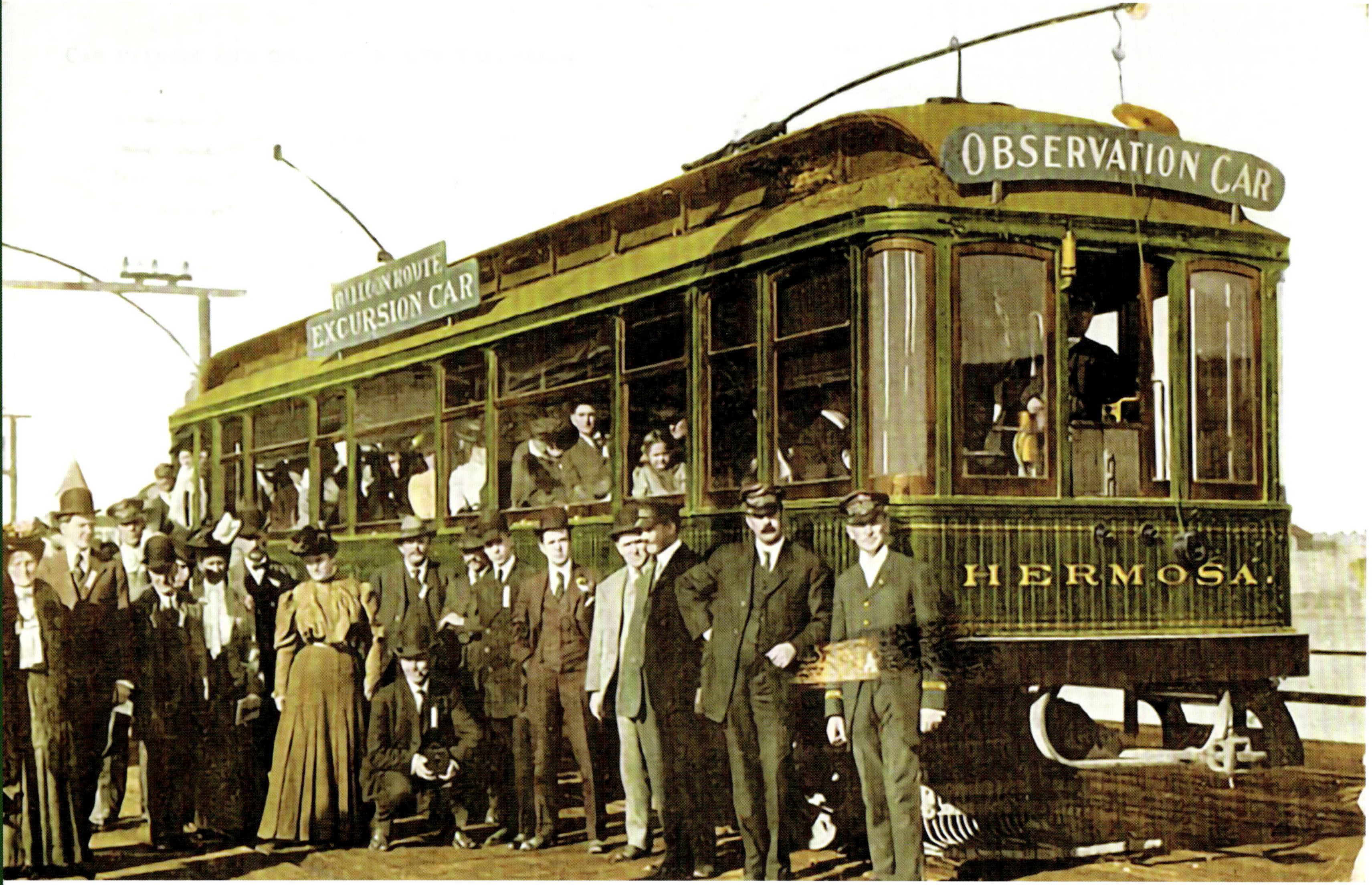

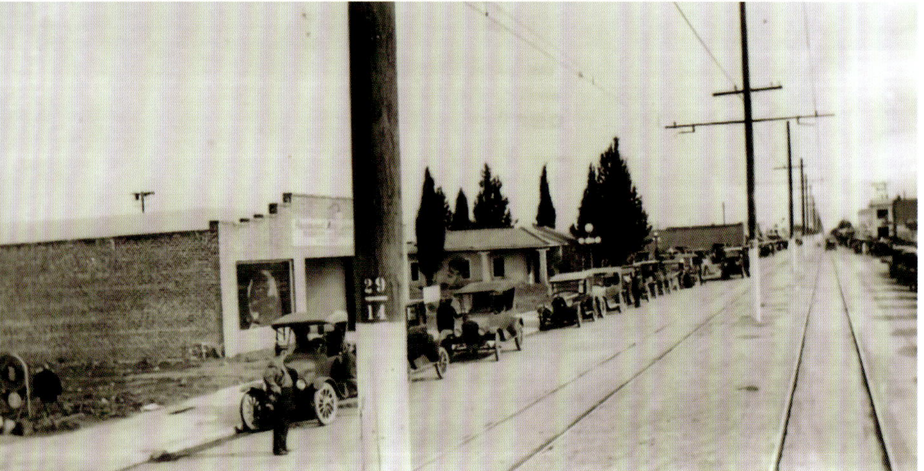

Top: A railcar of the Los Angeles Pacific Railroad's Balloon Route in 1905.

Bottom: Los Angeles Pacific Railroad tracks running down Sherman Way in Owensmouth, California, which is now incorporated into Los Angeles and named Canoga Park.

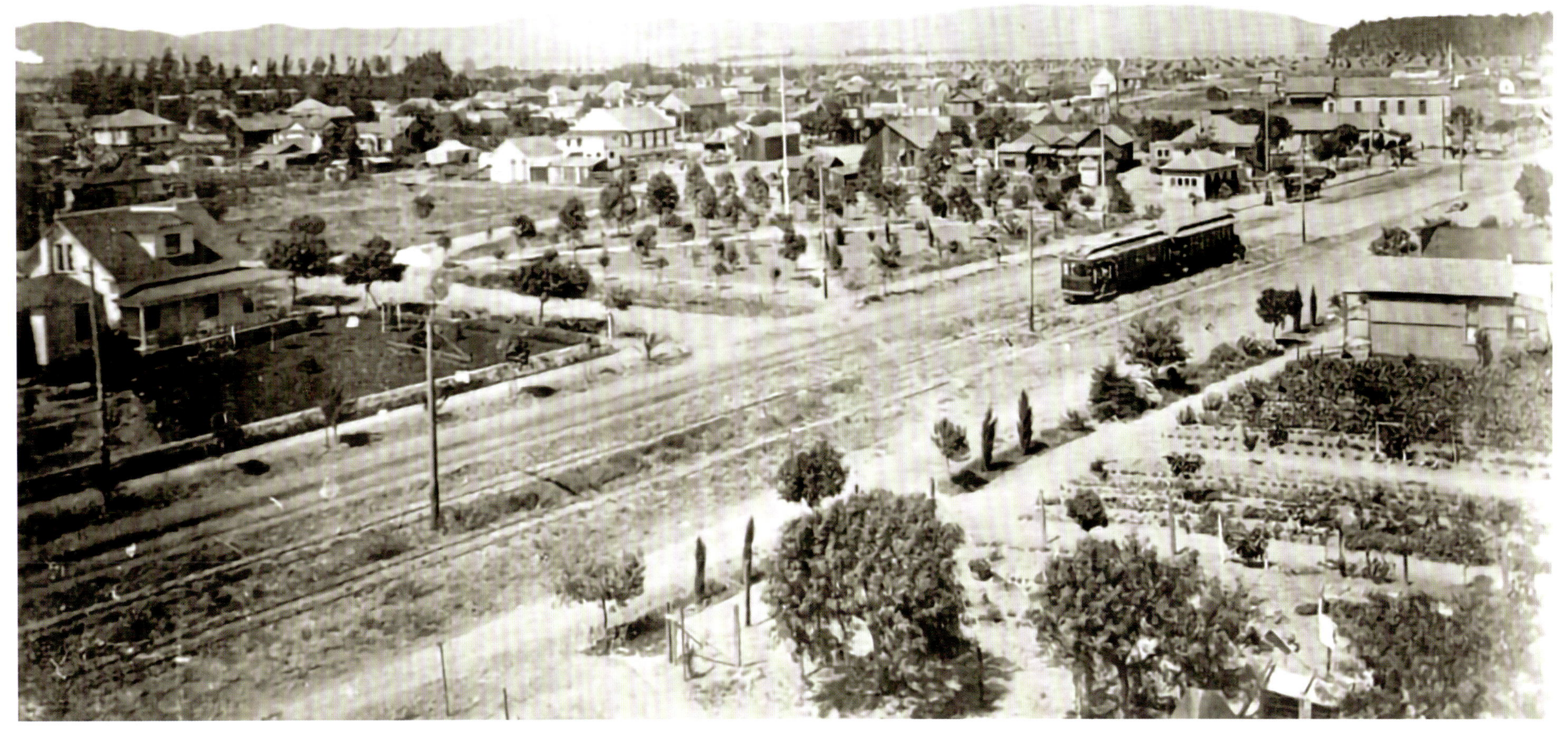

A railcar of the Los Angeles Pacific Railroad near the Sawtelle Veterans Home circa 1890.

Balloons, Hollywood, trolleys, and a "ten-dollar tour for just a buck." Those are some of the legacies of the Los Angeles Pacific Railroad, which existed formally for only seven years but left a lasting impact on Southern California transportation. Moses Sherman, a land developer who went by the nickname "General," and Eli Clark established the Los Angeles Pacific Railroad in 1899. Perhaps it's not surprising that both men were teachers earlier in their lives—Sherman in upstate New York and Clark in his native Iowa—because the men were about to give aspiring railway pioneers a lesson in not only railroading but also marketing and public relations.

With Clark serving as president, the two (who were also brothers-in-law) began acquiring smaller railways in order to run passenger cars from Pasadena through L.A. and Hollywood and out to "hot spots" like Santa Monica, Ocean Park, and other beach towns to the south. They merged with the Los Angeles–Santa Monica Railroad and Los Angeles, Hermosa Beach and Redondo Railway companies over the first few years of the 1900s, all while maintaining the Los Angeles Pacific name.

One of the earliest indications that those leading the railroad understood business was their handling of passenger traffic versus freight. During the waking hours, their focus was on the passenger experience. Overnight, though, their focus shifted to a lucrative freight business, a system designed to maximize their capacity along the rails. And that capacity was impressive. In the middle of the first decade of the twentieth century, they owned 405 cars: 144 passenger cars, 6 parlor cars, 17 electric locomotives, 221 freight cars, 5 mail and express cars, and 12 service cars. That rolling stock covered about 180 miles of rail at the height of the Los Angeles Pacific Railroad's heyday. Sherman and Clark made sure to increase the frequency of trains during commuting hours, and also on weekend routes out to the beaches.

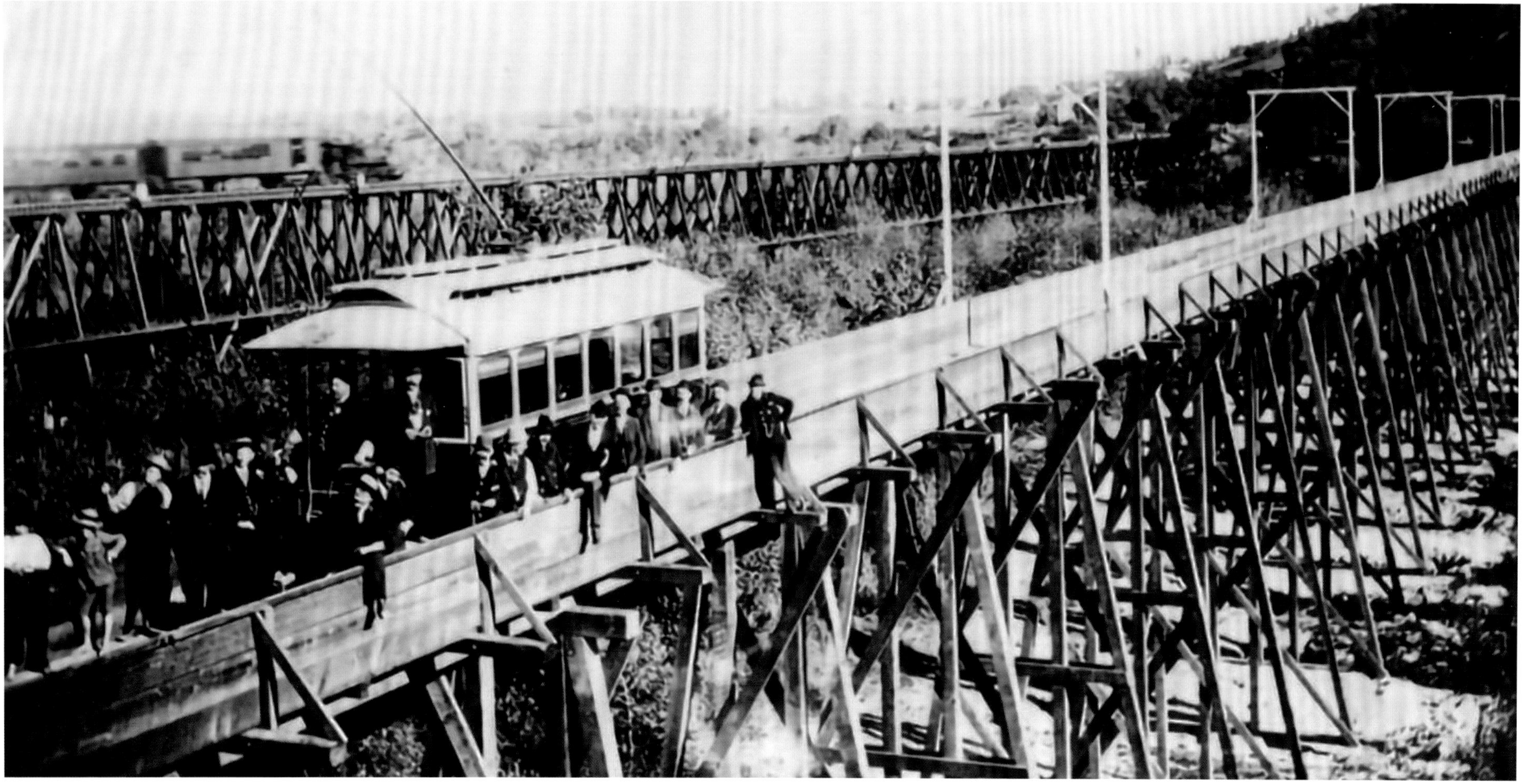

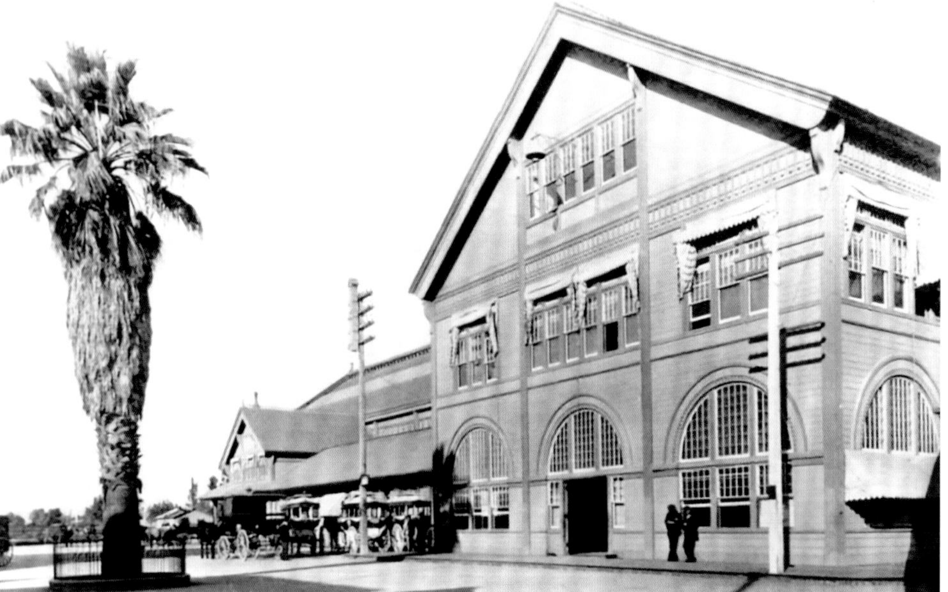

Top: These tracks running over Arroyo Seco in Highland Park, Los Angeles, were built by Pasadena and Los Angeles Railway, which was incorporated by the Los Angeles Pacific Railroad in 1895.

Bottom: The Los Angeles Pacific Railroad's Arcade Depot located at 6th and Central Streets in downtown Los Angeles. The railway built 47 miles of track that radiated around the Los Angeles area from this depot.

Their most talked-about line became known as the Balloon Route. Opening in 1901, a trolley took passengers for $1 from downtown L.A. through Hollywood, Santa Monica, Venice Beach, Redondo Beach, and then back downtown. It left at 9:30 a.m. and returned at 5:00 p.m. thanks to strategic stops along the way. Those included a Sunset Boulevard art studio, the bean fields of Beverly Hills, and a fish dinner at Playa del Ray. Another of the stops, the Streetcar Depot at the Sawtelle Veterans Home, is now on the National Register of Historic Places.

Sherman and Clark advertised the excursion all over Southern California. It was dubbed the "ten-dollar tour for just a buck." It used a fancy parlor car and had the exuberant backing of the Glen-Holly Hotel, the first hotel in Hollywood. "During the summer and winter tourist seasons an average of 10 thousand people per month are carried over the Balloon Route," wrote Los Angeles Pacific engineer P.H. Albright in the *Electric Railway Journal.* Another such daylong excursion brought passengers to the San Gabriel Mission, with interesting stops along the way. Between those daylong trips, beach runs, and commuter service, the line was the talk of the town. It was a short but successful run for Sherman and Clark, who in 1906 agreed to sell the company to Henry Huntington's Pacific Electric Railway for $6 million.

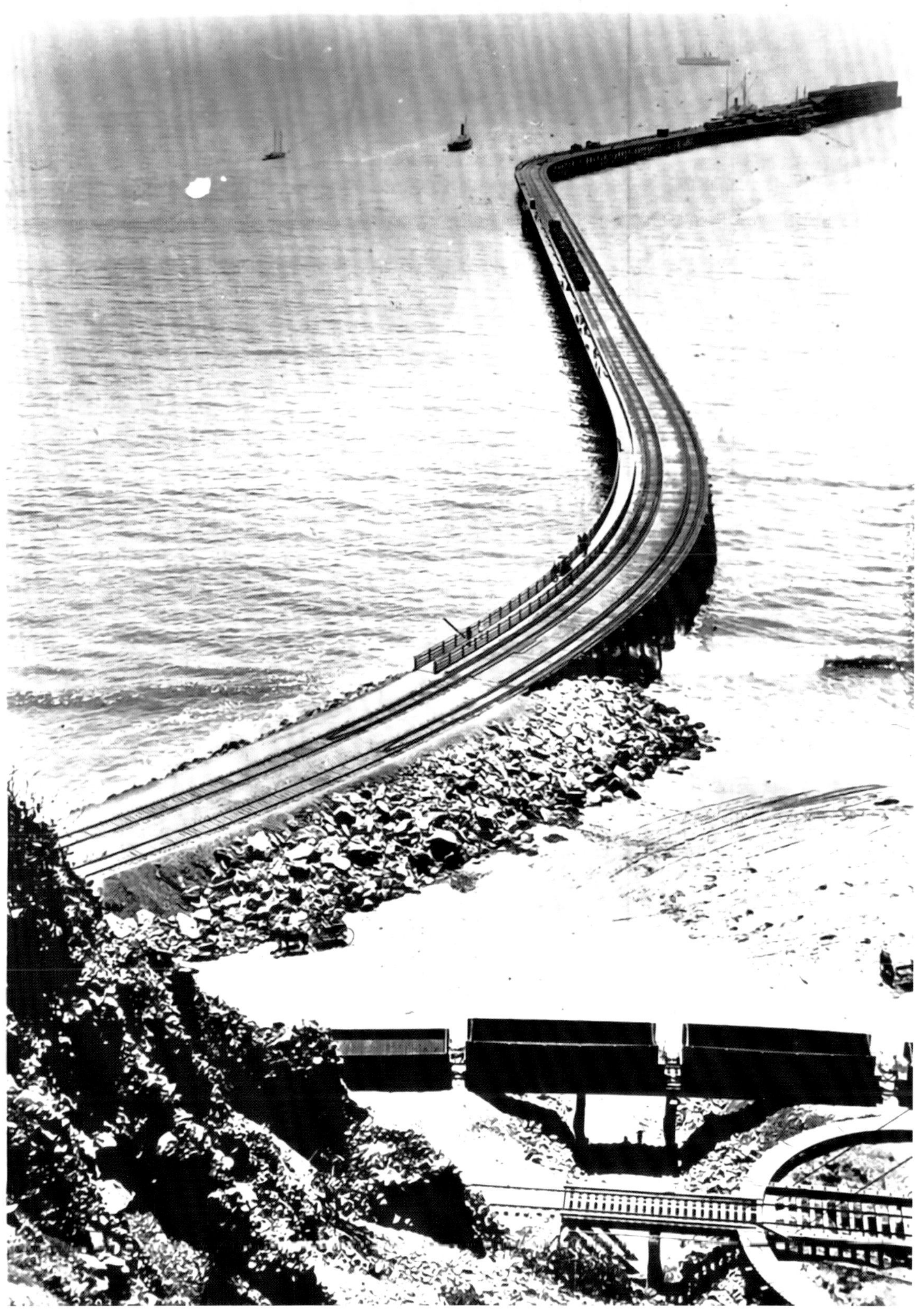

The Long Wharf in Santa Monica was used by the Los Angeles Pacific Railroad to move lumber along their freight lines.

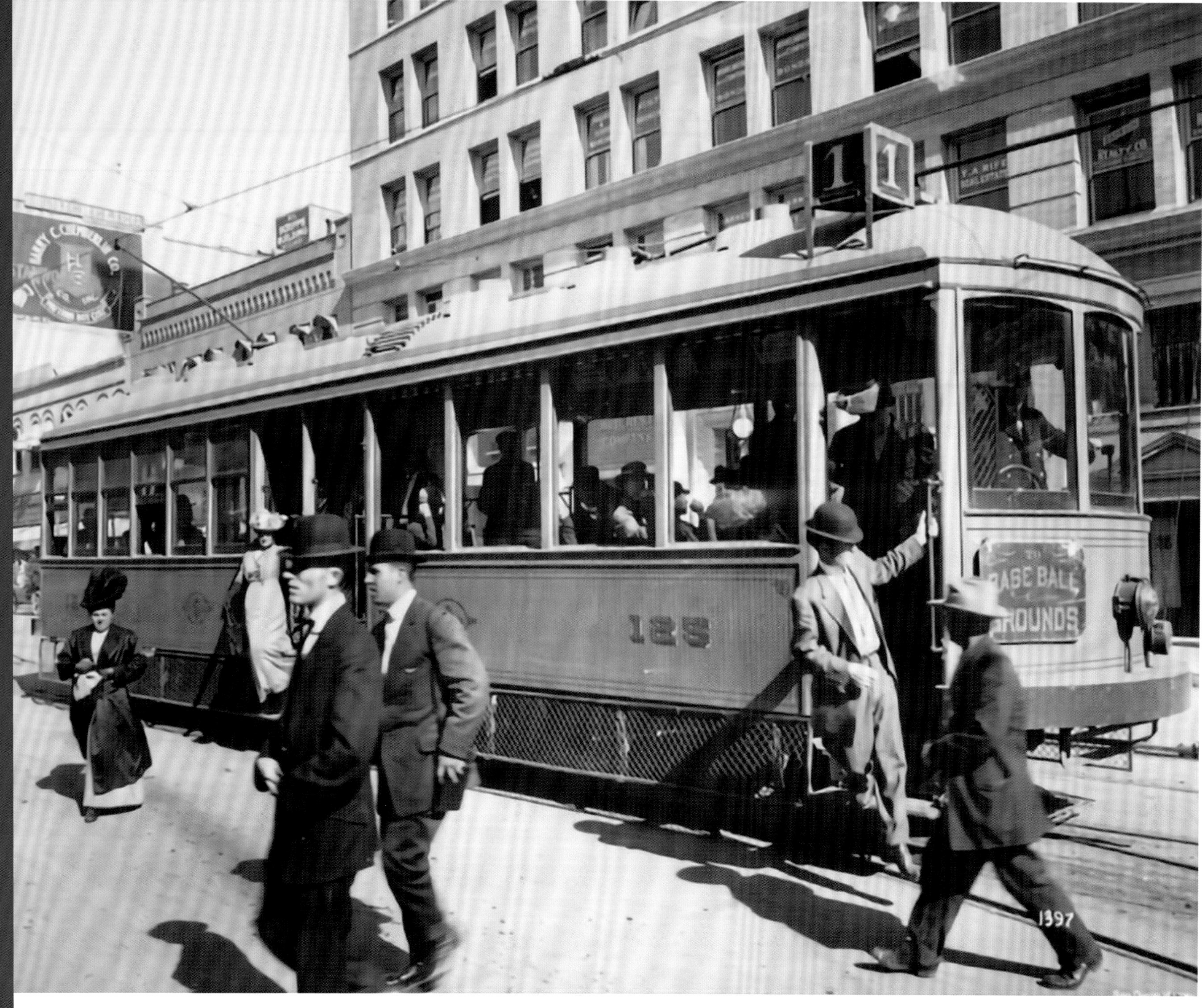

A photo of a San Diego Electric Railway streetcar at the intersection of 5th and Broadway around 1915.

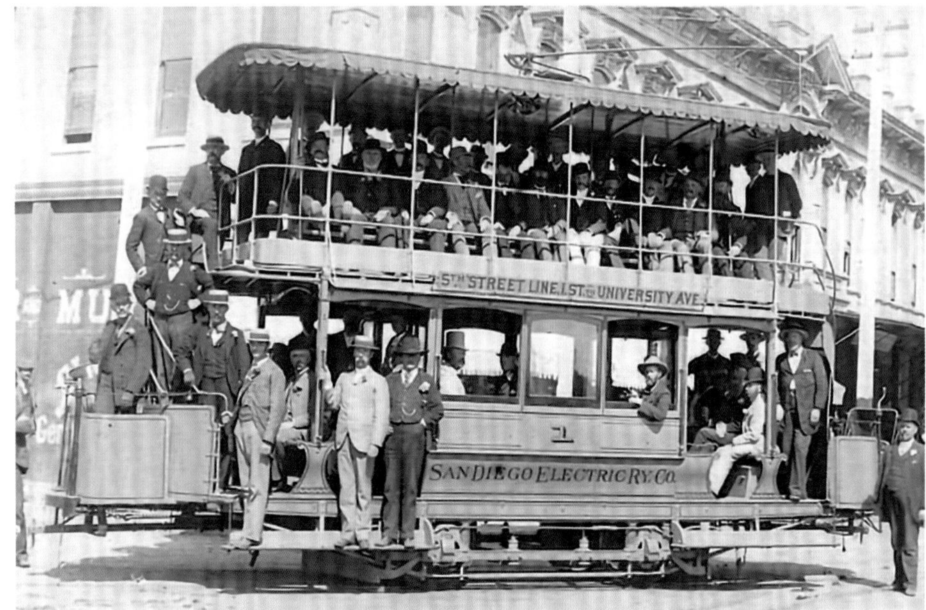

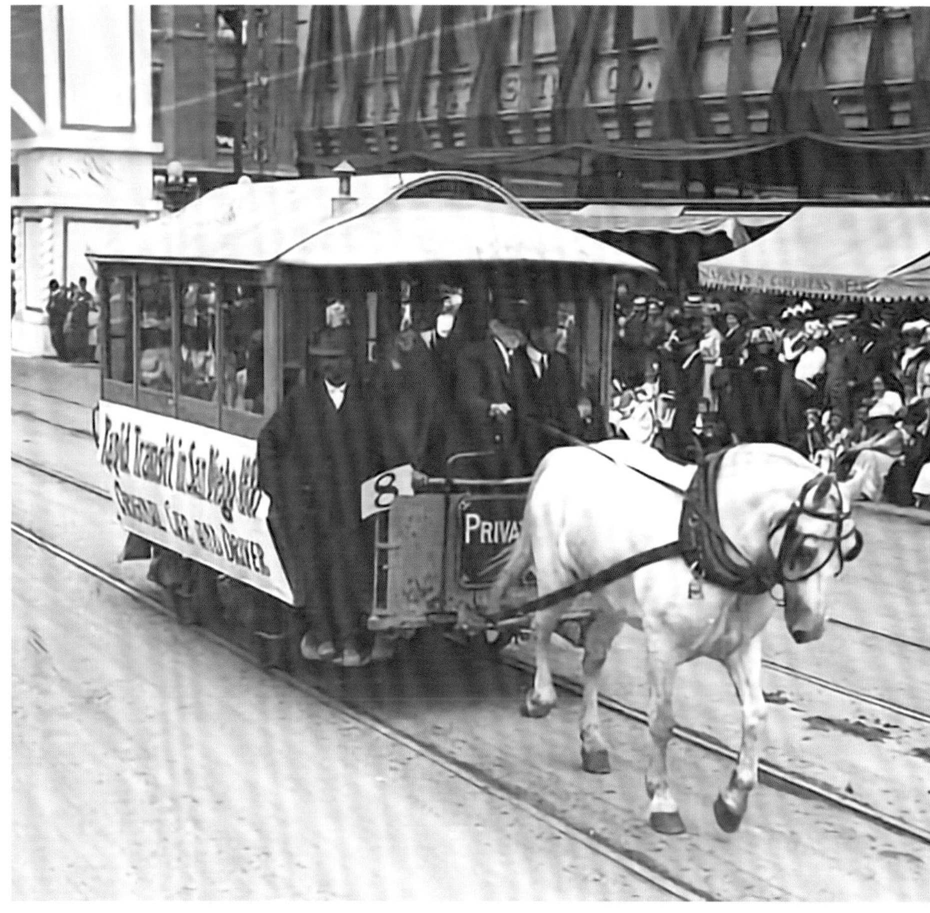

Left: The inaugural run of the double-decker Car No. 1 in 1892.

Right: A photo of a 1911 parade in San Diego where an 1866 horse-drawn trolley was a part of the procession.

Years before San Diego became a hot destination for those seeking surfing waves, year-round sunshine, or an ideal winter vacation spot, the city put some thought into how it might handle the growth that would inevitably arrive. Its first streetcars hit the roads in 1886 in the form of open-air coaches pulled by horse or mule through the town's central business district. The fare was a nickel. The West Coast's first electric streetcar connected downtown with "Old Town" the following year.

The electric streetcar was the precursor to the San Diego Electric Railway (SDERy), founded in 1891 when John D. Spreckels took control of the failing streetcar operation and turned it into a business that helped shape San Diego's growth. His company acquired the Belt Line and Ocean Beach railroads over the last decade of the nineteenth century with plans to turn them into single-track, electric trolley lines.

Spreckels went all out. The San Diego Electric Railway began with five routes and continued to stretch all over the San Diego area. The SDERy, as it became known, grew to serve more than 165 miles in its early-twentieth-century heyday. Spreckels built a new power plant in 1905 to handle the electricity required to run the growing streetcar business, and added motorized buses in the 1920s as the city continued to grow.

World War II brought a big boom to San Diego, a military town, that few cities could rival. The influx of residents certainly kept the public transit system hopping, but the growing popularity of the automobile that had begun in the 1930s had the opposite impact. With SDERy struggling, it was sold to the Western Transit Company for $5.5 million in 1948. Some of its cars and buses remain popular museum attractions in Southern California, commemorating a company that helped propel major growth in San Diego.

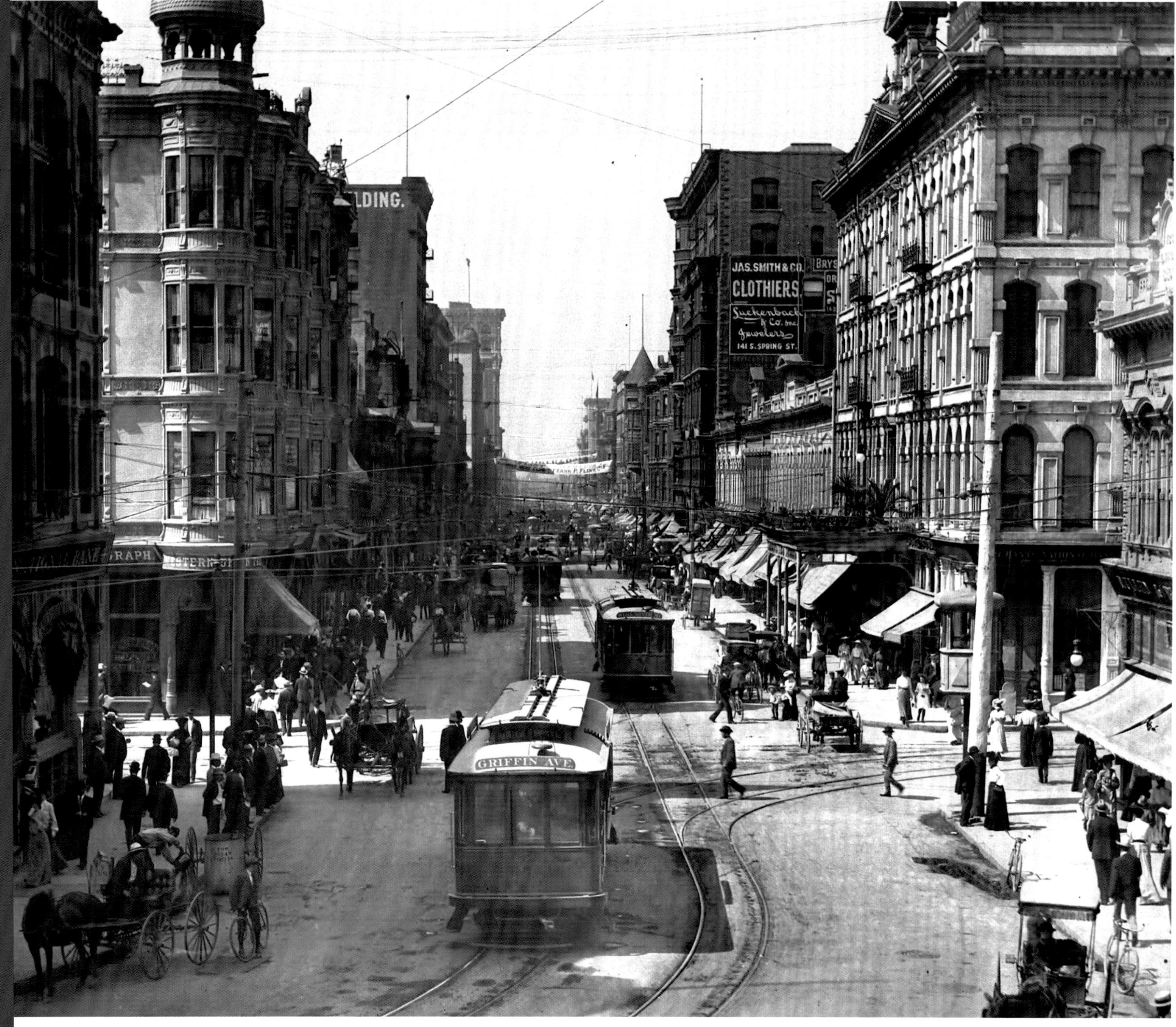

A photograph looking south down Spring Street from First Street in downtown Los Angeles, circa 1900.

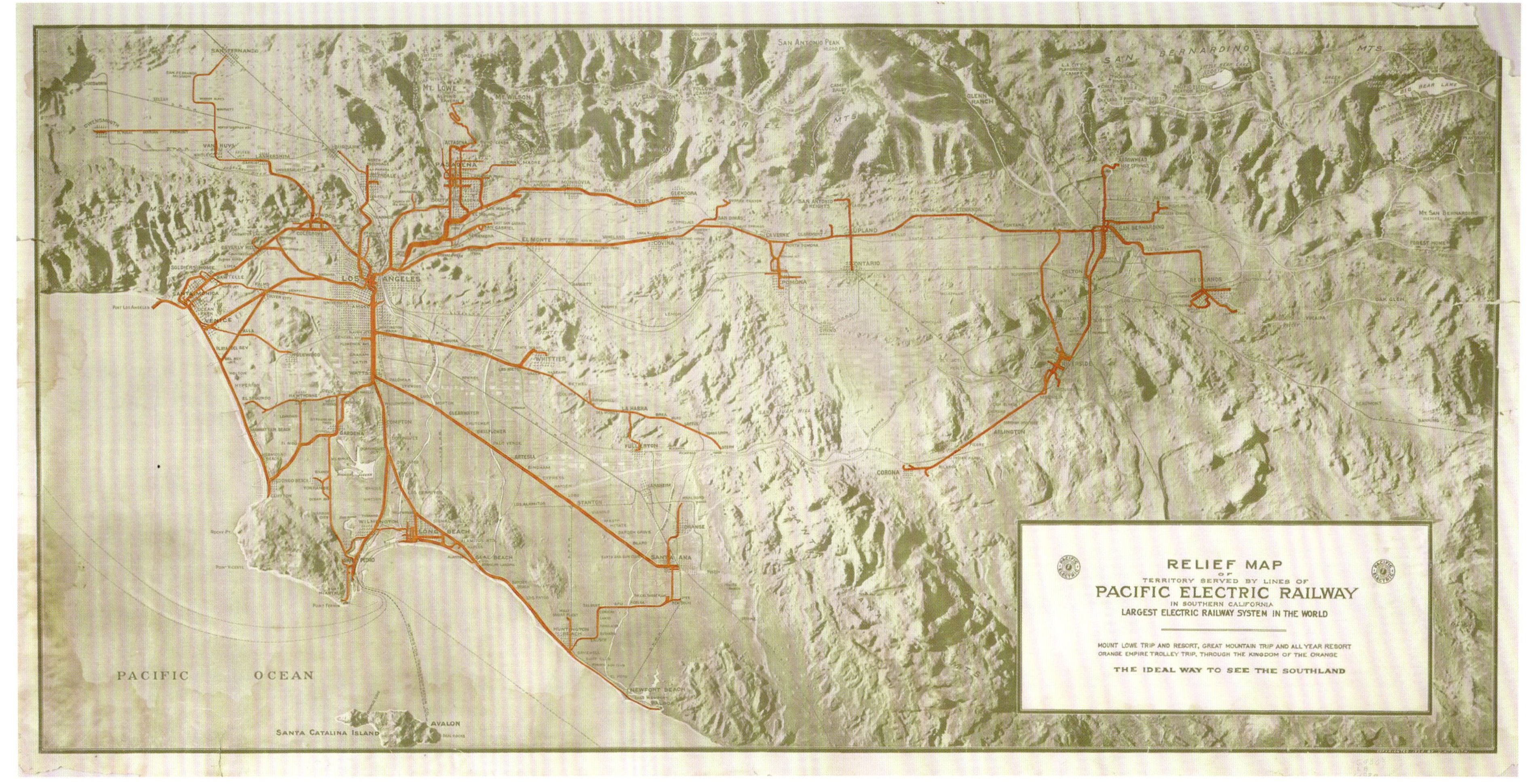

A relief map of routes provided by the Pacific Electric Railway in the 1920s.

Painting the town red was the goal of the Pacific Electric Railway (PE), with Los Angeles being the town. One look at a map of the company's "Red Line" trains shows that, for many years, it was a successful venture indeed. "There were eleven-hundred miles of track throughout Southern California that ran through Orange County, San Bernardino County, Riverside County, and of course Los Angeles County," said Michael Patris of the Pacific Electric Railway Historical Society. "So people, from as early as 1902 when Pacific Electric Railway was founded, were able to go pretty much anywhere. They could go to the beach, they could go to the mountains… they could go inland and see orange groves, they could even go to the San Gabriel Mission. It was pretty amazing."

Indeed, the Pacific Electric Railway was, by all accounts, the largest interurban rail system in the world. Passengers could use it to get virtually anywhere they wanted in the sprawling Los Angeles area. Rail and real-estate tycoon Henry Huntington founded the operation in 1901. Huntington had vast experience with electric trolley transportation, having served as vice president of Southern Pacific Railroad—owned by his uncle—for whom he oversaw an effort to merge several smaller, street railroads into a larger, organized network.

What he did with PE in Los Angeles, along with prominent L.A. banker Isaias W. Hellman, would dwarf the San Francisco operation. Acquiring several smaller interurban rail systems, PE spread its wings in every direction in Los Angeles. Huntington and Hellman bought up both rail companies and property at a breakneck pace and their trains were pulling into the suburbs long before they were called suburbs. The first big project was opening a line to Long Beach on July 4, 1902. It was only the beginning for their distinct, red-colored trains.

Newport Beach (1904), Glendale (1904), San Pedro (1904), Glendora (1907), Pomona (1911), Yorba Linda (1911), Burbank (1911), Van Nuys (1911) and San Bernardino (1915) followed. In addition to the lines themselves, PE was ahead of its time in focusing on keeping the customer happy. Service was fast and friendly, and its trains were impeccably clean. It opened a 400,000-square-foot office building at 6th and Main Streets. It launched an iconic, circular logo (mostly red, of course) that boasted "comfort, speed, safety." And by 1907, it declared itself the "greatest electric railway system on earth." Few debated the claim.

The Great Merger of 1911, in which PE absorbed seven other railways, made the company the largest interurban electric passenger railway in the world. It came at a time when the City of Los Angeles had just come off of a decade in which the population multiplied more than threefold. Now with more than 2,000 trains covering 1,000-plus miles, profits soared. Huntington invested much of the revenue back into the business, stretching lines west to the Pasadena area and hitting many of the beach towns along the coast. Between connecting L.A. neighborhoods and buying up real estate in those very towns, his empire grew. The Red Line became a Southern California institution, surging in popularity in the 1920s.

That was the heyday, however. Huntington's estate continued to run the company following his death in 1927, but the Great Depression and World War II made continued operation challenging. The growing popularity of automobiles and buses following the war did the same. PE had served more than 100 million passengers by 1945. Passenger service was discontinued, though, in 1953, as freight service had a better chance of paying the bills. Finally, on April 8, 1961, the last Red Car made its final run.

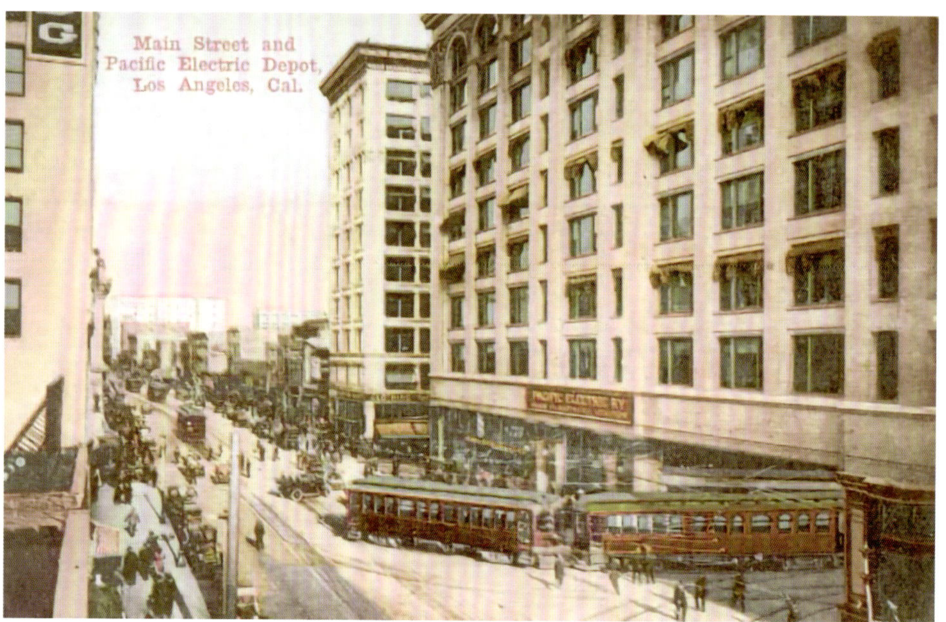

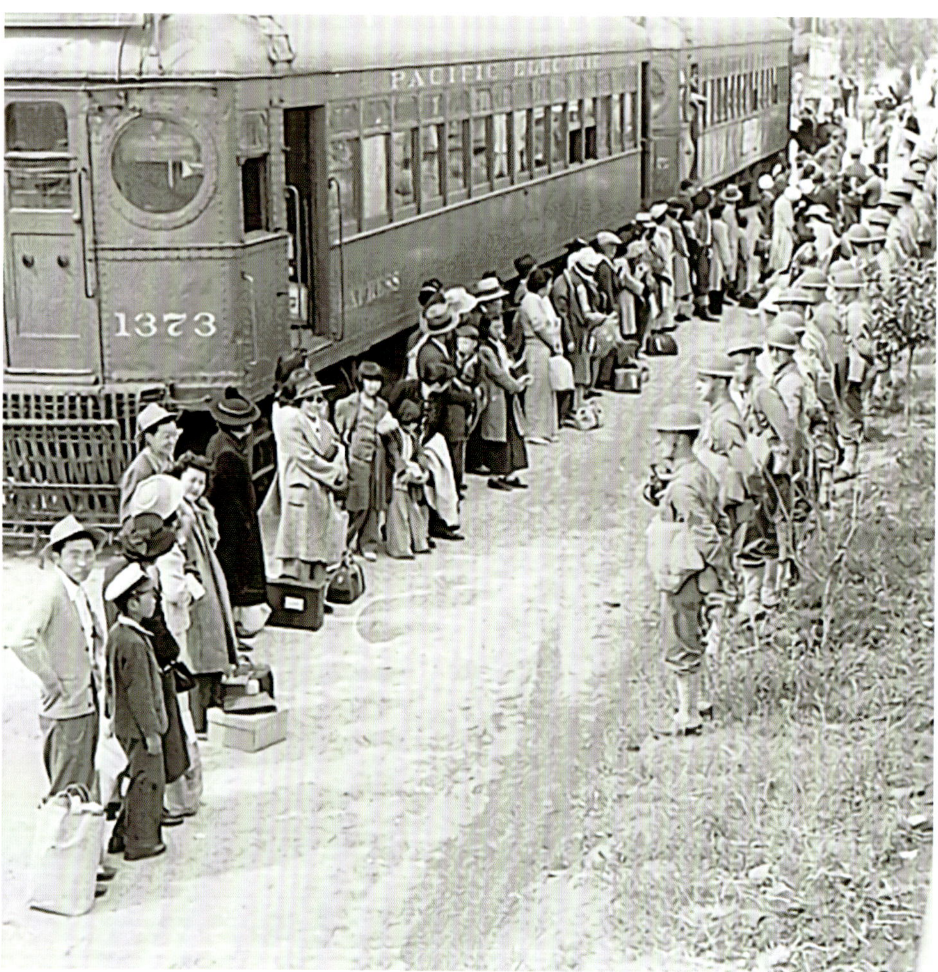

Top: The PE's main depot in downtown Los Angeles, in 1910.

Bottom: Japanese Americans from San Pedro arriving at the Santa Anita Race Track on April 4, 1942, by way of the PE, awaiting internment processing during World War II.

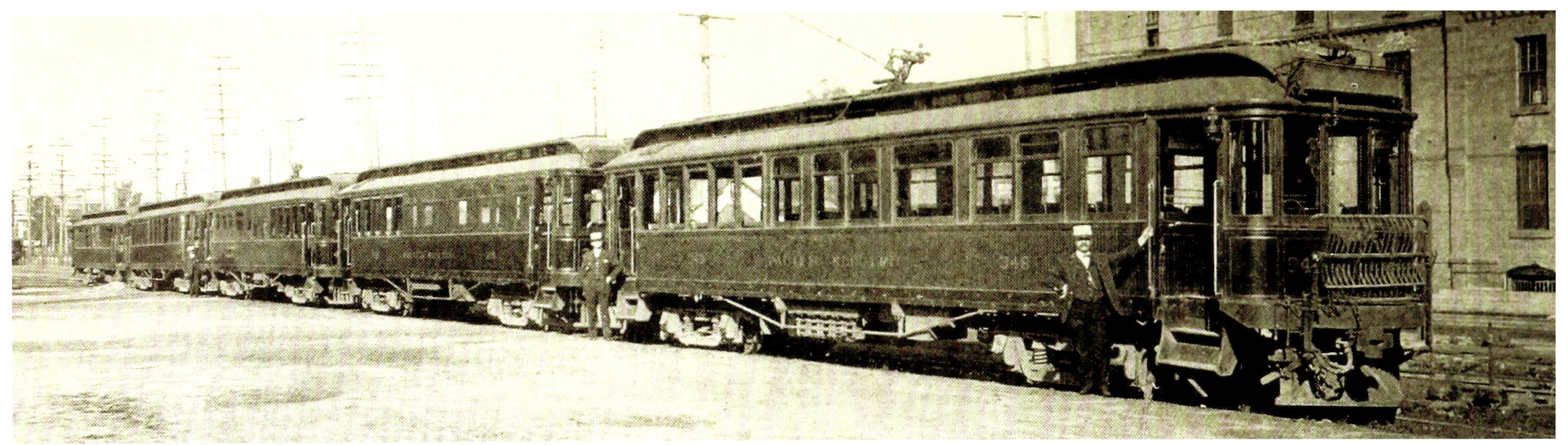

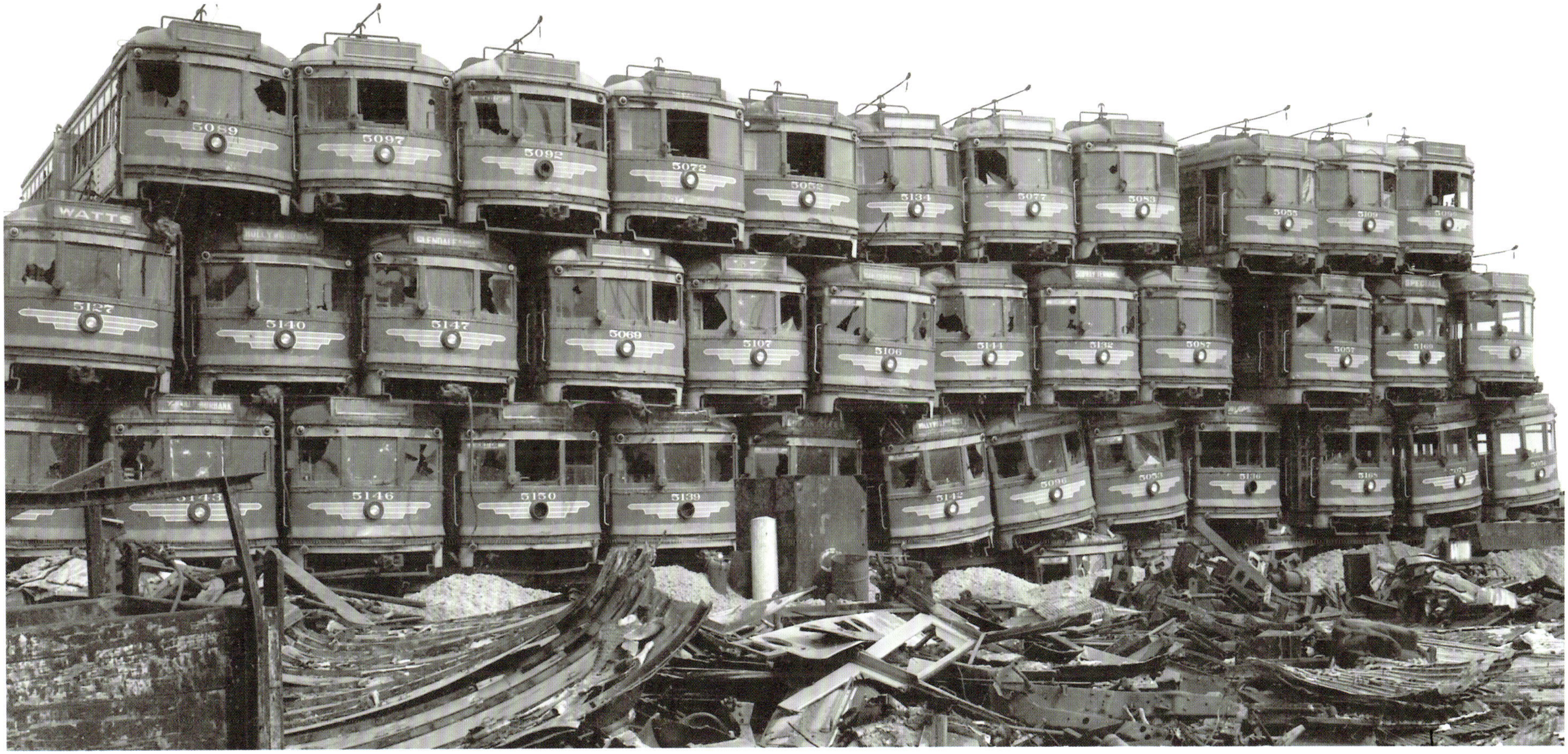

Top: A photo of a PE train, in 1908.

Bottom: A junkyard on Terminal Island with a stack of PE railcars. The demise of the railway and the fall in ridership coincided with the widespread use of the automobile in America and the construction of many freeways in the Los Angeles area during the 1950s.

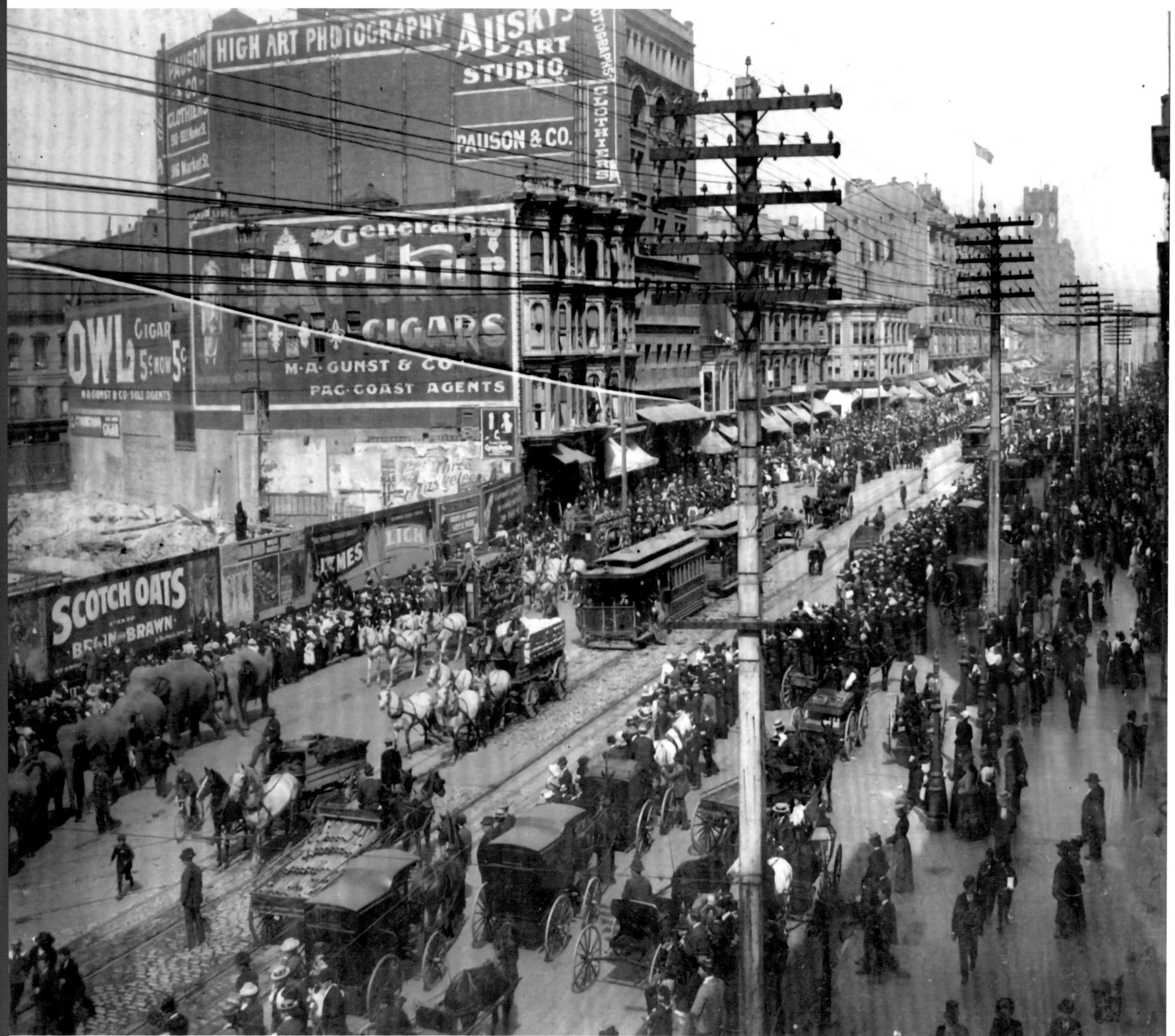

A parade for the Ringling Brothers running down Market Street in San Francisco.

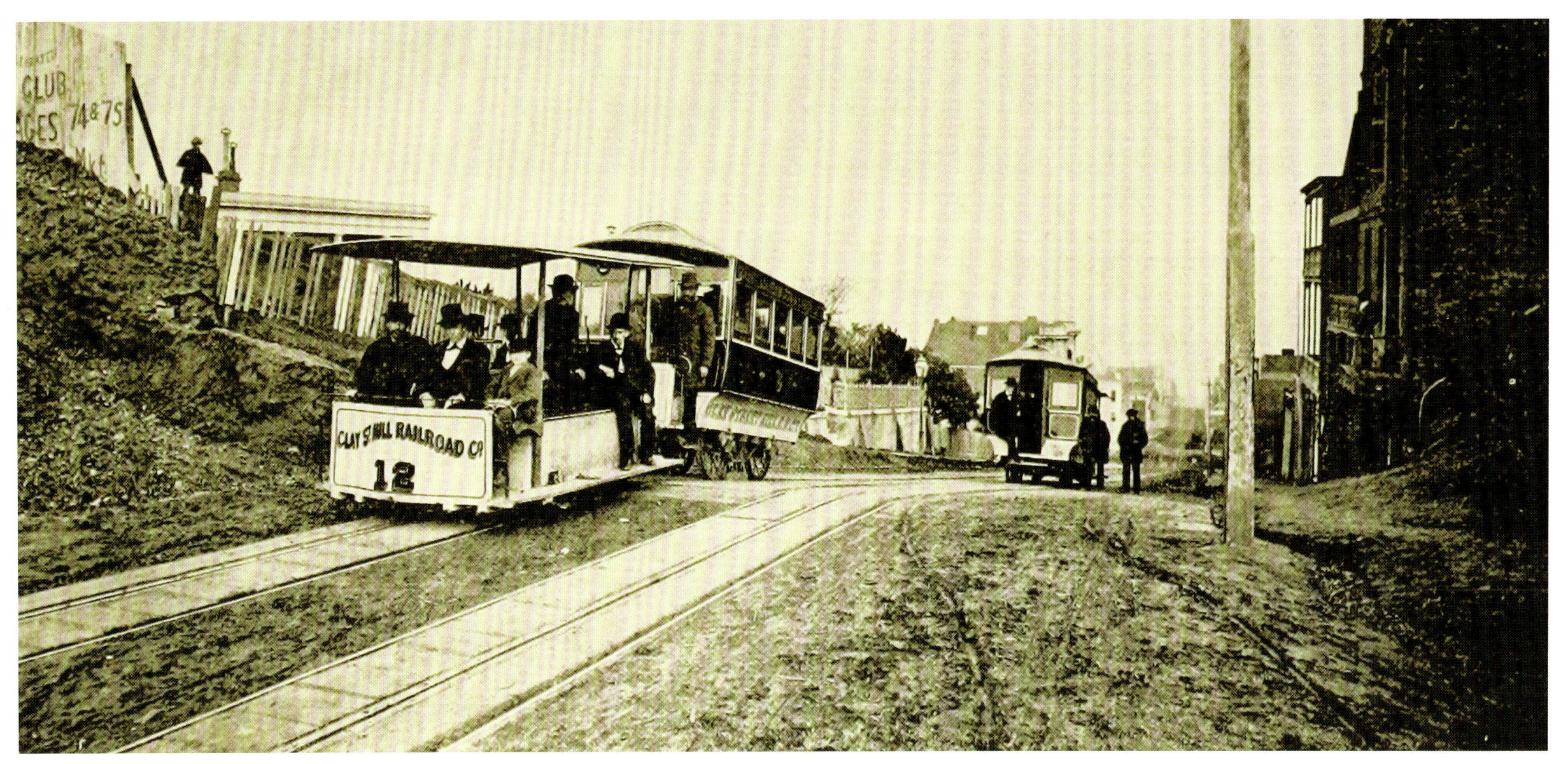

A photo of the first cable cars on Clay Street just days after San Francisco's Clay Street Railroad was opened.

Nowhere in the world does a city's public transit system define it quite like it does in San Francisco, California, home of the cable car. This iconic form of transportation was invented in San Francisco, and the "City by the Bay" still operates the last remaining public cable car system on the globe. It's no wonder the mention of a cable car brings to mind images of passengers waving out the windows while riding up and down the city's steep hills.

It was actually a tragic scene, however, that brought about this distinction in San Francisco. In 1869, on a wet summer day in the city, Andrew Smith Hallidie watched horses under a whip as they tried to pull a streetcar up a steep hill on the wet cobblestones of Jackson Street. Between the hill, the weight of the car, and the slick footing, five horses were dragged to their deaths.

It was then that the idea was sparked in Hallidie—a steam-powered, cable-driven rail system that could handle San Francisco's hills. His father had obtained a patent in his native England for a cable made of wire rope before Andrew came to the United States in 1852, in hopes of striking it rich in the Gold Rush. He first used the cable for suspension bridges and the hauling of ore from mines. His cable car idea proved to be the winner though.

Andrew, partnering with the Clay Street Railroad, worked diligently in 1873 to install a cable line on Clay Street. The first cable car test took place there on August 2, 1873, and on September 1 public service began. The project was so successful that several other companies moved to get in on the action. Over the next two decades, there were 23 different lines traversing some 53 miles of a city that, even today, is only 46 square miles in area.

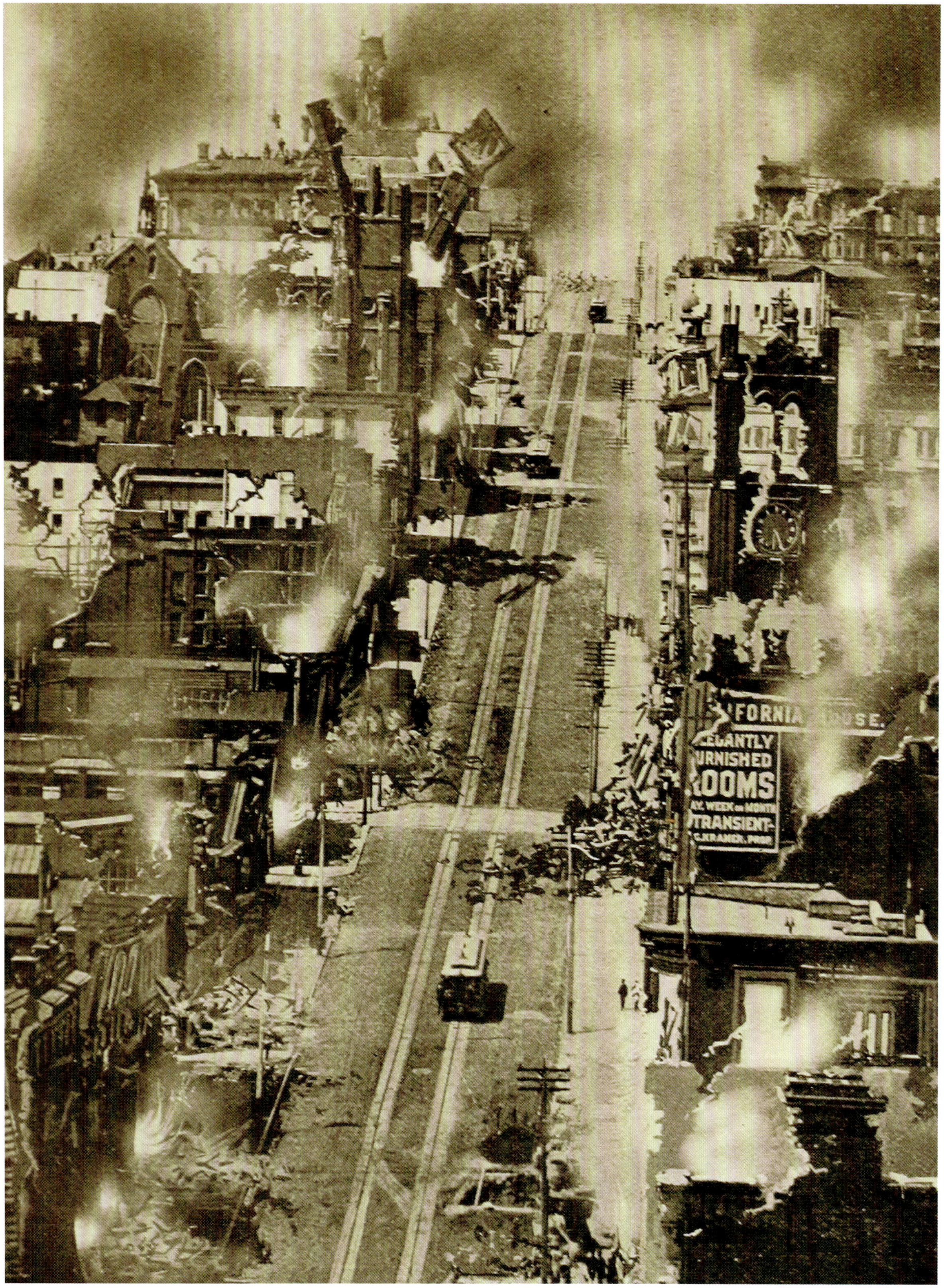

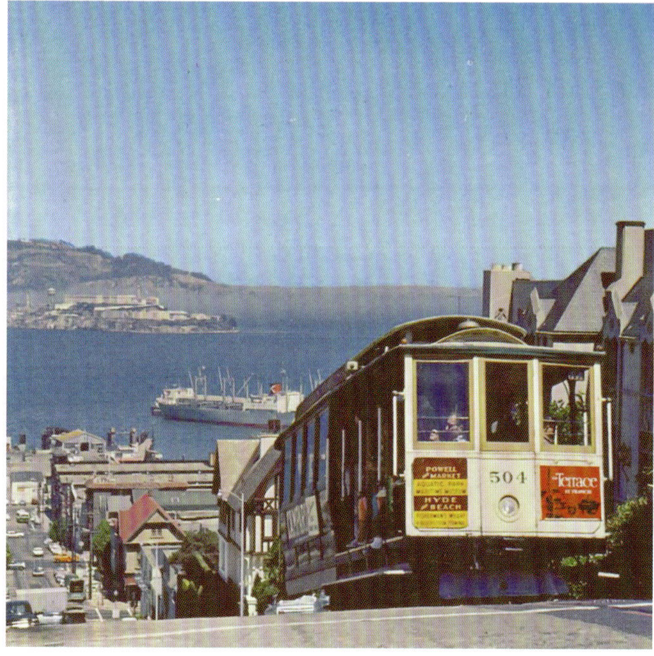

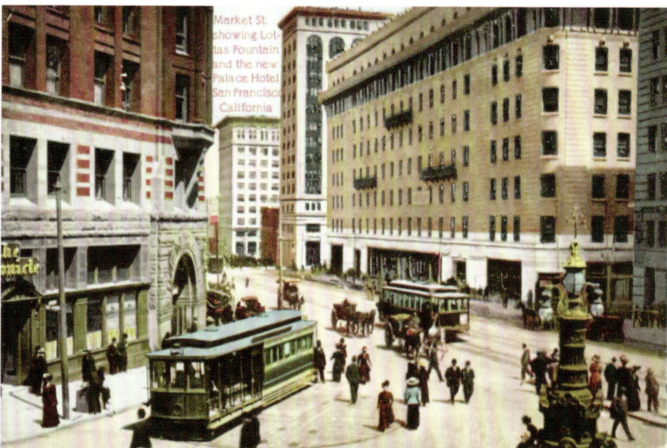

Left: Cable cars still operating after the 1906 San Francisco Earthquake.

Right Top: A cable car ascending the Hyde Street Hill from Lombard Street. Hyde Street is the steepest section of track in the present-day cable-car system with a 20 percent gradient.

Right Bottom: Cable cars in 1915 at the intersection of Market and Geary Streets.

By the turn of the century, electricity had replaced steam in powering the cars more efficiently. A setback was soon coming, however, in the form of a devastating earthquake on April 18, 1906. It took a while for the city—and its talked-about cable cars—to get back on the rails. By the middle of the twentieth century, higher operating costs were also threatening the transit system.

In response to a mayoral edict to scrap the cable car lines in 1947, local activist Friedel Klussmann founded the Citizens' Committee to Save the Cable Cars, forever penning her name in San Francisco lore. The committee launched a petition and public-relations drive to keep the cable cars running due to their value not only as a transit system but as a city treasure. Their Measure 10 won in a landslide vote later that year, and the cable cars kept running.

Faster, more modern transit systems caused the cable cars to shrink in quantity over the years, but they have never lost their place in the hearts of San Franciscans. The San Francisco cable car system was honored as a National Historic Landmark in 1964. The cars celebrated their 100-year anniversary with a ceremony on the Clay Street Hill in 1973. And after a complete refurbishing of the cars and a rebuilding of the system in the 1980s, a ribbon-cutting celebration at Union Square took place in 1984. Today, just three lines remain: two routes from downtown (near Union Square) to Fisherman's Wharf, and another on California Street. While it's mostly tourists rather than commuters who ride these days, the legacy of cable cars in San Francisco has been preserved. The San Francisco Cable Car Museum on Mason Street remains a popular attraction.

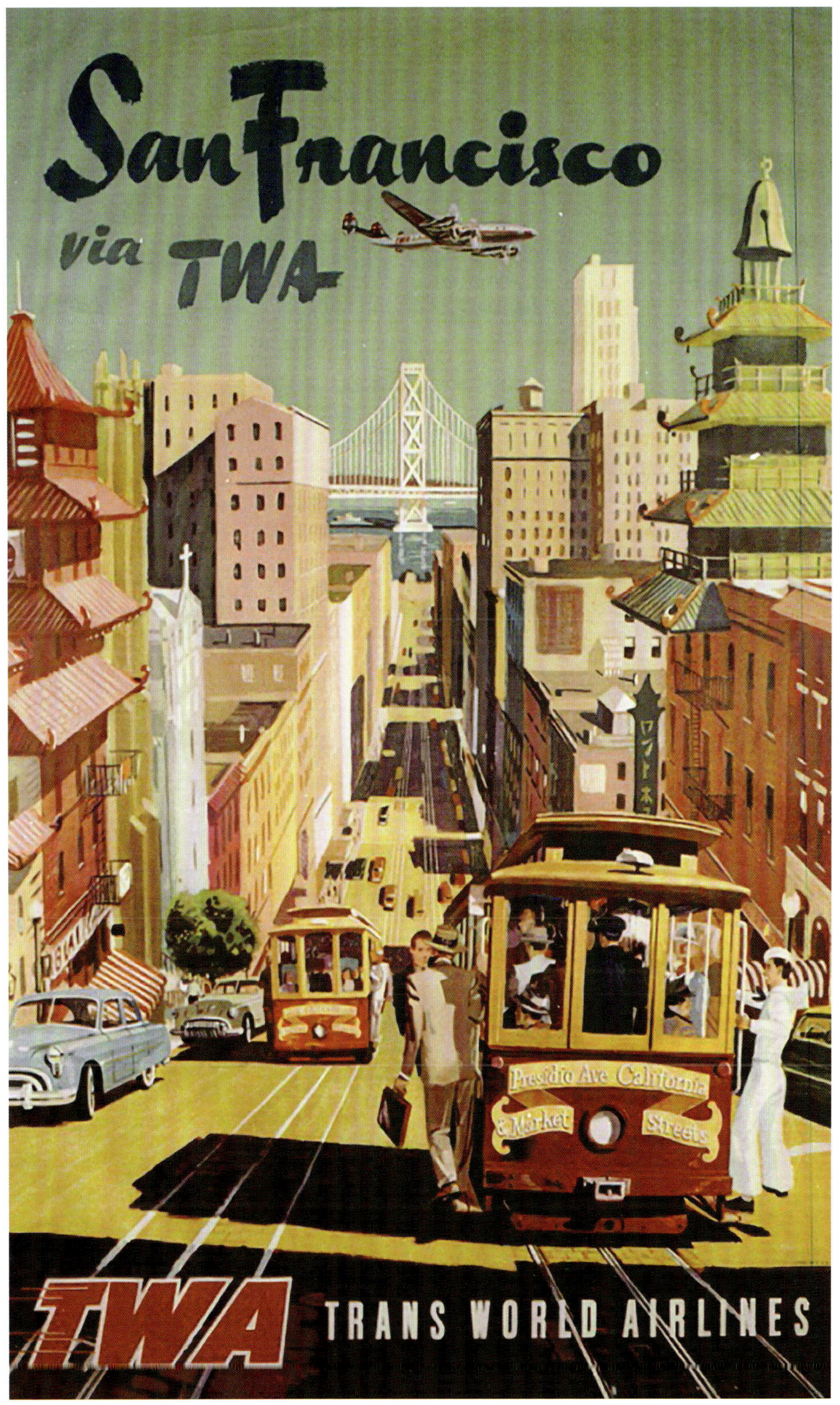

A TWA brochure advertising the highlights of San Francisco and showing cable-car passengers "surfing" on the sides of the cable cars.

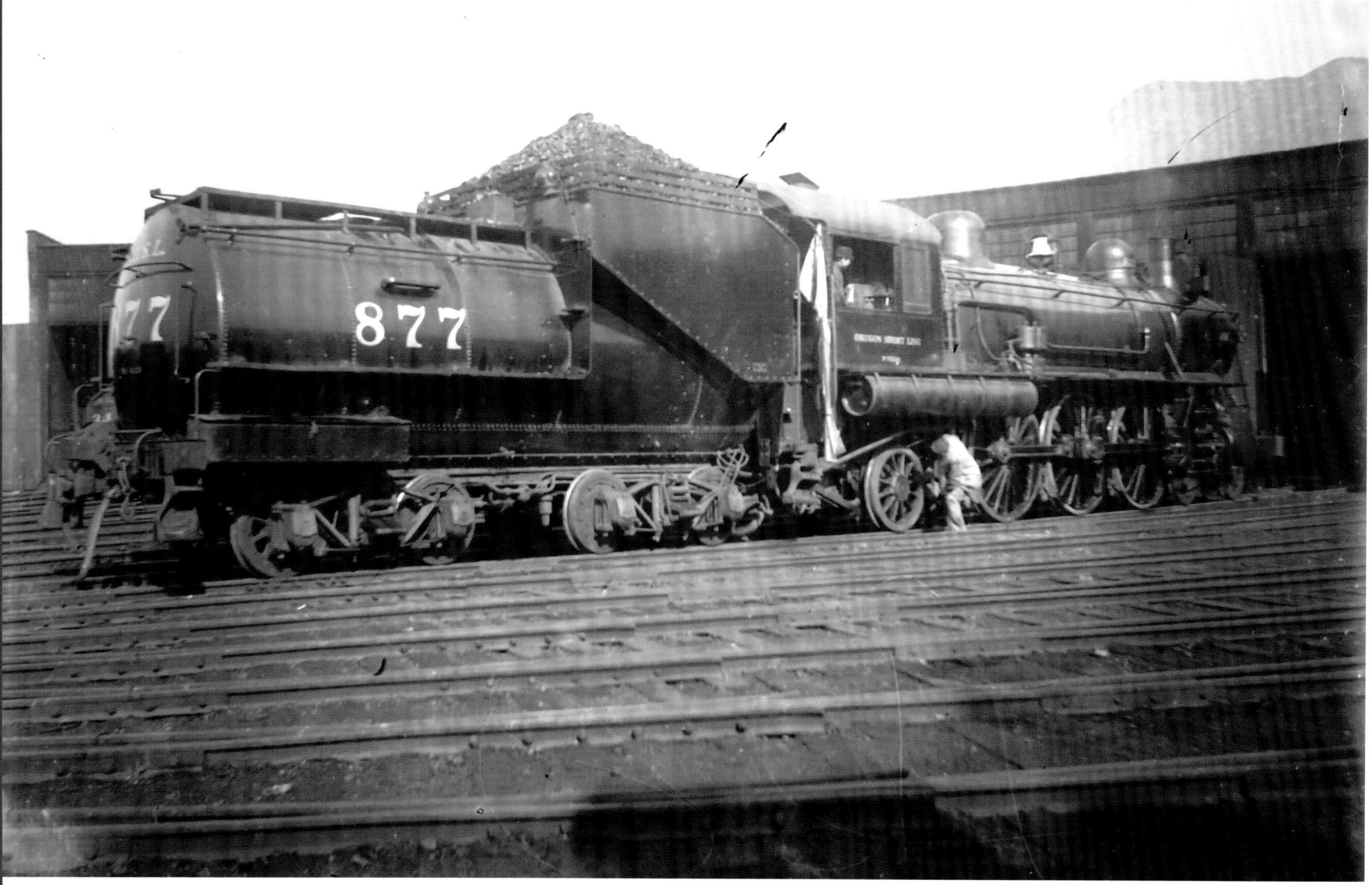

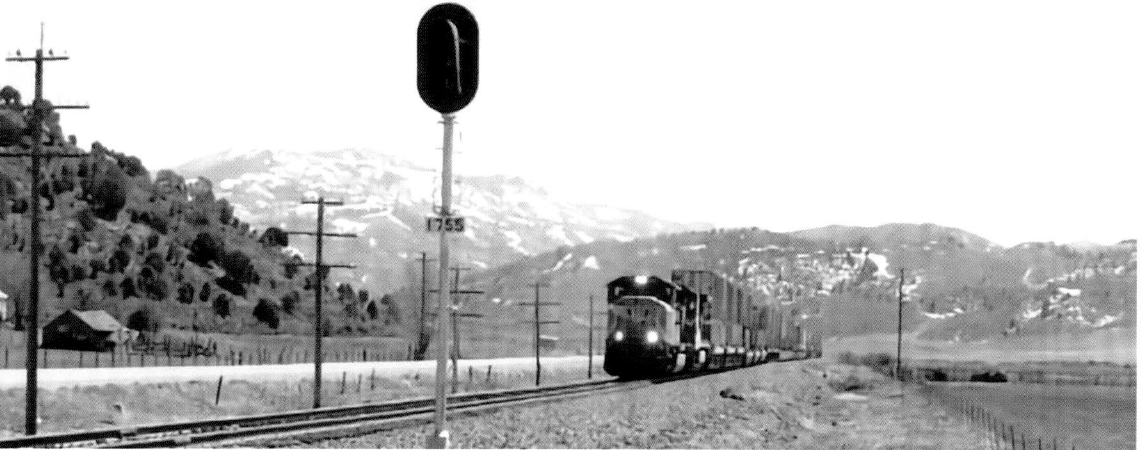

Top: The OSL was an independently financed railway for some time before the Union Pacific took control of the Oregon Short Line's board in 1898.

Bottom: The OSL running through Blaser Canyon in Idaho.

The Oregon Short Line Railroad (OSL) made a mighty big impact on at least two counts in the late nineteenth and early twentieth centuries. It was founded in 1881 as the Oregon Short Line Railway, a subsidiary of the mighty Union Pacific Railway (UP). It was named for being the shortest route from Wyoming to Oregon, though it also operated in Idaho, Utah, and Montana during its more than century of life. The railway's line started from UP's main line in Granger, Wyoming, and by 1882 the OSL had meandered into Idaho. It was extended to Huntington, Oregon, in 1884, and using leased track, the line terminated in Portland, Oregon.

A merger made it the Oregon Short Line and Utah Northern Railway for a while in the late 1800s, but it became the Oregon Short Line Railroad just before the turn of the twentieth century. By this time, it had already established its value as a route to the West. The Southern Pacific had cut off UP's progress due to its own transcontinental efforts, so UP needed a western route to take it past Utah. The OSL was the answer.

The little line's other claim to fame came in the early 1900s, when its parent company put it on the task of getting people to Yellowstone after UP president E. H. Harriman visited the national park in 1905. He ordered the construction of a line from the UP terminus in St. Anthony, Idaho, to Yellowstone's western border in Montana.

The OSL completed the job in late 1907, and passengers began to reach the park by train the following spring. Traffic through the west entrance of the park nearly doubled in that first year of rail service. A little more than 4,000 visitors used the west entrance in 1907. By 1915, that number swelled to more than 32,000—a majority of which were arriving by rail. Struggling in 1938 on the heels of the Great Depression, UP leased the OSL and other subsidiaries. The OSL continued to operate under that arrangement until 1987, when it was fully absorbed into the UP fleet.

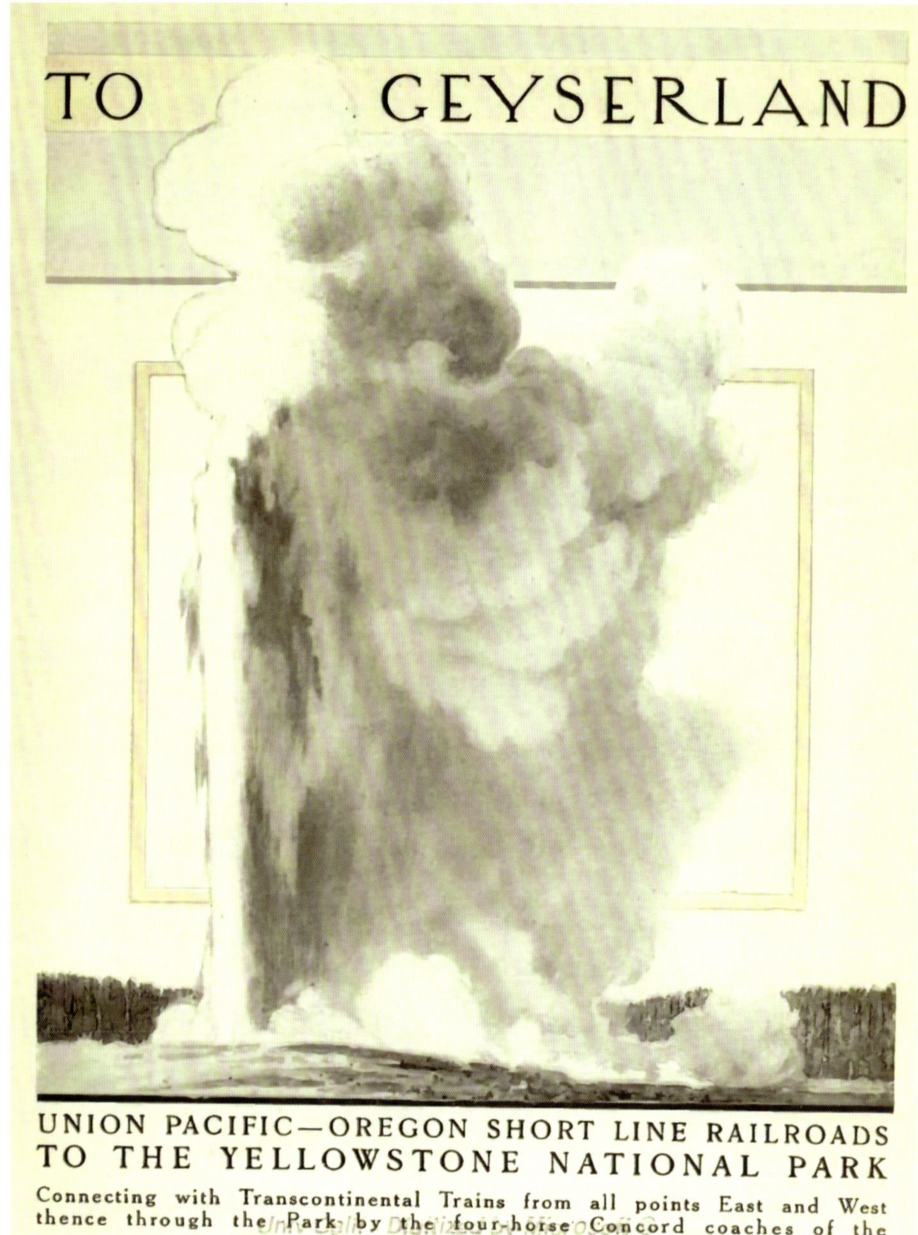

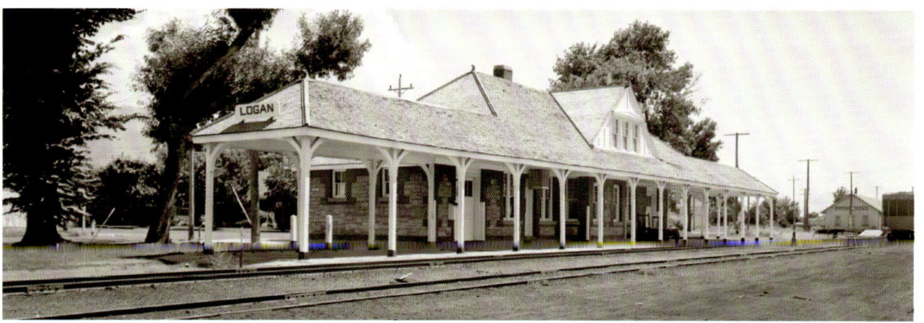

Top: The cover of a brochure advertising access to Yellowstone National Park.

Bottom: The OSL's station at Logan, Utah, in 1967. By the time this photo was taken, the Union Pacific had taken over operation of the track.

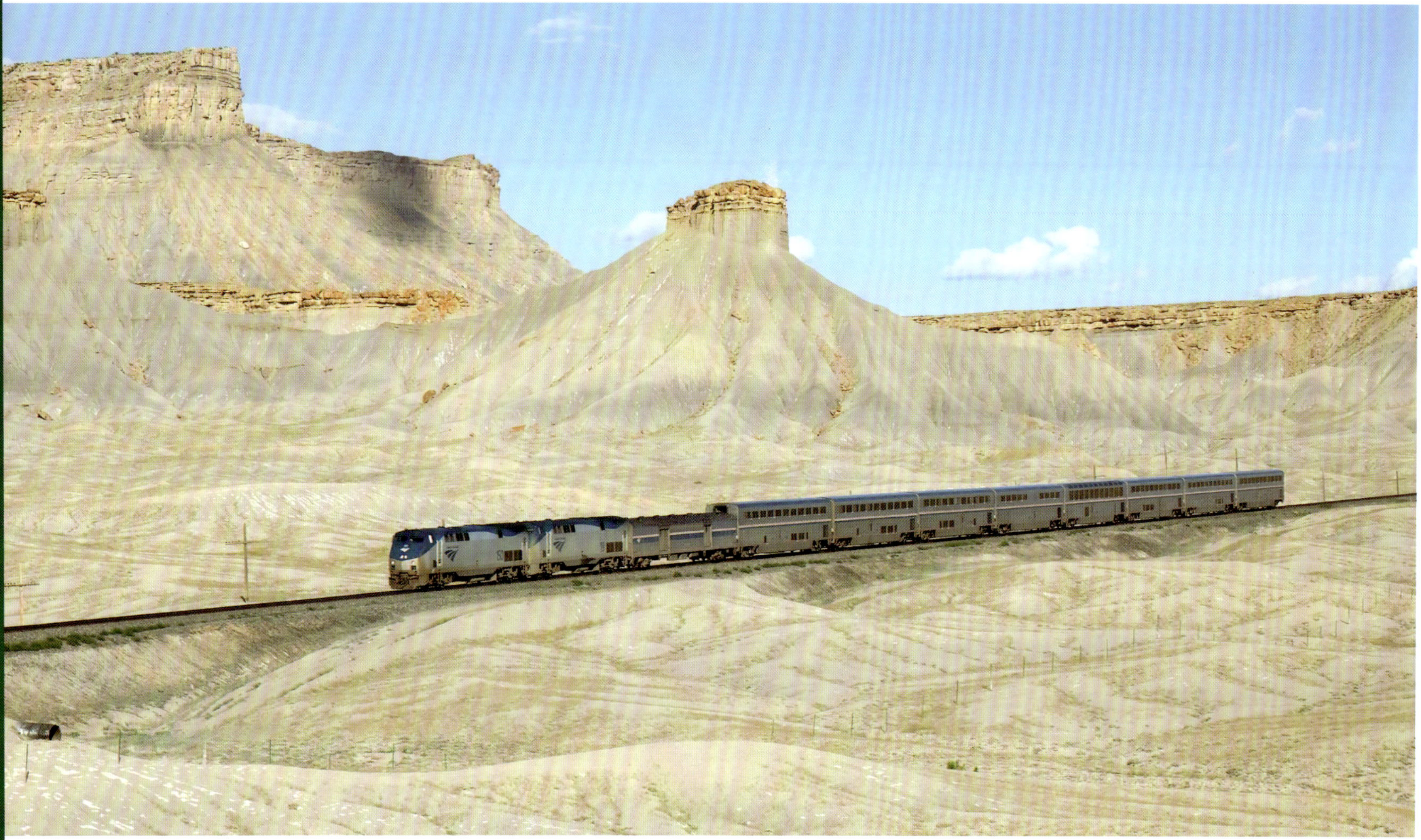

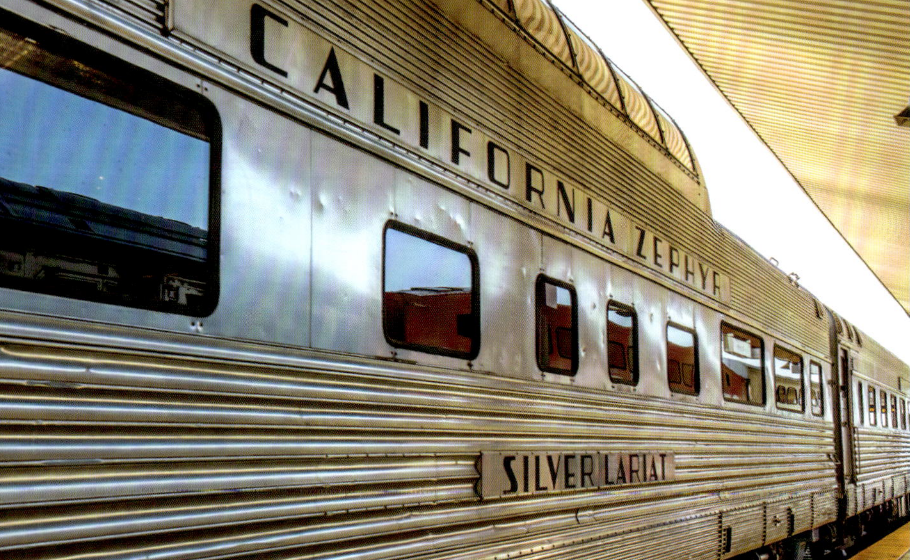

Top: The *California Zephyr* traveling in front of the Book Cliffs between Green River and Floy, Utah.

Bottom: Today, the *California Zephyr* operates with a baggage car, a transition sleeper, two sleeping cars, a dining car, three coaches, and a lounge car.

Amtrak's *California Zephyr*, following the second-longest route in America, provides service between Chicago and the San Francisco Bay Area through some of the nation's most spectacular scenery. Crossing the Great Plains, Rocky Mountains, and the Sierra Nevada, the train currently covers 33 stops, including Omaha, Nebraska; Denver, Colorado; Salt Lake City, Utah; and Reno, Nevada.

Before Amtrak took over, this epic passenger train was collectively operated by the Chicago, Burlington and Quincy (CB&Q), the Denver and Rio Grande Western (D&RGW), and the Western Pacific Railroads. Each claimed the *California Zephyr* was "the most talked about train in America." These operators wanted to provide passengers access to the Golden Gate International Exposition of 1939 and worked collectively to connect their railways, expanding on the route previously traveled by the *Exposition Flyer*.

The CB&Q provided service between Chicago and Denver, the D&RGW provided tracks between Denver and Salt Lake City, and the Western Pacific Railroad ran from Salt Lake City to Oakland. This conglomeration operated for 10 years, until they decided to streamline their service and rename it the *California Zephyr* in 1949.

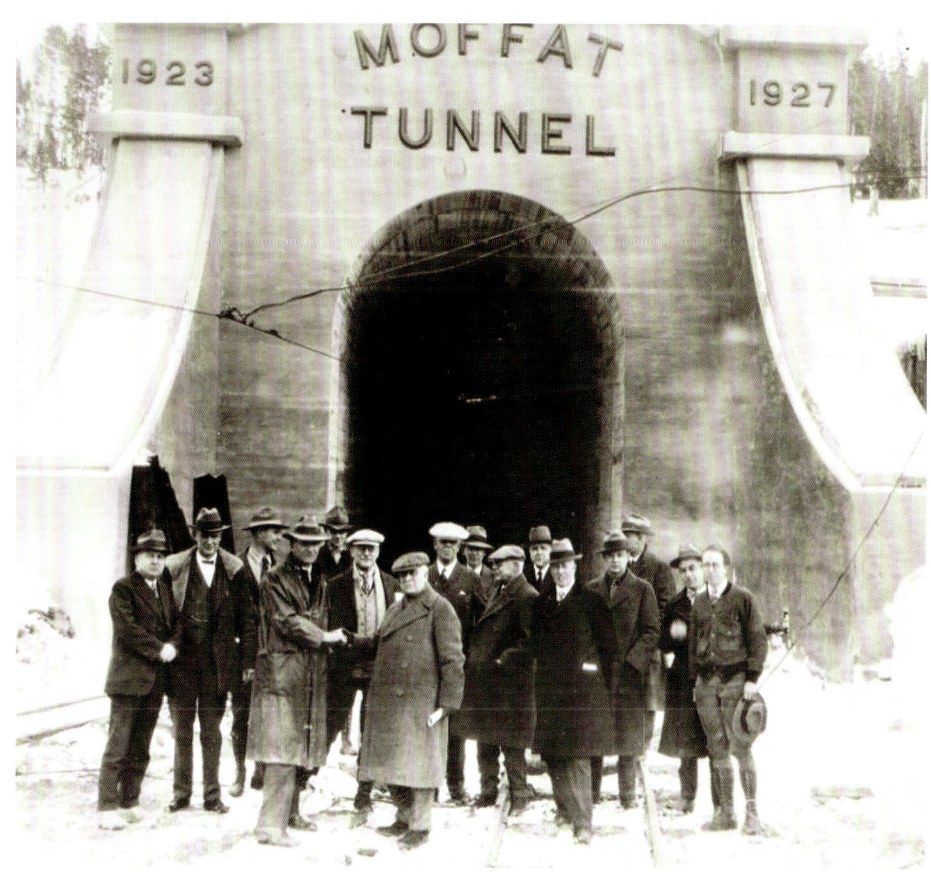

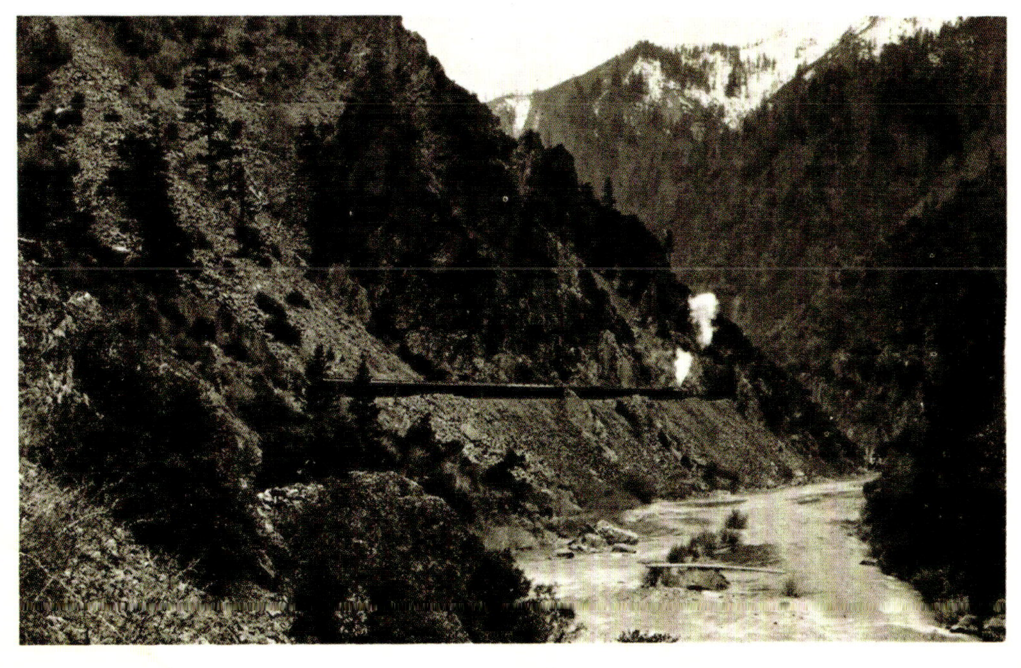

Left: A postcard circa 1939–1949 showing the *Exposition Flyer* in the Feather River Canyon before it was renamed to the *California Zephyr.*

Right: CB&Q officials at an inspection of the Moffat Tunnel in 1934. This tunnel cuts through the Continental Divide in Colorado, and the *California Zephyr* still passes through it today.

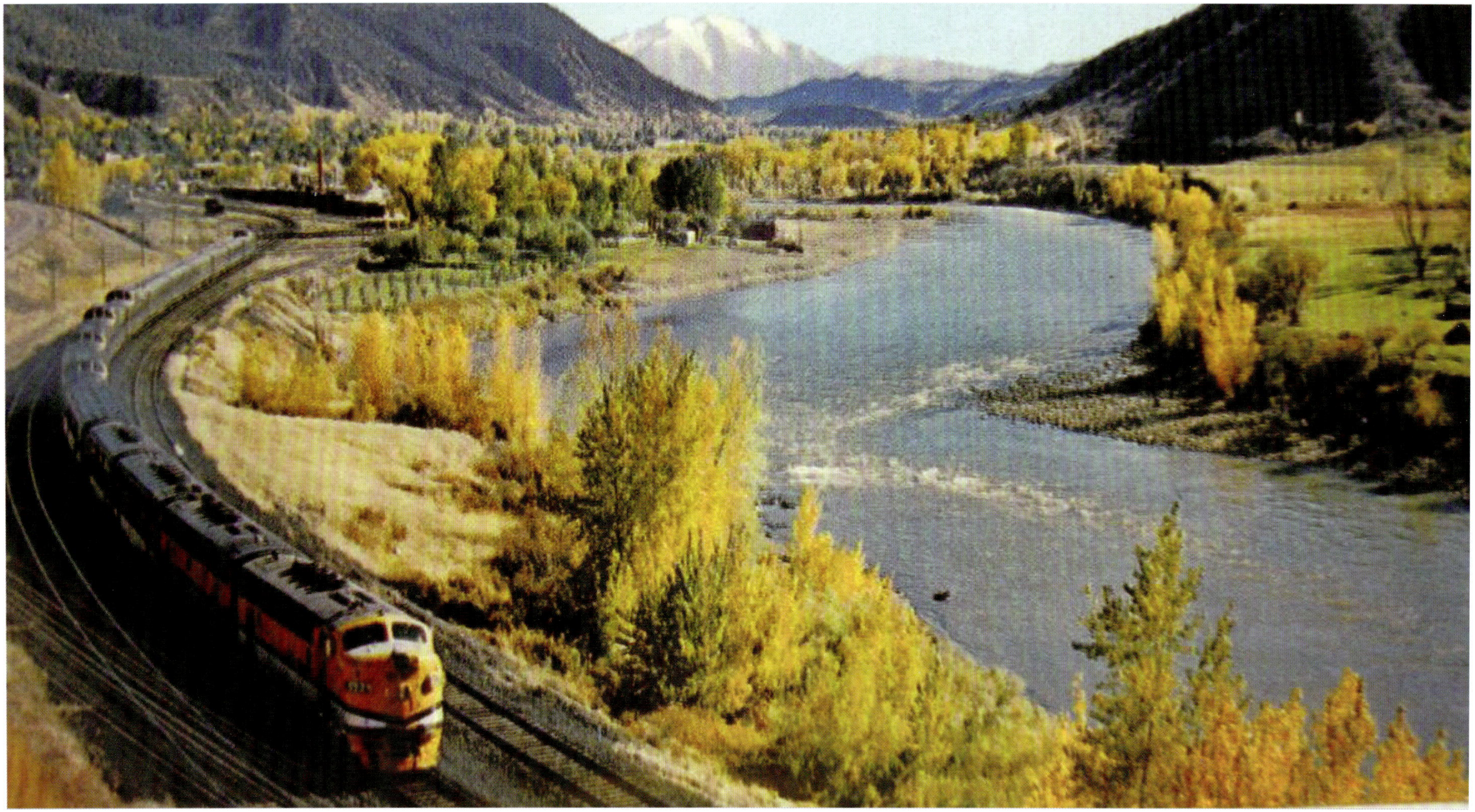

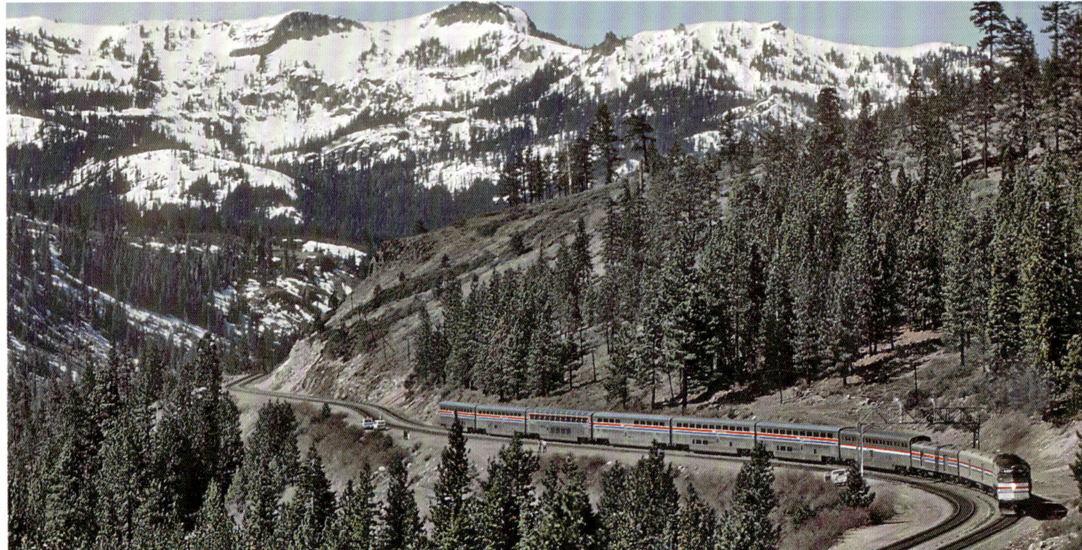

Top: A postcard showing the *California Zephyr* passing the Colorado River.

Bottom: The *San Francisco Zephyr* at Donner Pass in April 1981.

Despite its popularity, the *California Zephyr* was not immune to the falling passenger rates that struck the nation during the mid-twentieth century. The D&RGW and the Western Pacific Railroad applied to discontinue their service in 1969, and the Western Pacific was permitted to do so, provided that the CB&Q and the D&RGW continued to provide some semblance of the semi-transnational route. The last *California Zephyr* of the era ran on March 22, 1970.

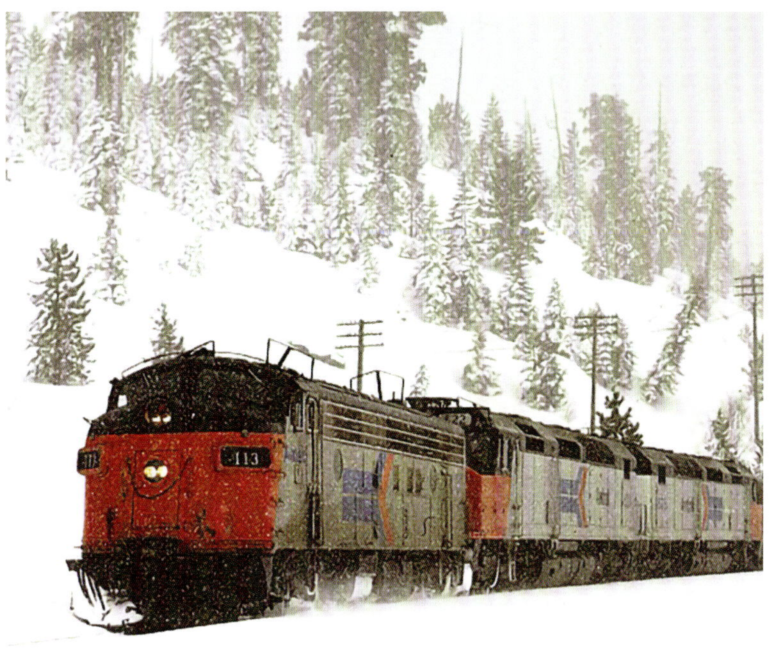

The *California Zephyr* revived the next year as Amtrak consolidated its own rail network across the nation, but the D&RGW refused to join Amtrak's rail system, leading Amtrak to re-route the train away from Colorado and onto the Union Pacific Overland Route through southern Wyoming. It wasn't until 1983, when D&RGW decided to join Amtrak, that the *California Zephyr*'s original track was reunited.

Its route no longer runs all the way to its Oakland terminus but now terminates at the train station in Emeryville, California. Today the train covers 2,438 miles, making it Amtrak's longest route with daily service. A one-way trip takes an average of 52 hours to complete, and it offers a variety of sleeper service arrangements for the long trip.

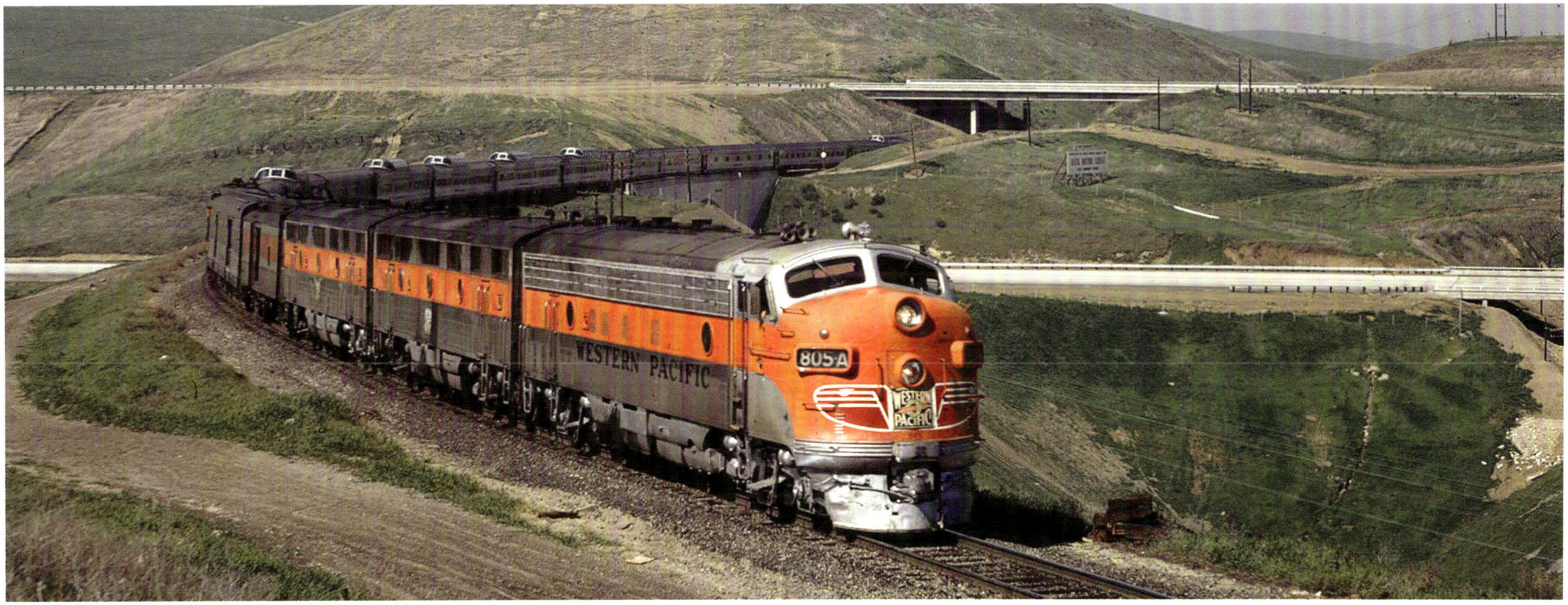

Top: The *California Zephyr* before it was owned by Amtrak, traveling through Feather River Canyon in northeastern California.

Bottom: The vista-dome coach *Silver Scout*.

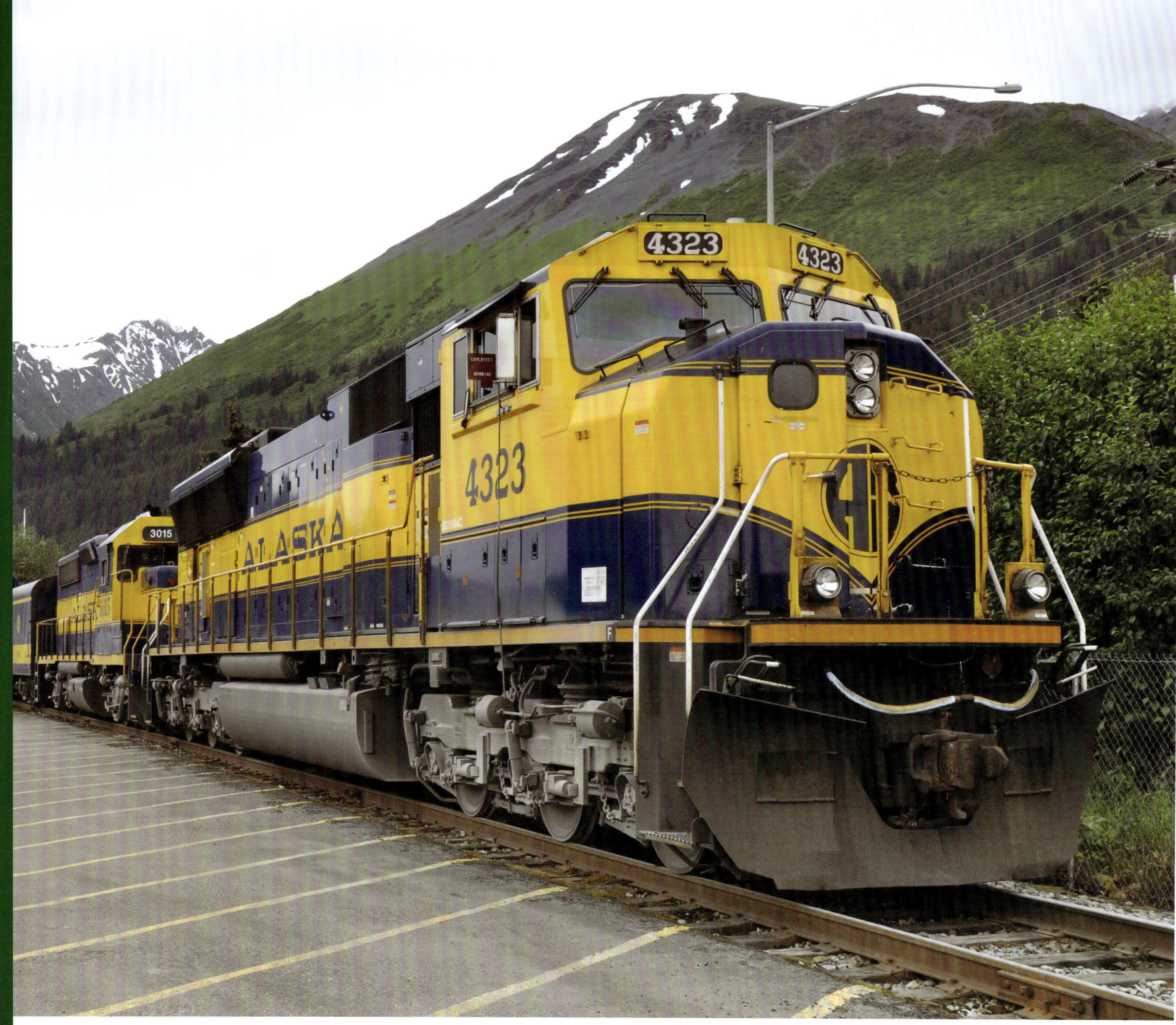

On clear days, the *Denali Star* provides tourists with spectacular views of the Alaska Range.

Trekking through more than 300 miles of Alaskan wilderness, the *Denali Star* provides seasonal rail service between Anchorage and Fairbanks along the Alaska Railroad. From May to September, the *Denali Star* is an easy way for tourists to get to Denali National Park, with dome cars for viewing the tremendous, snow-covered peaks of the Alaska Range. Daily service runs both directions toward Fairbanks and Anchorage, while the *Aurora Winter Train* operates for the rest of the year a few days out of the week on a flag-stop service.

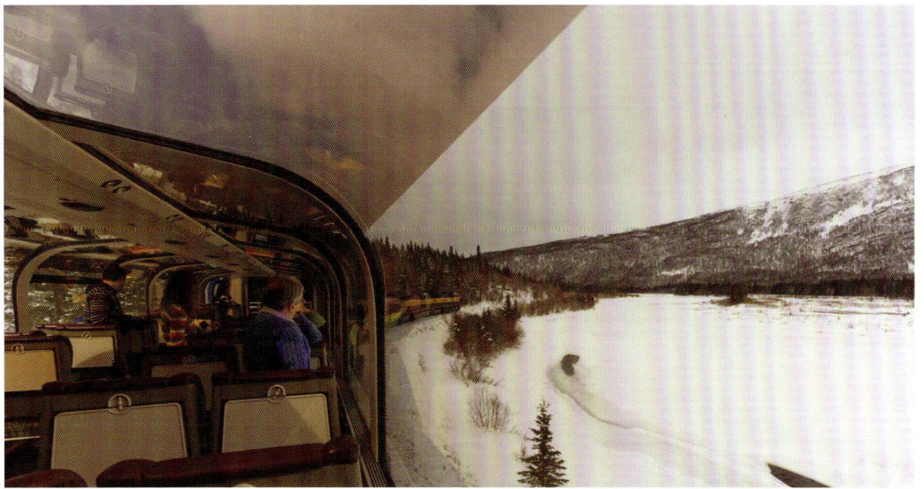

The Alaska Railroad system was built by the United States government between 1909 and 1914. Aside from passengers who want to get a glimpse of the Alaskan wilderness, large amounts of freight makes its way into the interior by way of this rail. Although the Alaska Railroad is not connected to the contiguous 48 states, freight is received from three rail barges that operate between Harbor Island, Washington, and Anchorage, Alaska. The railway also operates as a lifeline to residents who live in the Hurricane Area, which is not accessible by road. Residents of this area receive all of their food and goods by rail, and can enter or depart from the Alaksa Railroad by using the flag-stop service, which allows them to have a train stop anywhere they need without the presence of a depot or station. The Alaska Railroad is as wild as a line can get, taking passengers and residents of the Alaskan wilderness into truly remote areas that have no other regular access to the outside world.

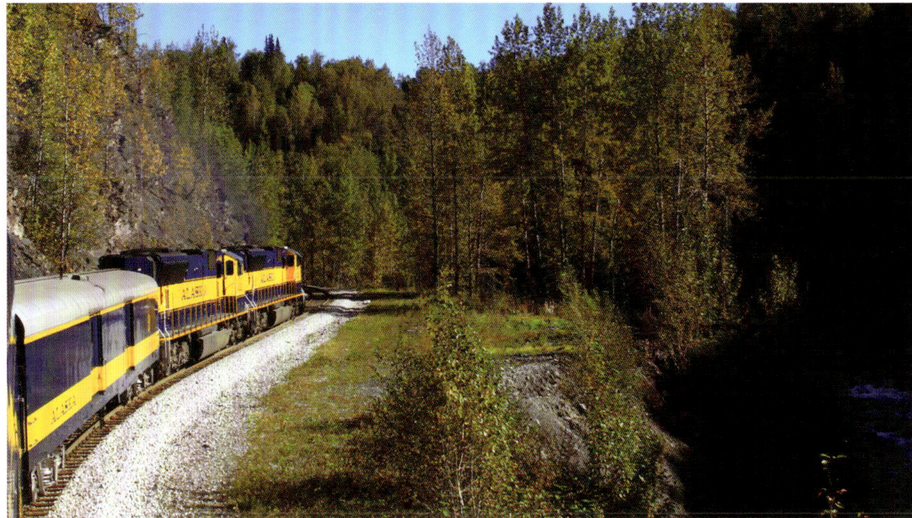

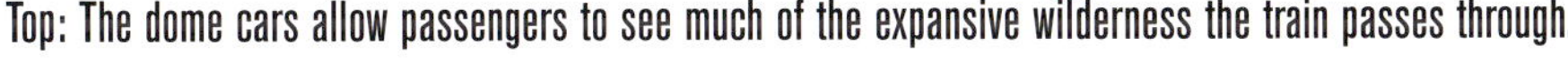

Top: The dome cars allow passengers to see much of the expansive wilderness the train passes through.

Middle: Diesel engines have been operating on the Alaska Railroad since 1944, while steam engines continued to operate until 1966.

Bottom: The *Aurora Winter Train* provides semi-weekly service during the winter months.

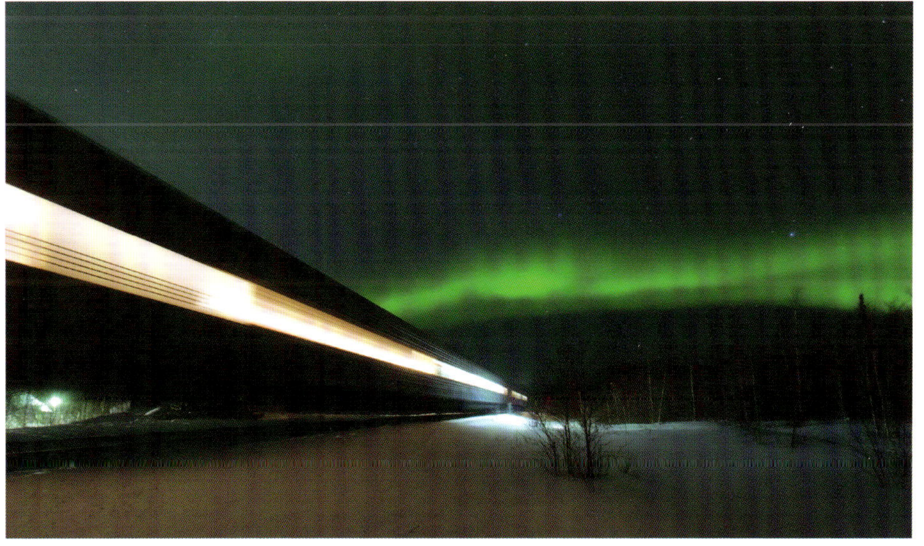

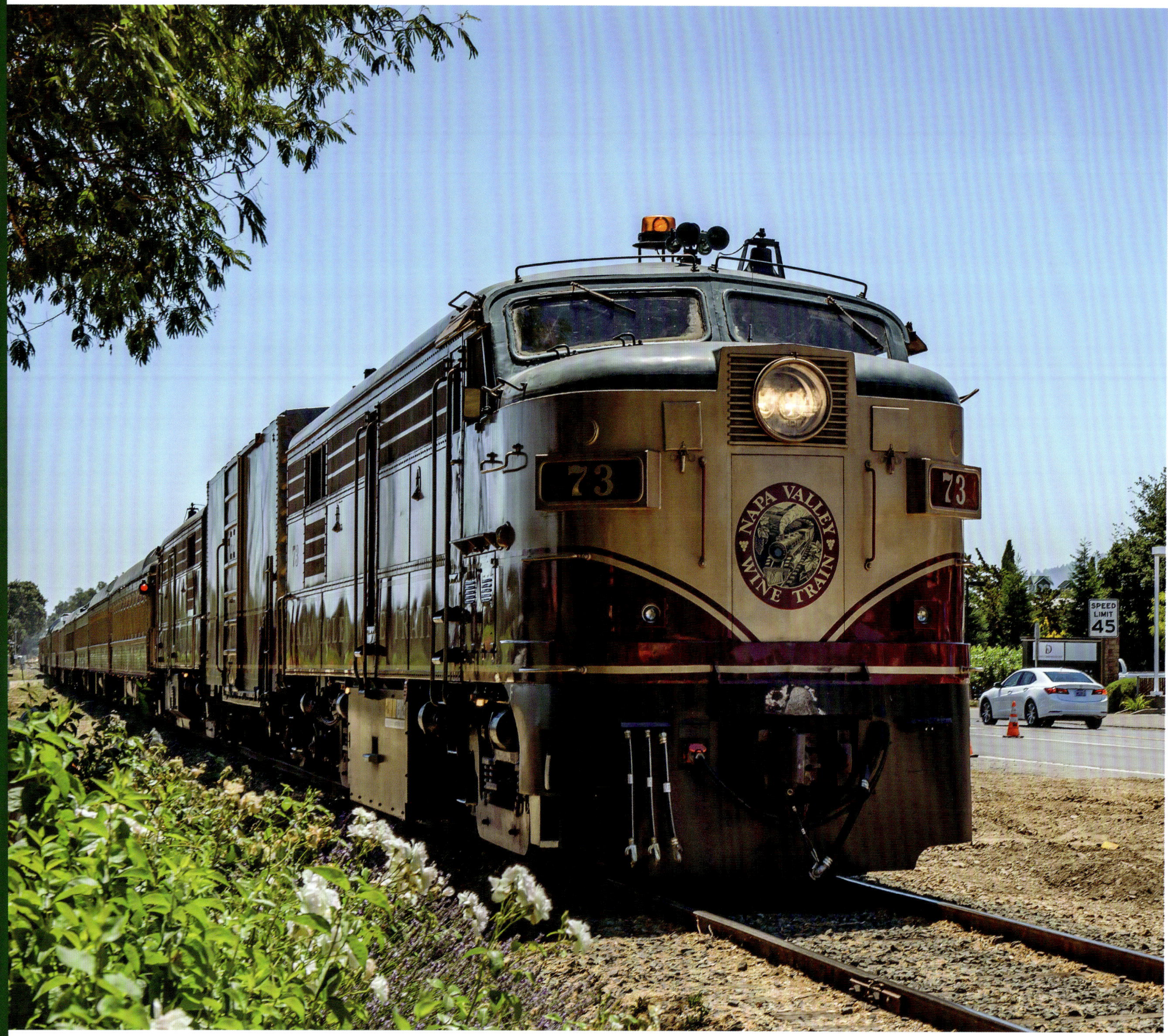

Locomotive #73 has been modified to burn a combination of natural gas and diesel as a cleaner fuel source. Locomotive #70 also underwent this conversion but has since been reverted back to strictly diesel.

Operating as a tourist expedition through Northern California's wine-rich Napa Valley since 1989, the Napa Valley Wine Train has transported nearly two million passengers between the four stations along the line. Following much of California State Route 29, the railway runs between Napa and St. Helena, California, for some 18 miles of the track's original 42 mile length. Built in 1864 by California pioneer Samuel Brannan, the train originally ran from Vallejo, California, to help bring tourists into the resort town of Calistoga, California. Going bankrupt only years after it opened, Brannan's railway was bought by various other railway companies throughout the next one hundred years for use as freight transportation in the area.

The new wine-centric focus of the railway came about in 1989 after the Southern Pacific Railroad notified local officials that it planned to abandon operating on the line. At that time, the Napa Valley Railroad was formed by a group of local entrepreneurs in order to provide a train service for wine enthusiasts. Although the new service was strongly opposed by locals, the Napa Valley Railroad was given right of way on the tracks and began operations September 16, 1989.

The tour is a three-hour journey through some of America's most luscious vineyards and expensive farmland, paired with a meal cooked on-site and local wines available in the lounge cars. Some excursions offer optional stops for passengers to get off and take tours at some of the region's numerous wineries. Or visitors can simply enjoy the ride as they pass the time with a fine glass of wine.

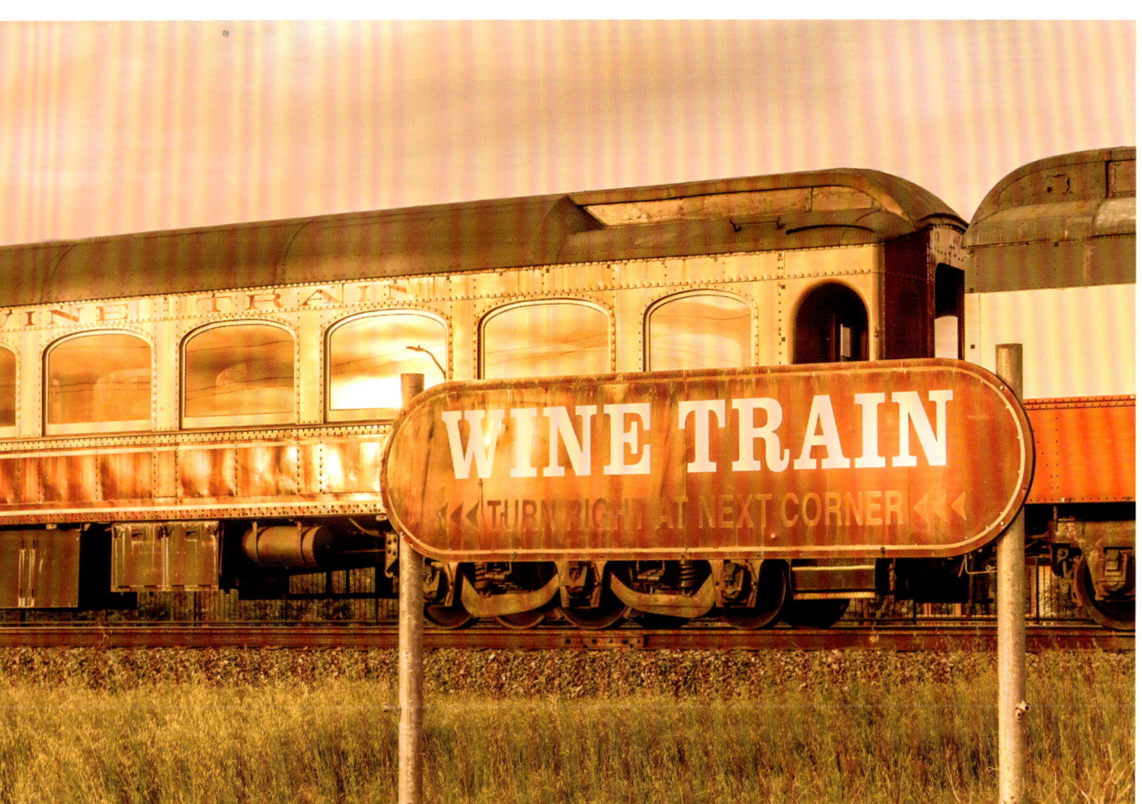

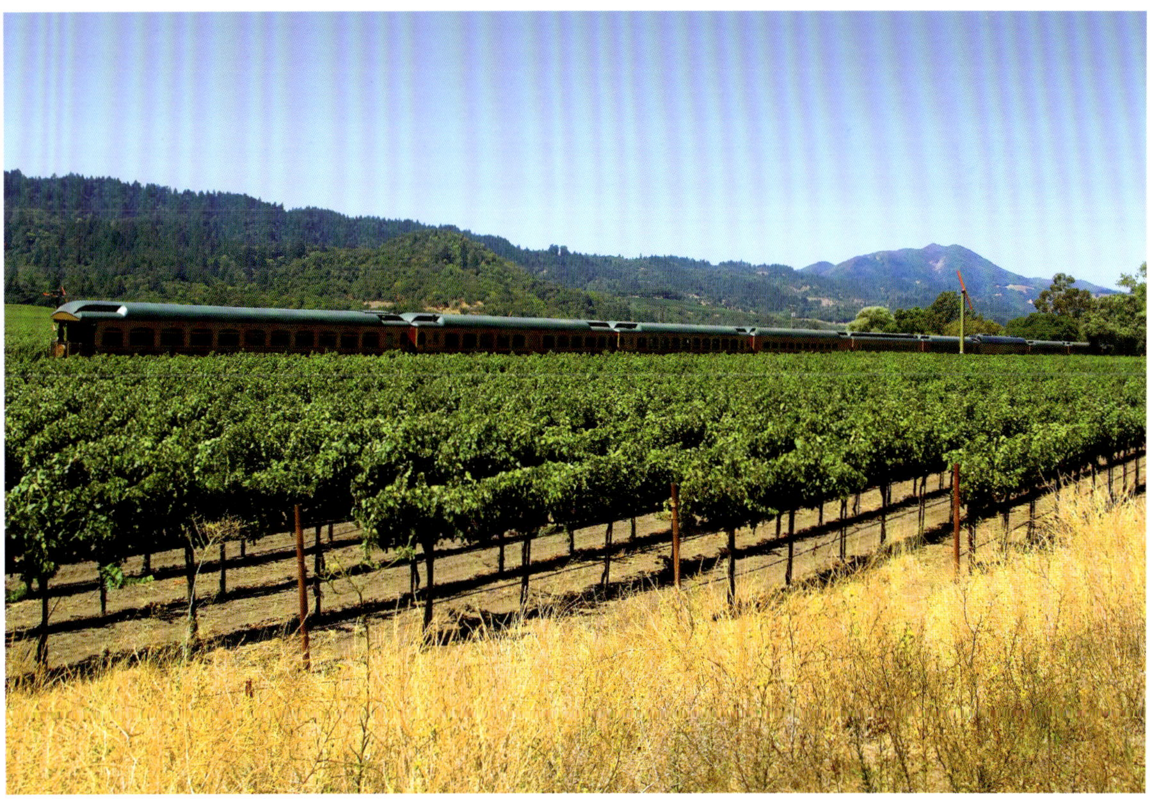

Top: The Napa Valley Wine Train operates a variety of daytime, evening, and seasonal experiences.

Bottom: By the end of the nineteenth century, Napa Valley had as many as 140 wineries in the area.

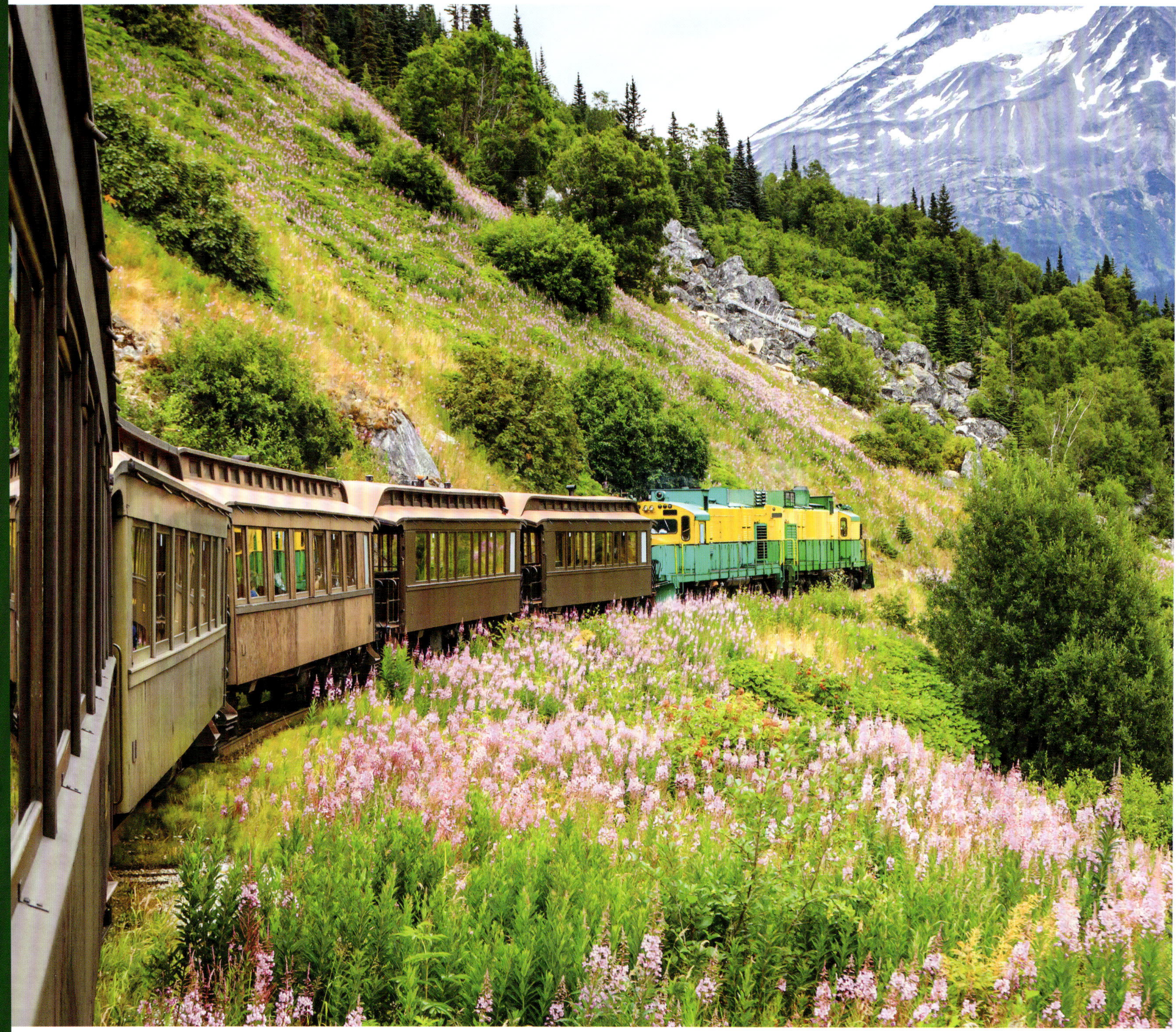

Tours on the White Pass and Yukon Route are available between May and October. A train service from Skagway to White Pass summit in Alaska does not require a passport, although two other routes which cross into Canada do.

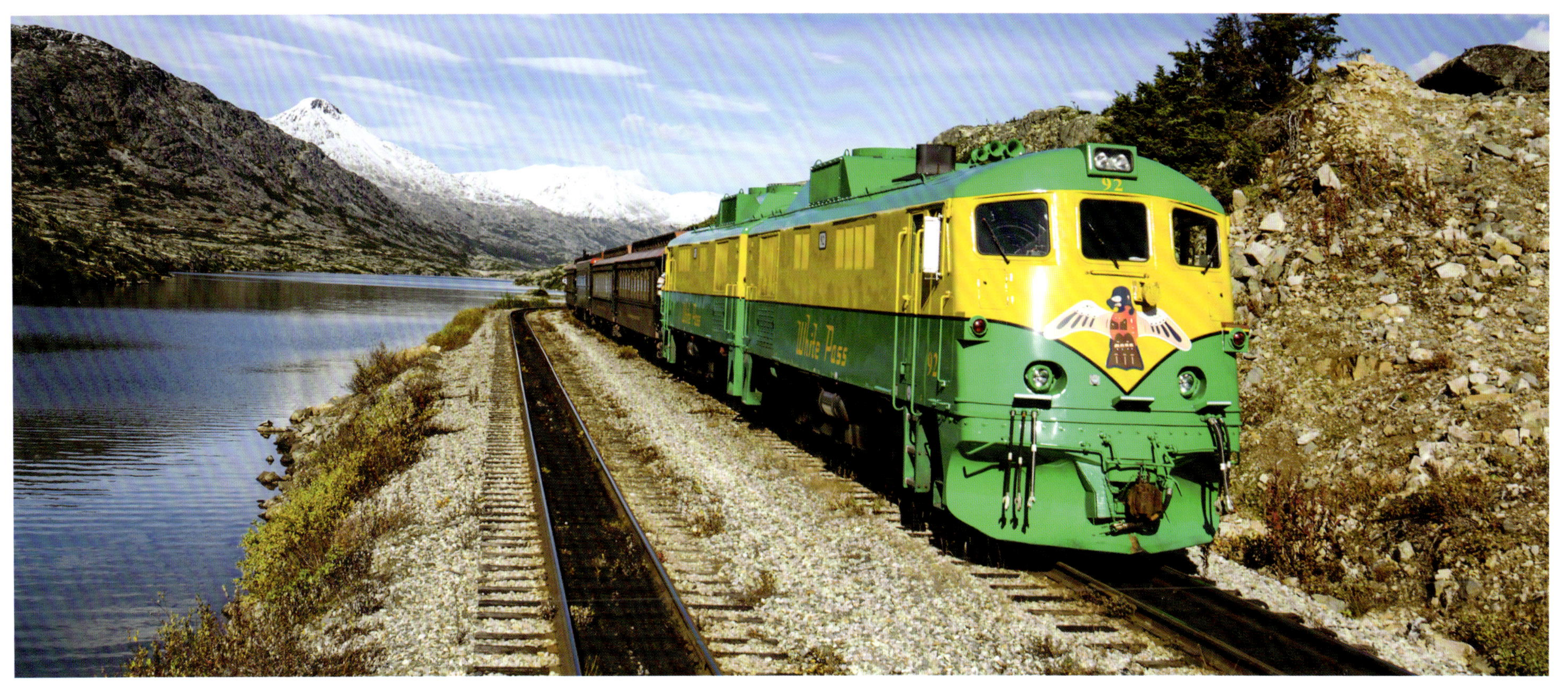

The track of the White Pass and Yukon Route is a narrow-gauge track at three feet in width. The narrow gauge allowed for construction costs to be much lower than standard guage because the smaller roadbed required less blasting.

The White Pass and Yukon Route is a vestige of the determination that fueled prospectors during the Klondike Gold Rush of 1897. Built in notoriously rough and mountainous terrain along the United States/Canada border, the White Pass and Yukon Route runs from Skagway, Alaska, through British Columbia, Canada, and then terminates at Carcross, in the Yukon Territory of Canada. Although the original train route made it as far north as Whitehorse, Yukon Territory, the tracks that lead to Whitehorse are no longer certified for use by the Canadian Transportation Agency.

Before the railway was built, Skagway was only accessible by sea, and the gold fields in the Yukon Territory were only accessible by foot from Skagway by traversing over the White Pass or the Chinook Trail, the only two passes through the Boundary Range of the Coastal Mountains. In those days, prospectors coming from America were often told to turn around by the Canadian authorities at the border if they did not have adequate supplies for the winter—often considered a ton's worth of goods, requiring multiple trips up and down the pass. The Klondike Gold Rush began in 1897, a year after George Carmack and Skookim Jim's discovery of gold in Bonanza Creek, and the construction for the White Pass and Yukon Route started in 1898 with the help of various companies and British investors.

By the time the railway had opened, the gold rush was bust, and many of the nearly 100,000 people who came running into the treacherous territory began to retreat home. The railway was used to haul copper, silver, lead, passengers, and other freight through the largely inaccessible areas of Alaska and Yukon Territory until 1982. As Alaskan tourism began to increase in the late twentieth century, the prospect of using the railway as a tourist attraction became more viable. Cruise ships often visited the port of Skagway, and the tracks of the White Pass and Yukon Route lie right at the edge of the port, creating a convenient way for passengers to gain access to the Alaskan wilderness. Today the railway operates as a tourist attraction, leading visitors into the vast and indomitable wilderness of Alaska.

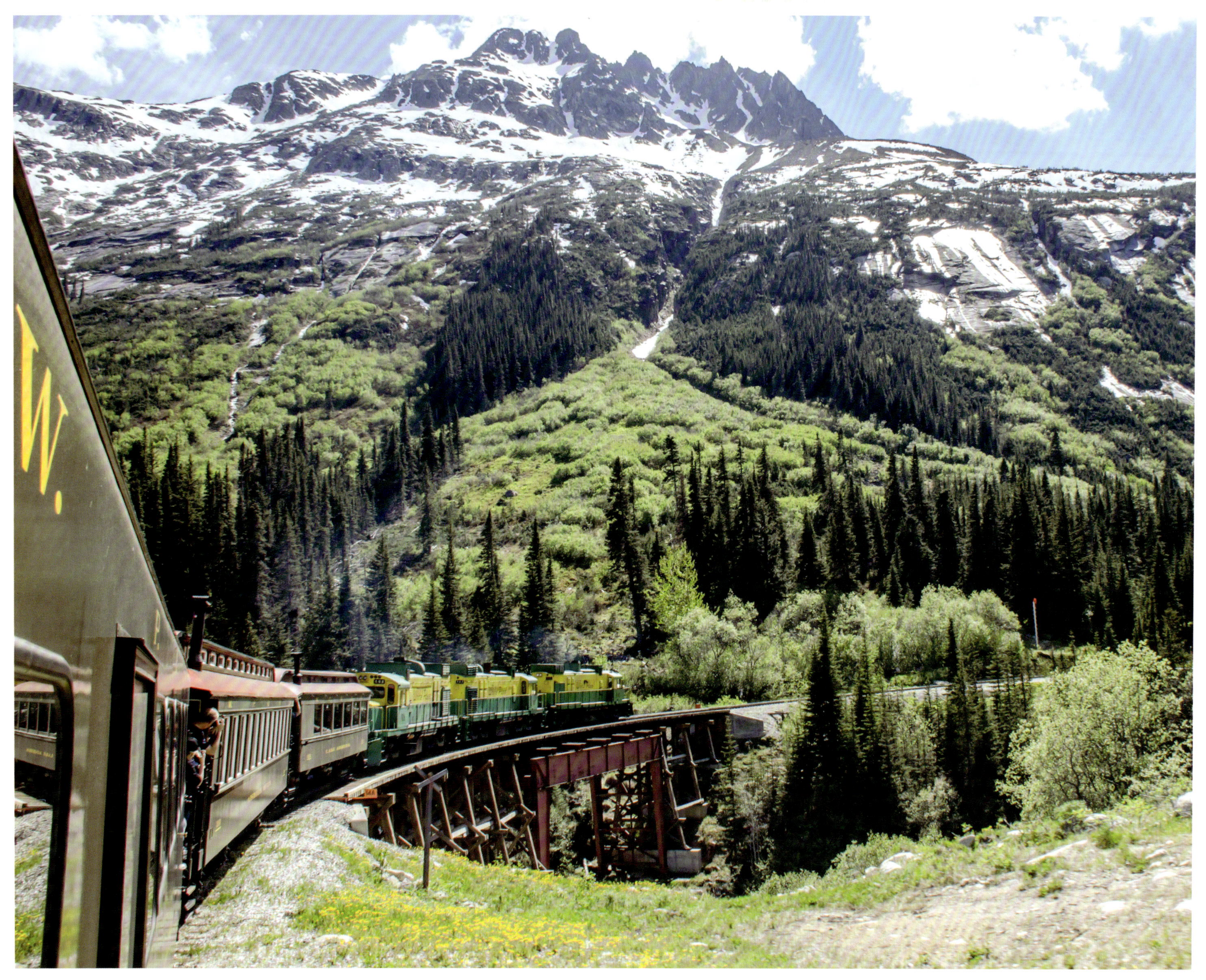

The Boundary Ranges contain many ice fields and act as a physical marker for the border between southeast Alaska and British Columbia.

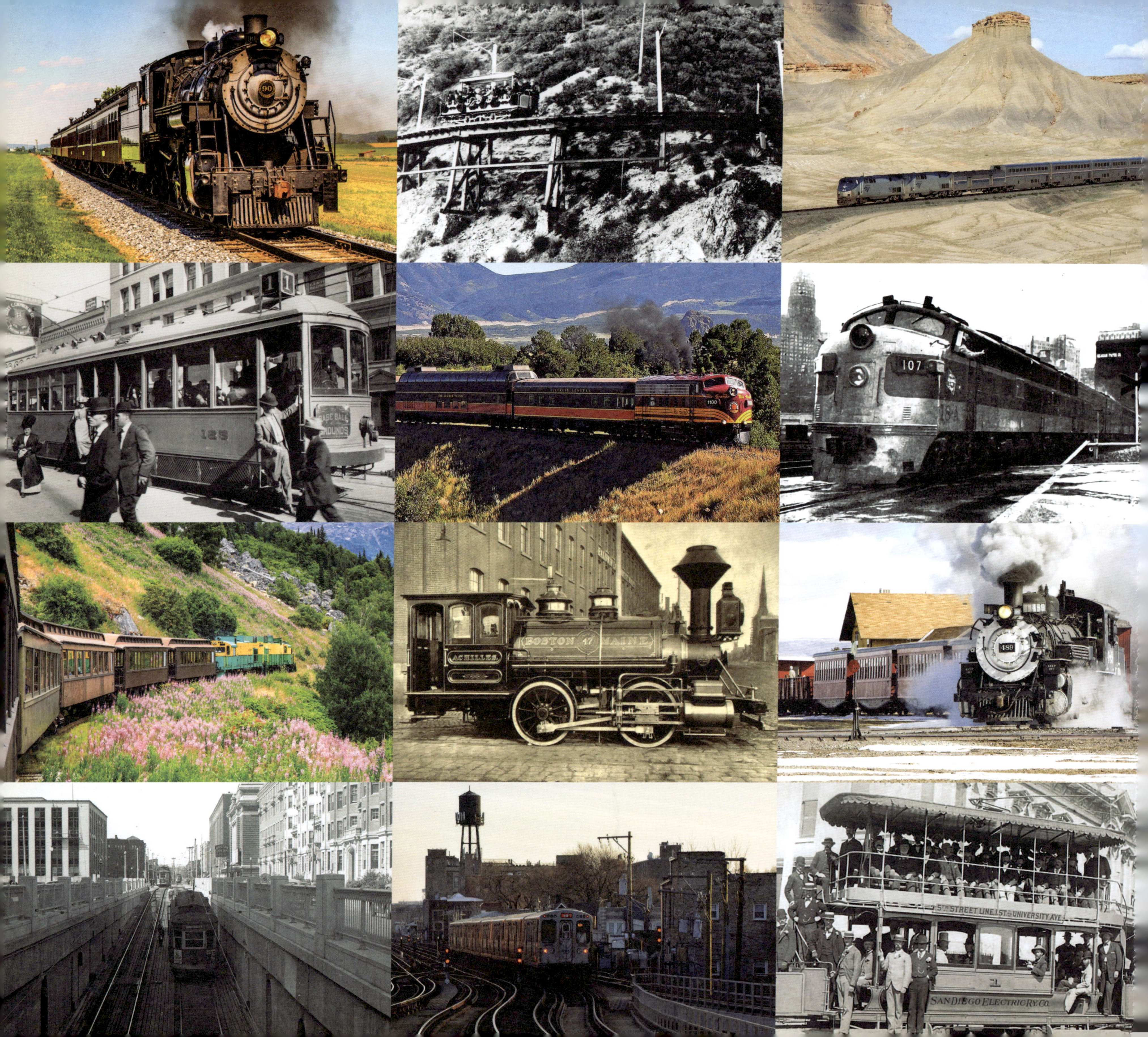
90
125
1100
107
BOSTON 47 MAINE
ACHILLES
480
5th STREET LINE 1st UNIVERSITY AVE
SAN DIEGO ELECTRIC RY CO.